METHODS IN MOLECULAR BIOLOGY™

Series Editor
John M. Walker
School of Life Sciences
University of Hertfordshire
Hatfield, Hertfordshire, AL10 9AB, UK

For further volumes:
http://www.springer.com/series/7651

Mucins

Methods and Protocols

Edited by

Michael A. McGuckin

Immunity, Infection and Inflammation Program, Mater Medical Research Institute, South Brisbane, QLD, Australia

David J. Thornton

Wellcome Trust Centre for Cell-Matrix Research, Faculty of Life Sciences, University of Manchester, Manchester, UK

Humana Press

Editors
Michael A. McGuckin
Immunity, Infection and Inflammation Program
Mater Medical Research Institute
South Brisbane, QLD, Australia
mmcguckin@mmri.mater.org.au

David J. Thornton
Wellcome Trust Centre for Cell-Matrix Research
Faculty of Life Sciences
University of Manchester
Manchester, UK
dave.thornton@manchester.ac.uk

ISSN 1064-3745 e-ISSN 1940-6029
ISBN 978-1-4939-5886-3 ISBN 978-1-61779-513-8 (eBook)
DOI 10.1007/978-1-61779-513-8
Springer New York Dordrecht Heidelberg London

Printed on acid-free paper

Humana Press is part of Springer Science+Business Media (www.springer.com)

Preface

Introduction to Mucin Biology and Technical Challenges of Mucin Research

Epithelial mucins are large complex cell surface and secreted glycoproteins produced by mucosal epithelial cells. Mucins are a major component of the interface between the external world and mucosal tissues, where they provide lubrication, hydration, and a biological and physical barrier to potential toxins, particles, and pathogens. Mucins provide many challenges to researchers due to their large size, complex biochemical nature, and the viscous gels that they form when secreted. Overcoming these challenges is centrally important to a full understanding of mucosal biology and the contribution of mucins to normal human physiology and disease. In this volume of the *Methods in Molecular Biology* series, we have highlighted the technical challenges while describing procedures that are specifically relevant to the analysis of mucins and their contribution to mucosal biology. We have gathered a group of experts together to overview the best approaches to analysing each specific area of mucin biochemistry, physiology, and biophysics before providing individual detailed experimental protocols together with troubleshooting and interpretation tips. We have avoided detailing methods where the analysis of mucins is consistent with standard approaches for other proteins. The volume is designed to be a useful resource for those entering the mucin field and to facilitate those already studying mucins to broaden their experimental approaches to understanding mucosal biology.

The initial three chapters deal with the complexities of working with mucin genes, the challenges of the isolation and biochemical analysis of mucin glycoproteins and methods for detecting and quantifying mucins. The next two chapters concern detection of mucin core proteins by mass spectrometry and techniques for identifying sites of *O*-glycosylation on the mucin core proteins. These are followed by two chapters concerning the analysis of the biosynthesis of secreted mucins and the synthesis and intracellular trafficking of the cell-surface mucins. Then, there are three chapters that focus on the use of mass spectrometry-based methodologies to analyze the complex and diverse *O*-glycans present on mucins. The book then changes focus to methods used to assess mucus and mucin physiology and pathophysiology beginning with a chapter detailing methods for analyzing degradation of mucins. Then, there are three chapters concerned with assessing mucus in situ, including in vivo measurement of mucus thickness and production. This is followed by chapters describing the culture of mucus-producing human bronchial epithelial cells and techniques for assessing mucus production and secretion by those cultures. The last three chapters describe methods for assessing mucins in vitro and in vivo in the context of pathophysiology including infection.

South Brisbane, QLD, Australia — *Michael A. McGuckin*
Manchester, UK — *David J. Thornton*

Contents

Contributors

LUBNA H. ABDULLAH • *Cystic Fibrosis/Pulmonary Research and Treatment Center, University of North Carolina, Chapel Hill, NC, USA*

PABLO ARGÜESO • *Harvard Medical School, Schepens Eye Research Institute, Boston, MA, USA*

ISABELLE BRELOY • *Medical Faculty, Institute of Biochemistry II, University of Cologne, Cologne, Germany*

BRIAN BUTTON • *Department of Medicine, University of North Carolina, Chapel Hill, NC, USA*

STEPHEN D. CARRINGTON • *Veterinary Science Centre, University College Dublin, Belfield, Dublin, Ireland*

ANTHONY P. CORFIELD • *School of Clinical Sciences, Bristol Royal Infirmary, Bristol, UK*

JULIA R. DAVIES • *Department of Oral Biology, Faculty of Odontology, Malmö University, Malmö, SE, Sweden*

C. WILLIAM DAVIS • *Cystic Fibrosis/Pulmonary Research and Treatment Center, University of North Carolina, Chapel Hill, NC, USA*

BURTON F. DICKEY • *Department of Pulmonary Medicine, The University of Texas M.D. Anderson Cancer Center, Houston, TX, USA*

CHRISTOPHER M. EVANS • *Department of Pulmonary Medicine, The University of Texas M.D. Anderson Cancer Center, Houston, TX, USA*

THOMAS A. GERKEN • *Department of Pediatrics and Biochemistry, Case Western Reserve University, School of Medicine, Cleveland, OH, USA*

ILENE K. GIPSON • *Harvard Medical School, Schepens Eye Research Institute, Boston, MA, USA*

JENNY K. GUSTAFSSON • *Department of Medical Biochemistry, Mucin Biology Group, University of Gothenburg, Gothenburg, Sweden*

FRANZ-GEORG HANISCH • *Institute of Biochemistry II, Medical Faulty, and Center for Molecular Medicine Cologne, University of Cologne, Köln, Germany*

GUNNAR C. HANSSON • *Department of Medical Biochemistry, Mucin Biology Group, University of Gothenburg, Gothenburg, Sweden*

CERI A. HARROP • *Wellcome Trust Centre for Cell-Matrix Research, Faculty of Life Sciences, University of Manchester, Manchester, UK*

SUMAIRA Z. HASNAIN • *Immunity, Infection and Inflammation Program, Mater Medical Research Institute, South Brisbane, QLD, Australia*

CATHERINE A. HAYES • *Medical Biochemistry, University of Gothenburg, Gothenburg, Sweden*

DAVID B. HILL • *Department of Medicine, University of North Carolina, Chapel Hill, NC, USA*

LENA HOLM • *Department of Medical Cell Biology, Uppsala University, Uppsala, Sweden*

REBECCA P. HUGHEY • *Department of Medicine, University of Pittsburgh School of Medicine, Pittsburgh, PA, USA*
JANE A. IRWIN • *Veterinary Science Centre, University College Dublin, Dublin, Ireland*
SAMAH ISSA • *Medical Biochemistry, University of Gothenburg, Gothenburg, Sweden*
CHUNSHENG JIN • *Medical Biochemistry, University of Gothenburg, Gothenburg, Sweden*
MALIN E.V. JOHANSSON • *Department of Medical Biochemistry, Mucin Biology Group, University of Gothenburg, Gothenburg, Sweden*
NICLAS G. KARLSSON • *Medical Biochemistry, University of Gothenburg, Gothenburg, Sweden*
MEHMET KESIMER • *Department of Biochemistry and Biophysics Cystic Fibrosis/Pulmonary Research Center, University of North Carolina, 4021 Thurston Bowles Bldg. CB#7248, Chapel Hill, NC, USA*
CAROL L. KINLOUGH • *Renal Electrolyte Division, Department of Medicine, University of Pittsburgh School of Medicine, Pittsburgh, PA, USA*
SARA K. LINDÉN • *Mucosal Immunobiology and Vaccine Center, University of Gothenburg, Gothenburg, Sweden*
LI LIU • *NIBRT, Fosters Avenue, Mount Merrion, Blackrock, Dublin, Ireland*
ELIZABETH MATTHEWS • *Veterinary Science Centre, University College Dublin, Dublin, Ireland*
MICHAEL A. MCGUCKIN • *Immunity, Infection and Inflammation Program, Mater Medical Research Institute, South Brisbane, QLD, Australia*
SZILARD NEMES • *Medical Biochemistry, University of Gothenburg, Gothenburg, Sweden*
MIA PHILLIPSON • *Department of Medical Cell Biology, Uppsala University, Uppsala, Sweden*
RAY PICKLES • *Pulmonary Diseases and Critical Care Medicine, Department of Medicine, University of North Carolina, Chapel Hill, NC, USA*
LUCIA PICCOTTI • *Department of Pulmonary Medicine, The University of Texas M.D. Anderson Cancer Center, Houston, TX, USA*
CHIN WEN PNG • *Immunity, Infection and Inflammation Program, Mater Medical Research Institute, South Brisbane, QLD, Australia*
KARINE ROUSSEAU • *Wellcome Trust Centre for Cell-Matrix Research, Faculty of Life Sciences, University of Manchester, Manchester, UK*
PAULINE M. RUDD • *NIBRT, Fosters Avenue, Mount Merrion, Blackrock, Dublin, Ireland*
JOHN K. SHEEHAN • *Department of Biochemistry and Biophysics, Cystic Fibrosis/Pulmonary Research Center, University of North Carolina, Chapel Hill, NC, USA*
YONG H. SHENG • *Immunity, Infection and Inflammation Program, Mater Medical Research Institute, South Brisbane, QLD, Australia*
HENRIK SJÖVALL • *Department of Medical Biochemistry, Mucin Biology Group, University of Gothenburg, Gothenburg, Sweden*
SIMON STAUBACH • *Institute of Biochemistry II, Center of Molecular Medicine, University of Cologne, Cologne, Germany*
DALLAS M. SWALLOW • *Research Department of Genetics, Evolution and Environment, University College London, London*

DAVID J. THORNTON • *Wellcome Trust Centre for Cell-Matrix Research, Faculty of Life Sciences, University of Manchester, Manchester, UK*
CLAES WICKSTRÖM • *Department of Oral Biology, Faculty of Odontology, Malmö University, Malmö, SE, Sweden*
CÉDRIC WOLBER • *Cystic Fibrosis/Pulmonary Research and Treatment Center, University of North Carolina, Chapel Hill, NC, USA*

Chapter 1

Mucin Methods: Genes Encoding Mucins and Their Genetic Variation with a Focus on Gel-Forming Mucins

Karine Rousseau and Dallas M. Swallow

Abstract

Mucin genes encode the polypeptide backbone of the mucin glycoproteins which are expressed on all epithelial surfaces and are major constituents of the mucus layer. Mucins are, thus, expressed at the interface between the external and the internal environment of the organism, and represent the first line of defence of our body. These genes often have an extensive region of repetitive exonic sequence which codes for the heavily glycosylated domain, whose roles include bacterial interactions and gel hydration. This region shows, in several of the genes, considerable inter-individual variation in repeat number and sequence. Because of their site of expression and their high variability in this important domain, mucin genes are good candidates for conferring differences in genetic susceptibility to multifactorial epithelial and inflammatory disease. However, progress in characterizing the genes has been considerably slower than the rest of the genome because of their size and the GC-rich content of the large, repetitive variable region. Some of the issues relating to the study of these genes are discussed in this chapter. In addition, methods and approaches that have been used successfully are described.

Key words: MUC gene, Tandem repeat domain, Polymorphism, SNP, Disease association

1. Introduction

As is seen elsewhere in this volume, mucins are extracellular proteins containing large domains that are rich in serine and threonine residues and are heavily O-glycosylated, and they are mainly expressed by epithelial cells. Apart from these general properties, however, they have a variety of other different features reflecting a number of diverse functions and they are not all closely related. They can, for example, be attached to the membrane or secreted. However, their complete cloning and protein characterization has been slow, which has made their gene nomenclature difficult, and has led to the use of a single set of gene symbols (*MUC*) for genes that are not necessarily evolutionarily related.

Michael A. McGuckin and David J. Thornton (eds.), *Mucins: Methods and Protocols*, Methods in Molecular Biology, vol. 842, DOI 10.1007/978-1-61779-513-8_1, © Springer Science+Business Media, LLC 2012

Since the renaming of the first gene identified to encode a mucin-type protein, to *MUC1* (in the early 1990s), the number of *MUC* genes has increased to 18 (see Note 1). Of these, only 5 code for proteins which are secreted and involved in gel formation, and which some would argue were the only true mucins (i.e. critical to the formation of mucus gels). Four of these, *MUC6*, *MUC2*, *MUC5AC*, and *MUC5B*, are located on chromosome 11p15.5 and form a gene complex while the fifth mucin gene, *MUC19*, is located on chromosome 12q12 (1, 2). The four 11p15.5 gel-forming mucins are closely related and all five share common structural and functional characteristics (reviewed in ref. 3). The genes that encode the 11p15.5 mucins are thought to have evolved by duplication, accounting for their high level of similarity. For example, the exon/intron boundaries are highly conserved between the *MUC* genes on chromosome 11, as are the exon sizes.

In this chapter, we review the methodologies and approaches used to study the mucin genes and the difficulties that have been encountered, focusing on those encoding the gel-forming mucins, but refer to the genes encoding the other small and membrane-associated proteins where they provide good examples.

Although there are claims that the human and several other genomes are fully sequenced, this is *not* true for mucin genes and the sequences reported in some cases are not real and/or incomplete, mostly as a result of automated sequence assembly and incorrect annotation. This is misunderstood, even sometimes in the mucin field, and researchers can be totally misled by incorrect annotations and the fact that the Refseq (NCBI reference Sequence) entries are not fully correct. This is unlikely to be resolved by high-throughput re-sequencing which suffers from even more severe problems resulting from computational assembly.

Historically, the *MUC* genes were first of particular interest because of the extent of genetic polymorphism found at the gene and protein levels. This was due to the existence of a tandemly repetitive central region which codes for the heavily glycosylated domain that in many cases shows "variable number tandem repeat (VNTR) polymorphism," leading the genes to be considered as expressed "minisatellite" sequences (4). Of the genes encoding the secreted mucins, *MUC2* shows the largest range of relative allele sizes ranging from 40 to 185 repeats (Table 1 and Fig. 1), though *MUC6* shows the greatest heterozygosity of VNTR length alleles, and MUC5B lacks common VNTR length variants. Since mucins are in the first line of defence of our innate immune system, they represent the direct link between the outside environment and the inside of the organism. In addition, the existence of a high level of inter-individual variation has led to the suggestion that this variation has an impact on susceptibility to inflammatory disease, and to an array of studies to examine allelic association with inflammatory and epithelial disorders (Table 2). However, while there are many, now standard, tools for studying genes and their expression, the

Table 1
Tandem repeat characteristics of the secreted gel-forming mucins

Mucin gene	Size of the TR unit in bp	Size of the TR unit in aa	Range or size of the TR
MUC2	69	23	3.3–11.4 kb (40–185 repeats)
MUC5AC	24	8	6.5–7.5 kb
MUC5B	87	29	10 kb
MUC6	507	169	8–13.5 kb (15–25 repeats)
MUC19	Variable	Variable	ND

ND indicates not determined
MUC5B and *MUC5AC* also show allelic length variation but to a lesser extent, these have been described in detail by Vinall et al. (43) (see Notes 5 and 7). MUC19 was recently characterized by Zhu et al. (46)

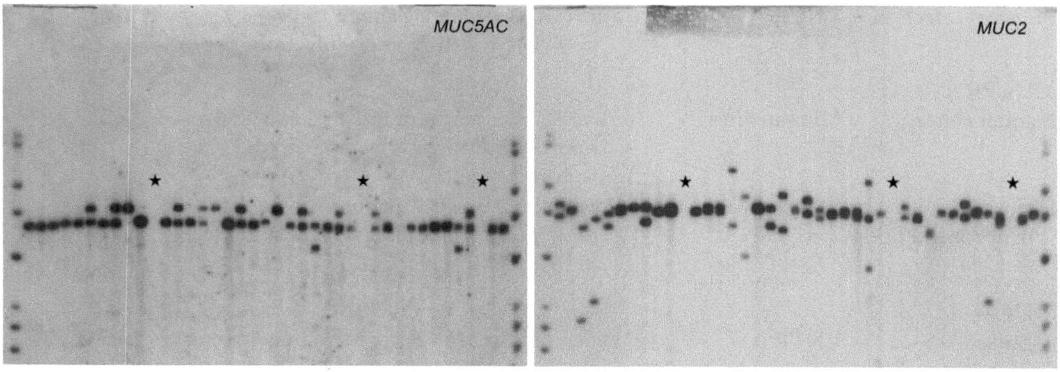

Fig. 1. Southern blots of genomic DNA for the same set of individuals hybridized with the *MUC5AC* and *MUC2* probes. Genomic DNAs were digested with *HinfI*, the Raoul molecular weight marker was electrophoresed in the first and last lane on both gels, a mix of two DNA of known genotype were applied to lanes 27. Lanes 12, 29, and 39 are shown with a star and were left as blank to orientate the gel. It is noteworthy that we have shown a statistically significant difference in the *MUC2* allele distribution between individuals of the three main *MUC5AC* TR genotypes (18), which is attributable to linkage disequilibrium but this correlation between the band sizes for the two genes is not obvious from these gels.

repetitive nature of the sequence corresponding to the glycosylated domain of mucins has led to a variety of difficulties, both practical and bioinformatic. Subheadings 3.2 and 3.3 cover these aspects.

Subheading 3.4 suggests a strategy for disease association studies. Different types of genetic variations in the mucin genes can influence their function. VNTR length variations have the potential to influence the properties of the mucus layer, since this domain carries most of the carbohydrate side chains which are involved in binding to microbes and other proteins, and are also involved in water retention in the mucus layer (3). VNTR length association

Table 2
Published studies in which allelic association of genes encoding gel-forming mucins is reported

Disease	Variation studied	Finding	Reference
MUC5AC			
Gastric cancer	SNPs	*MUC5AC** SNP association with risk of stomach cancer	(47)
Otitis media	VNTR	*MUC5AC* large alleles claimed to be more frequent in otitis media patients	(48)
MUC6			
Gastric cancer	Minisatellites	Rare short *MUC6* intronic minisatellite alleles claimed to influence expression and susceptibility to gastric carcinoma	(49)
	VNTR	Small *MUC6* VNTR alleles are more frequent in gastric cancer patients than in healthy individuals	(50)
	SNPs	No association between *MUC6* and risk of stomach cancer	(47)
H. pylori infection	VNTR	Short *MUC6* alleles claimed to be associated with *H pylori* infection	(51)
MUC5B			
Bladder cancer	Minisatellites	Possible association of intronic *MUC5B* minisatellite variants and susceptibility to bladder cancer	(52)
Diffuse panbronchiolitis	SNPs	Promoter analysis, aberrant expression of *MUC5B**, and disease association in diffuse panbronchiolitis	(15)
MUC2			
Asthma	VNTR	Differences in *MUC2* allele length between topic individuals with and without asthma	(53)
Gastric cancer	Variability of the first TR domain	Rare alleles associated with altered susceptibility to gastric carcinoma	(54)
Inflammatory bowel disease	SNP	Aberrant intestinal expression and allelic variants of *MUC2* associated with Crohn's disease	(55)
	VNTR	Ulcerative colitis is not associated with differences in *MUC2* mucin allele length	(56)
Gallstone disease	SNPs	*MUC2* SNP association with risk of gallstone disease in Chinese males	(57)
MUC19			
Inflammatory bowel disease		Genome-wide association defines more than 30 distinct susceptibility loci for Crohn's disease	(58)

*Since this article went to press two important papers have been published (59, 60, 62, 63)

has been well-studied for *MUC1*, where several studies have shown an association with gastric cancer (5–7). This has usually been done by Southern blot analyses, which remains the most effective method. Despite the progress of long-range Taq polymerase mixes, there is still a risk of not detecting extremely long alleles, although some investigators have succeeded in producing large fragments spanning the VNTR region in a few samples (8–10) (Burgess and Swallow 2006, unpublished). Amino acid substitutions occur within the tandem repeats and are also variable in different people (8, 11, 12) and can affect conformational flexibility (13) (see Note 2) but the extent of this variation has been barely investigated because the technique (12) is even more labour intensive and difficult than the Southern blots used for VNTR analysis. Outside the VNTR domain, there are rather few known coding single-nucleotide polymorphisms (SNPs) or rare variants in the human *MUC* genes that have clear functional consequences (see Note 3). One exception is the *MUC1* exon 2 SNP rs4072037 that alters splicing (14). Another likely important source of functional variation is within regulatory regions. There is an example of this in *MUC5B*, where one particular allelic combination of the promoter sequence is associated with and probably directly causal of higher expression than others (15, 16). As with other genetic association studies, variants of unknown function are often tested, usually being selected to "tag" the variability of the region, by exploiting observed patterns of allelic association. In the case of *MUC* genes, it has however been difficult to find suitable markers because of gaps in the human genome sequence and erroneous SNP entries. While there is a good tagging SNP for the *MUC7* VNTR (17) and there is evidence of LD stretching across the TR domains, in no other case have we noted a SNP with near 100% association with VNTR alleles ((18) and Swallow et al. unpublished). There are several hints in publications and databases that the 11p15.5 *MUC* gene region is subject to copy number variation (CNV), but although our own attempts to verify this for *MUC5AC* were initially suggestive of CNV, replication was unsuccessful. In some of the reported cases, the signal probably arises from the VNTR domains and the difficulty of working with GC-rich sequences.

The technological advances in SNP analyses now allow the genotyping of a large number of variations in very little time, and there has been increasing use of genome-wide association studies (GWAs), but until recently these have also suffered from gaps in coverage, and there are limitations to the methods of analysis because of the requirement to correct for multiple testing and also loss of information relating to rare variants.

Although secreted gel-forming mucin proteins in other species have been studied for a long time (19–21), until recently there has been little gene sequence information in non-human species apart from murine and bovine (2, 22–28). The recent explosion of genome sequencing provides us with the opportunity to predict the protein sequence of the homologous mucin genes for a number of species using the high degree of conservation observed between human and mouse (29, 30). This information which is essential for the understanding of their function or the development of new model systems is addressed in Subheading 3.5.

2. Materials

2.1. DNA Extraction from Whole Blood and Other Sources of Human DNA

1. Puregene Blood Kit (Qiagen-Gentra) for genomic DNA preparation.
2. Sample spectrophotometer by Nanodrop Technologies (ND-8000 from Thermo Scientific).
3. 3 mL of whole blood or other source of DNA, such as buccal swabs.

2.2. Southern Blot

1. Restriction enzymes: see Notes 4–7.
2. TBE buffer (1× = 0.89 M Tris–HCl, 0.1 M borate, 0.002 M EDTA buffer, pH 8.3): Prepared as a 10× or 5× stock (see Note 8).
3. For agarose electrophoresis: Horizontal gel tank 20×25-cm apparatus, and a 10×7-cm horizontal gel tank or equivalent.
4. Agarose, analysis grade, broad separation range for DNA/RNA.
5. Loading buffer for agarose gels: 0.25% (w/v) bromophenol blue, 0.25% (v/v) xylene cyanol, 40% (w/v) sucrose in water.
6. Stock solution of 2.5 mg/mL ethidium bromide (see Note 9).
7. Transilluminator.
8. Hybond N+ membrane (GE Healthcare).
9. Vacuum blotter (VacuGene XL, GE Healthcare).
10. Megaprime™ DNA Labeling System (GE Healthcare).
11. Sodium chloride/sodium citrate (SSC)-containing solutions: Prepare from a stock of 20× SSC (3 M NaCl, 0.3 M trisodium citrate) (see Note 10).
12. Denhardt's solution: Make as a 100× stock (2% (w/v) Ficoll, 2% (w/v) polyvinylpyrrolidone, 2% (w/v) bovine serum albumin, pH 7.2, and filter sterilized).
13. Sonicated Herring sperm DNA.
14. Molecular weight markers for agarose electrophoresis: 1-kb ladder, λ*HindIII*, and control genomic DNA samples containing alleles of known length.
15. Shaking water bath at 65°C.

16. Cling film.
17. Luminescent marking solution- Glo-bug X-ray marking solution (Radleys).

2.3. PCR

1. Oligonucleotide primers at 10× stock solution (5 pmol/μL).
2. PCR machine.
3. *Taq* polymerase and its reaction buffer (for long-range PCR, use specialized polymerase enzyme, such as Fermentas long PCR enzyme mix, Finnzymes DyNAzyme™ EXT DNA polymerase, from Thermo Scientific, or TaKaRa LA Taq from Lonza).
4. Deoxynucleotides (2 mM stock of each or a mix of each dNTP).
5. Agarose gels prepared using TBE (1–3% gel according to the size of the fragment).
6. Loading buffer: 0.25% (w/v) bromophenol blue, 0.25% (v/v) xylene cyanol, 15% (w/v) Ficoll.

2.4. Sequencing

1. ABI BigDye Terminator v3.1 Cycle Sequencing Kit (cat no. 4336917) (Applied Biosystems).
2. Cleanup solution (stock solution: 40% (w/v) PEG-8000, 1 M NaCl, 2 mM Tris–HCl (pH 7.5), 0.2 mM EDTA, 3.5 mM $MgCl_2$, working solution: 2 parts stock to 1 part water).
3. 5× SEQ buffer (400 mM Tris–HCl, pH9, 10 mM $MgCl_2$) or 5× Sequencing buffer supplied with BigDye Terminator v1.1 and v3.1 (kit, cat no. 4336697).
4. Between 20 and 100 ng of cleaned up PCR product.
5. DMSO.

2.5. Bioinformatics

UCSC Genome Browser Web site: http://genome.ucsc.edu/
ExPASy Proteomics Server: http://ca.expasy.org/
National Center for Biotechnology Information (NCBI): http://preview.ncbi.nlm.nih.gov/guide/

3. Methods

3.1. DNA Extraction

1. Prepare genomic DNA samples from whole blood or another convenient source using and following the instructions of the appropriate Puregene kit (see Note 11).
2. Quantify 1 μL of the DNA by using a Nanodrop or by measurement of the optical density at 260 nm after dilution (approx 1/100) and extensive mixing using a conventional spectrophotometer. For the latter, multiply by the dilution factor and the conversion factor of 50 to convert OD to micrograms per mL.

3. Check the integrity of the DNA by agarose electrophoresis of 1 μL of each sample plus 2 μL of loading buffer on small gels (0.8% (w/v) agarose gel in 1× TBE) in the presence of 50 ng/mL ethidium bromide, and inspection under ultraviolet (UV) light using a transilluminator (see Note 12).

3.2. Southern Blot Analysis

1. Treat 5–7 μg of DNA with the appropriate restriction enzymes (see Notes 4–7) in a final volume of 25 μL (with the buffer provided and as recommended by the manufacturer).
2. Check digestion of the DNA by electrophoresis of 3 μL of each sample plus 2 μL of loading buffer on small gels (0.8% in 1× TBE) in the presence of 50 ng/mL of ethidium bromide, and inspection under UV light.
3. For analysis of *MUC2* and *MUC5AC,* separate the *Hinfl* fragments (22 μL digest plus 7 μL of loading buffer) by electrophoresis using 0.8% (w/v) 20×25-cm agarose gels in 1× TBE, for 24 h at 2 V/cm.
4. For analysis of *MUC6,* separate the *PvuII* fragments (22 μL digest plus 7 μL of loading buffer) by electrophoresis using 0.5% (w/v) 20×25-cm agarose gels in IX TBE, at 2 V/cm for 24 h, followed by a complete change of the tank buffer and continued electrophoresis at 1.2 V/cm for a further 19 h.
5. Apply several kinds of markers to each gel: 1-kb ladder, λ *HindIII,* and DNA samples with alleles of known size.
6. Following electrophoresis, visualize the markers by post-staining with 0.4 mg/mL ethidium bromide in distilled water for 20 min (see Note 13).
7. Record the migration of the marker bands by making a photographic record, including a clear ruler aligned to the leading edge of the wells.
8. Depurinate the DNA with 0.25 M HCl for 30 min, with occasional gentle agitation.
9. Denature with 1.5 M NaCl and 0.5 M NaOH for 30 min, with occasional gentle agitation.
10. Neutralize with 0.5 M Tris–HCl, 1.5 M NaC'l, and 0.001 M EDTA, pH 7.2, for 30 min, with occasional gentle agitation (see Note 14).
11. Transfer the digested DNA onto Hybond N+ membranes by capillary blotting overnight or vacuum blotting for 2 h, both as recommended by the manufacturers, aligning the top of the membrane accurately.
12. Fix the DNA onto the filters by baking at 80°C for 2 h.
13. Detect the *MUC* genes using TR cDNA probes: SMUC41 for *MUC2* (31), JER58 for *MUC5AC* (32), and the cDNA reported in 33 for *MUC6,* and, when used, JER57 for *MUC5B*

(34). Label 25 ng by random primed labelling utilizing [α-^{32}P] dCTP and the Amersham Megaprime™ DNA Labeling System using the solutions and protocol provided (GE Healthcare).

14. Prehybridize the filters in a plastic box in 200 mL of 6× SCC, 5× Denhardt's, and 0.5% (w/v) SDS in a shaking water bath at 65°C (see Note 15).
15. After approx 4 h, prepare the hybridization solution. Add 500 μg of sonicated Herring sperm DNA to the labelled probe and boil for 5 min.
16. Add to the prehybridization solution and agitate the box to ensure that the probe is dispersed evenly.
17. Hybridize the filters overnight in the shaking water bath.
18. Wash the filters in several changes of SSC, with a final stringent wash of 0.1× SSC and 0.1% SDS at 65°C for 10 min.
19. Cover the wet filters with cling film, fix the filter into the cassette using tape, mark the filter position by using Glo-bug X-ray solution, and conduct autoradiography using X-ray film.
20. Determine the relative sizes of the fragments by plotting a standard curve using the control *MUC* alleles (detected after transfer by autoradiography) as well as the commercial size markers (see Note 16). Carefully transfer the position of the top of the filter onto the autoradiograph after development by using luminescent Glo-bug marks to reposition the autoradiograph in the cassette. Measure all distances from this start line.
21. For the allele length distribution studies, you can display the results in histogram form grouping the fragment size in 500-bp steps (see Note 17). For *MUC5AC*, report the variation as two-size classes as indicated, and "other" for unusual sizes (Fig. 1) (see Note 5).

3.3. Standard and Long-Range PCR

1. To each 2 μL DNA sample (2–10 ng of DNA), add the following PCR reagents: 1 μL of ABgene 10× buffer IV containing $MgCl_2$ [750 mM Tris–HCl (pH 8.8 at 25°C), 200 mM $(NH_4)_2SO_4$, 0.1% (v/v) Tween 20, 15 mM $MgCl_2$], 1 μL of each of dATP, dCTP, dGTP, dTTP, at 2 mM, 2.5 pmol of the forward primer, and 2.5 pmol of the reverse primer. Add distilled water to make a final reaction volume of 10 μL (see Note 18, and Subheading 2.3).
2. Initiate thermal cycling by denaturation at 95°C for 5 min, followed by cycling of 30 s at 95°C, 30 s at the optimal annealing temperature, and 1 min at 72°C or 0.5 kb/min at 70°C (see Note 19). Add a final elongation step of 72°C for 5 min to the end of the thermal program.
3. Visualize PCR products by agarose gel electrophoresis (1–3% gels as appropriate).

3.4. Single-Nucleotide Polymorphisms

Many commercial companies now provide a rapid high-quality sequencing service, but although this does save time, data analysis is still the most time-consuming step (see Note 20).

3.4.1. Sequencing

1. Purify template by adding 3× volume (30 μL) "cleanup" solution to each PCR reaction. Mix.
2. Centrifuge the PCR plate at 1,500 × *g* for 60 min.
3. Remove the lids, invert the plate, and place back in the centrifuge on a piece of tissue paper. Centrifuge at low speed (<20 × *g*) for 30 s (see Note 21).
4. Add 150 μL of 70% ethanol to each sample and centrifuge at 1,500 × *g* for 10 min (do not mix).
5. Remove the lids, invert the plate, and place back in the centrifuge on a piece of tissue paper. Centrifuge at low speed (<20 × *g*) and stop immediately the centrifuge reaches speed.
6. Dry the samples for 15 min at room temperature or 5 min at 65°C.
7. Add 10 μL of water to each sample and leave for 15 min to re-suspend.
8. Run 2 μL of this on an agarose gel in order to check for the presence of a product after cleaning.
9. Prepare enough sequencing reaction mix for the number of samples with 2.15 μL of 5× ABI or 5× HM-SEQ buffer, 0.35 μL of Big Dye v3.1, 1 μL of primer (see Note 22) at 1.6 μM, 4.25 μL of distilled water, and 0.25 μL of DMSO per sequencing reaction.
10. Mix with 2 μL of cleaned up PCR product [equivalent to 20–50 ng DNA (see Note 23)].
11. Initiate thermal cycling by denaturation at 95°C for 10 min, followed by 25 cycles of 45 s at 96°C, 30 s at 50°C, and 4 min at 60°C.
12. After cycling, centrifuge the plate at 100 × *g* for 1 min.
13. For each plate, prepare 290 μL 125 mM EDTA + 3,500 μL 100% ethanol in a dispensing trough.
14. Dispense 33 μL to each sample and as quickly as possible.
15. Mix by vortexing and centrifuge at 1,500 × *g* for 60 min.
16. Remove the lids, invert the plate, and centrifuge at low speed (<20 × *g*) for a few seconds.
17. Add 30 μL of 70% ethanol to each sample and centrifuge at 1,500 × *g* or 10 min. Do not mix.
18. Remove the lids, invert the plate, and centrifuge at low speed (<20 × *g*) for a few seconds.
19. Remove the lids and allow the samples to air dry for 15 min at room temperature or for 5 min at 65°C.

20. Add 6 μL HiDi loading buffer (Applied Biosystems).
21. Analyze sequencing reaction on one of the ABI applied Biosystems sequencing machines.

3.4.2. Restriction Fragment Length Polymorphism

1. Design oligonucleotide primers to obtain a fragment between 300 and 600 bp with the variable restriction site located approximately 1/3 along the fragment to result in two fragments of different length (see Note 24).
2. Amplify the genomic DNA as described above, and digest 3 μL of PCR product (in a total volume of 15 μL) with the appropriate restriction enzyme, following the manufacturer's instructions (see Note 25).
3. Separate the DNA fragments by agarose gel electrophoresis (see Subheading 3.3, step 3).

3.4.3. Allele-Specific Methods

There are a number of other methods of genotyping that depend on allele-specific reactions or hybridization that can be used "in-house" or commercially (see Note 26). This can range from designing the whole assay in-house or having an assay designed commercially but performed in-house or entirely performed commercially. One such example is the TaqMan technology (Applied Biosystems, Foster City, CA). TaqMan probes are designed by Applied Biosystems for the SNPs selected and polymerase chain reactions (PCRs) are performed in preferably 384-well microplates using a "real-time" PCR machine (see Note 27). Fluorescence is then measured using an Applied Biosystems® Real-Time PCR Systems and data analyzed using TaqMan® Genotyper Software. Other companies which provide SNP genotyping services are Illumina (http://www.illumina.com/) and K Biosciences (http://www.kbioscience.co.uk/).

An amenable allele-specific method that can be used in-house is TETRA ARMS (Amplification Refractory Mutation System (35, 36)), which can be entirely performed with a single standard PCR machine followed by agarose gel electrophoresis and without fluorescent dyes.

1. Design four primers, two "external," one forward, one reverse, to form a product of approximately 300–500 bp and two internal allele-specific primers, one forward and one reverse, situated on each side of the polymorphic site, with the appropriate mutations to obtain specificity (see (35, 36) and Note 28).
2. Use a Thermostart buffer (without magnesium chloride) with a Thermostart Taq and add magnesium chloride separately. Determine by titration the most effective magnesium chloride concentration for your assay (the standard recommended final concentration is 2.5 mM). Use the internal primers at a higher final concentration (1 μM) than the external primers (0.2 μM).

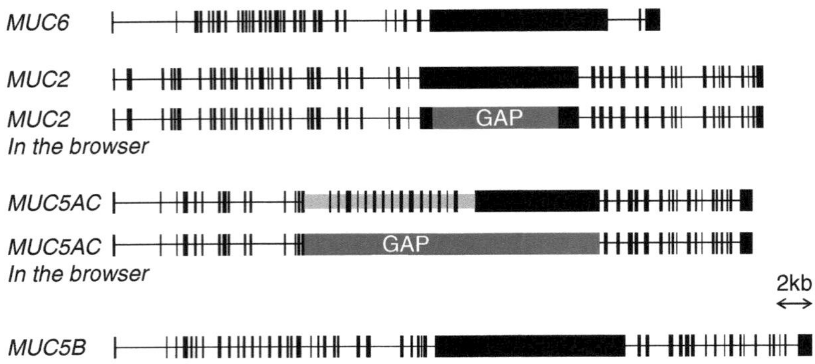

Fig. 2. Scaled representation of the four gel-forming mucin genes located on chromosome 11p15.5. The thin lines represent the introns and the boxes represent the exons. *MUC2* and *MUC5AC* are represented twice since their structure is incorrect in the human genome browser database. The grey boxes represent sequences missing in the human genome sequence. For *MUC5AC*, the complete genomic sequence is not known; therefore, the sequence and size of the intron from exon 15 to the tandem repeat are missing but the mRNA sequence is known and is shown in light grey. Considering the conservation of exon/intron boundaries between the genes, we can assume that at least 15 exons and 14 introns are missing in this region.

3.5. Bioinformatics

3.5.1. The Genome Browser (http://genome.ucsc.edu/)

1. Use the genome browser to find out background information about your gene of interest; reference sequences, working draft of several genomes, SNP data. In the current human genome database (GRCh37, February 2009) produced by the Genome Reference Consortium, the mucin genes are well-represented but are some errors and some missing domains (see Notes 29–33 and Fig. 2).
2. Find your gene using the genome browser by searching using a gene name from the genome page (http://genome.ucsc.edu/cgi-bin/hgGateway?org=Human&db=hg19&hgsid=168916865) or by using the program Blat supplied to submit a sequence (DNA, mRNA, or protein) (http://genome.ucsc.edu/cgi-bin/hgBlat?command=start).
3. Use the browser to scroll along as well zoom in and out of a chromosomal region.
4. Use the selection boxes below to select which track you wish to visualize and in which format, and click refresh. Tracks are organized by categories: mapping and sequencing tracks, phenotypes and disease association, gene and gene prediction tracks, mRNA and EST tracks, expression, regulation, comparative genomics, and variation and repeats. (Not all tracks are available for all genomes, since they are added as data

become available). Move the tracks into the order you wish to view them by clicking on the grey vertical bar and dragging and dropping into position.

5. Click on the link to each annotation to obtain detailed information.

3.5.2. Protein Database

Mucin protein sequences can be obtained from the Expasy Web site (http://ca.expasy.org/tools/dna.html) (see Note 34). In addition, Professor GC Hansson has compiled a useful database of mucin sequences on the following Web site: http://www.medkem.gu.se/mucinbiology/databases/.

3.6. Gel-Forming Mucin Genes and Disease Association

1. Select the approach according to the availability of patient material. If families are available, examine the pattern of inheritance and conduct linkage analysis (37), if parents and siblings are available conduct a transmission disequilibrium test (TDT test (38)) by comparing the numbers of transmitted and non-transmitted alleles in the affected and unaffected progeny; for unrelated patients and controls, design as case–control study or take advantage of an existing longitudinal cohort (see Note 35).
2. Target your study to known or putative functional variation, e.g.: tandem repeat length or use an SNP tagging approach or extensive re-sequencing (see Notes 36 and 37).

3.7. Mucin Genes in Other Species: In Silico Deduction from Whole-Genome Sequence Databases

Here, we describe a method for predicting the sequence of homologous mucin genes. This is limited to the non-TR domains since in most cases these central regions are absent or incomplete.

Predicted sequences do not replace experimentally obtained mRNA sequences but can be used to add to mass spectrometry databases to study conservation/evolution of sequences and to design oligonucleotide primers for sequencing and/or for real-time PCR quantification. Here, we take the example of predicting the sequence of the equine *Muc5b* gene.

1. From the UCSC genome gateway page (http://genome.ucsc.edu/cgi-bin/hgGateway), select the desired species, and under the position or search term option enter MUC5B and submit.
2. The resulting page shows that there are no equine mRNA identified for *MUC5B* but that human and mouse *Muc5b* mRNAs do align to the equine genome. Follow the link for the AF086604 sequence (longest 5′ human *MUC5B* sequence (39)).
3. This page shows how many times this sequence aligns in the equine genome (see Note 38). Follow the link to enter the Equine genome browser. Alternatively, a human MUC5B

sequence can be submitted to Blat on the appropriate page. There are two links per result, one to the browser and the other to the alignment. Click on the link to the browser. Select the tracks you would like to see from the menus below: e.g. non-horse Refseq genes.

4. Zoom out using the buttons labelled 3× or 10× to show other genes upstream and downstream of the alignment. In the case of *Muc5b*, a 10× zoom shows that the other MUC genes *Muc2* and *Muc5ac* can be found upstream of *Muc5b* and that the *tollip* gene is located downstream. This is in accordance to the gene order in the human genome.
5. Click on the human RefSeq annotation to view the alignment between the two species. On the genomic alignment, the potential equine exons corresponding to the human *MUC5B* mRNA are shown in capitals and blue. Copy and paste into a word document, and remove all paragraph marks, spaces, and numbers.
6. Translate the sequence covering each potential exon using the translation interface of the Expasy Web site and use the option to visualize the nucleic acid sequence (http://ca.expasy.org/tools/dna.html).
7. Identify the coding sequence using the high level of conservation of the location of the exon/intron boundaries and following the rules for end and start of intron sequences ag/gt. Figure 3 shows the amino acid sequence of human MUC2, MUC5AC, MUC6, and MUC5B mucins highlighting the position of the introns.
8. Inspect and check all the exons.
9. Copy and paste the entire sequence, delete all intronic sequences, and translate.
10. Submit this translation to the NCBI program Blast, which should result in the first hit being to the mucin of interest, in our case Muc5b. Figure 4 shows a small segment of the alignment of Muc2 amino acid sequence from several species predicted using the method described above.

a

```
MUC2    MGLP---------LARLAAVCLALSLAGGSELQTEGRTRYHGRN-----------------------------------
MUC5AC  MSVGRRKLALLWALA-LALACTRHTGHAQDGSSESSYKHHPALSPIARGPSGVPLRGATVFPSLRTIPVVRASNPAHNGR
MUC5B   MGAPSACRTLV-LALAAMLVVPQAETQGPVEPSWGNAGHTMDGGA    PTSSPTRRVSFVPPVTVFPSLSPLNPAHNGR
MUC6    MVQRWLLLSCCGALLSAGLANTSYTSPGLQRLKDSPQTAPDKG-----------------------------------

MUC2    VCSTWGNFHYKTFDGDVFRFPGLCDYNFASDCRGSYKEFAVHLKRGPGQAEAPAGVESILLTIKDDTIYLTRHLAVLNGA
MUC5AC  VCSTWGSFHYKTFDGDVFRFPGLCNYVFSEHCGAAYEDFNIQLRR--SQESAAPTLSRVLMKVDGVVIQLTKGSVLVNGH
MUC5B   VCSTWGDFHYKTFDGDVFRFPGLCNYVFSEHCRAAYEDFNVQLRRGLVG--SRPVVTRVVIKAQGLVLEASNGSVLINGQ
MUC6    QCSTWGAGHFSTFDHHVYDFSGTCNYIFAATCKDAFPSFSVQLRRGPDGSISRIIVELGASVVTVSEAII---SVKDIG-

MUC2    VVSTPHYSPGLLIEKSDAYTKVYSRAGLTL----MWNREDALMLELDTKFRNHTCGLCGDYNGLQSYSEFLS-DGVLFSP
MUC5AC  RVLLPFSQSGVLIQQSSSYTKVEARLGLVL----MWNHDDSLLLELDTKYANKTCGLCGDFNGMPVVRELLS-HNTKLTP
MUC5B   REELPYSRTGLLVEQSGDYIKVSIRLVLTF----LWNGEDSALLELDPKYANQTCGLCGDFNGLPAFNEFYA-HNARLTP
MUC6    -VISLPYTSNGLQITPFGQSVRLVAKQLELELEVVWGPDSHLMVLVERKYMGQMCGLCGNFDGKVT-NEFVSEEGKFLEP

MUC2    LEFGNMQKINQPDVVCEDPEEEVAPASCSEHRAECERLLTAEAFADCQDLVPLEPYLRACQQDRCRCPGGDT--CVCSTV
MUC5AC  MEFGNLQKMDDPTEQCQDPVPEPPR-NCSTGEGICEELLHGQLFSGCVALVDVGSYLEACRQDLCFCEDTDLLSCVCHTL
MUC5B   LQFGNLQKLDGPTEQCPDPLPLPAG-NCTDEEGICHRTLLGPAFAECHALVDSTAYLAACAQDLCRCPT-----CPCATF
MUC6    HKFAALQKLDDPGEICTFQDIPSTHVRQAQHARGCTQLLTLVAP-ECSVSKEPFVLS--CQADVAAAPQPGPQNSSCATL

MUC2    AEFSRQCSHAGGRPGNWRTATLCPKT-CPGNLVYLESGSPCMDTCSHLEVSSLCEEHRMDGCFCPEGTVYDDIGDS-GCV
MUC5AC  AEYSRQCTHAGGLPQDWRGPDFCPQK-CPNNMQYHECRSPCADTCSNQEHSRACEDHCVAGCFCPEGTVLDDIGQT-GCV
MUC5B   VEYSRQCAHAGGQPRNWRCPELCPRT-CPLNMQHQECGSPCTDTCSNPQRAQLCEDHCVDGCFCPPGTVLDDITHS-GCL
MUC6    SEYSRQCSMVGQPVRRWRSPGLCSVGQCPANQVYQECGSACVKTCSNSEHS--CSSSCTFGCFCPEGTDLNDLSNNHTCV

MUC2    PVSQCHCRLHGHLYTPGQEITNDCEQCVCNAGRWVCKDLPCPGTCALEGGSHITTFDGKTYTFHGDCYYVLAKGDHNDSY
MUC5AC  PVSKCACVYNGAAYAPGATYSTDCTNCTCSGGRWSCQEVPCPGTCSVLGGAHFSTFDGKQYTVHGDCSYVLTKPCDSSAF
MUC5B   PLGQCPCTHGGRTYSPGTSFNTTCSSCTCSGGLWQCQDLPCPGTCSVQGGAHISTYDEKLYDLHGDCSYVLSKKCADSSF
MUC6    PVTQCPCVLHGAMYAPGEVTIAACQTCRCTLGRWVCTERPCPGHCSLEGGSFVTTFDARPYRFHGTCTYILLQSPQLPED

MUC2    ALLGELAPCGSTDKQTCLKTVVLLADKKNAVVFKSDGSVLLNQLQV-NLPHVTASFSVFRPSSYHIMVSMAIGVRLQVQ
MUC5AC  TVLAELRRCGLTDSETCLKSVTLSLDGAQTVVVIKASGEVFLNQIYT-QLPISAANVTIFRPSTFFIIAQTSLGLQLNLQ
MUC5B   TVLAELRKCGLTDNENCLKAVTLSLDGGDTAIRVQADGGVFLNSIYT-QLPLSAANITLFTPSSFFIVVQTGLGLQLLVQ
MUC6    GALMAVYDKSGVSHSETS--LVAVVYLSRQDKIVISQDEVVTNNGEAKWLPYKTRNITVFRQTSTHLQMATSFGLELVVQ

MUC2    LAPVMQLFVTLDQASQGQVQGLCGNFNGLEGDDFKTASGLVEATGAG-FANTWKAQSTCHDKLDWLDDPCSLNIESANYA
MUC5AC  LVPTMQLFMQLAPKLRGQTCGLCGNFNSIQADDFRTLSGVVEATAAA-FFNTFKTQAACPNIRNSFEDPCSLSVENEKYA
MUC5B   LVPLMQVFVRLDPAHQGQMCGLCGNFNQNQADDFTALSGVVEAT-AH-FANTWKAQAACANSRNSFEDPCSLSVENENYA
MUC6    LRPIFQAYVTVGPQFRGQTRGLCGNFNGDTTDDFTTSMGIAEGT-ASLFVDSWRAGN-CPDALERETDPCSMSQLNKVCA

MUC2    -EHWCSLLKKTETPFGRCHSAVDPAEYYKRCKYDTCNCQNNEDCLCAALSSYARA-CTAKGVMLWGWREHV--CNKDVGS
MUC5AC  -QHWCSQLTDADGPFGRCHAAVKPGTYYSNCMFDTCNCERSEDCLVRRAVLLRARLCA-KGVQLGGWRDGV--CTKPMIT
MUC5B   -RHWCSRLTDPNSAFSRCHSIINPKPFHSNCMFDTCNCERSEDCLCAALSSYVHA-CAPKGVQLSDWRDGV--CTKYMQN
MUC6    ETH-CSMLLRTGTVFERCHATVNPAPIYKRCMYQACNYEETFPHICAALGDYVHA-CSLRGVLLWGWRSSVDNCTIP---

MUC2    CPNSQVFLYNLTTCQQTCRSLSEADSHCLEGFAPVDGCGCPDHTFLDEKGRCVPLAKCSCYHRGLYLEAGDVVVRQEER-
MUC5AC  CPKSMTYHYHVSACQPTCRSLSEGDITCSVGFIPVDGCICPKGTFLDDTGKCVQASNCPCYHRGSMIPNGESVHDSGAI-
MUC5B   CPKSQPYAYVVDACQPTCRGLSEADVTCSVSFVPVDGCTCPAGTFLNDAGACVPAQECPCYAHGTVLAPGEVVHDEGAV-
MUC6    CTGNTTFSYNSQACERTCLSLSDRATECHHSAVPVDGCNCPDGTYLNQKGECVRKAQCPCILEGYKFILAEQSTVINGIT

MUC2    CVCRDGRLHCRQIRLIGQ-SCTAPKIHMDCSNLTALATSKPRALSCQTLAAG--YYHTECVSGCVCPDGLMDDGRGGCVV
MUC5AC  CTCTHGKLSCIGGQAPAP-VCAAPMVFFDCRNATPRGTGAGCQKSCHTLDMT--CYSPQCVPGCVCPDGLVADGEGGCIT
MUC5B   CSCTGGKLSCLGASLQKSTGCAAPMVYLDCSNSSAGTPGAECLRSCHTLDVG--CFSTHCVSGCVCPPGLVSDGSGGCIA
MUC6    CHCINGRLSCPQRLQMFLASCQAPKTFKSCSQSSENKFGAACAPTCQMLATGVACVPTKCEPGCVCAEGLYENAYGQCVP

MUC2    EKECPCVHNNDLYSSGAKIKVDCNTCTCKRGRWVCTQA-VCHGTCSIYGSGHYITFDGKYYDFDGHCSYVAVQDYCGQNS
MUC5AC  AEDCPCVHNKASYRAGQTIRVGCNTCTCDSRMWRCTDD-PCLATCAVYGDGHYLTFDGQSYSFNGDCEYTLVQNHCGGKD
MUC5B   EEDCPCVHNEATYKPGETIRVDCNTCTCRNRRWECSHG-LCLGTCVAYGDGHFITFDGDRYSFEGSCEYILAQDYCGDNT
MUC6    PEECPCEFSGVSYPGGAELHTDCRTCSCSRGRWACQQGTHCPSTCTLYGEGHVITFDGQRFVFDGNCEYILATDVCGVNY

MUC2    S-LGSFSIITENVPCGTTGVTCSKAIKIFMGRTELKLED-KHRVVIQRDEGHHVAYTTREVGQYLVVESST-G---IIVI
MUC5AC  STQDSFRVVTENVPCGTTGTTCSKAIKIFLGGFELKLSHRK-VEVIGTDESQEVPYTIRQMGIYLVVDTDI-G---LVLL
MUC5B   T-HGTFRIVTENIPCGTTGTTCSKAIKLFVESYELILQEGTFKAVARGPGGDPPYKIRYMGIFLVIE-TH--G---MAVS
MUC6    SQPT-FKILTENVICGNSGVTCSRAIKIFLGGLSVVLADRNYTVTGEEPHVQLGVTPGALS---LVVDISIPGRYNLTLI

MUC2    WDKRTTVFIKLAPSYKGTVCGLCGNFDHRSNNDFTTRDHMVVSSELDFGNSWKEAPTCPDVSTNPEPCSLNPHRRSWAEK
MUC5AC  WDKKTSIFINLSPEFKGRVCGLCGNFDDIAVNDFATRSRSVVGDVLEFGNSWKLSPSCPDALAPKDPCTANPFRKSWAQK
MUC5B   WDRKTSVFIRLHQDYKGRVCGLCGNFDDNAINDFATRSRSVVGDALEFGNSWKLSPSCPDALAPKDPCTANPFRKSWAQK
MUC6    WNRHMTILIRIARASQDPLCGLCGNFNGNMKDDFETRSRYVASSELELVNSWKESPLCGDVSFVTDPCSLNAFRRSWAER

MUC2    QCSILKSSVFSICHSKVDPKPFYEACVHDSCSCDTGGDCECFCSAVASYAQECTKEGACVFWRTPDLCPIFCDYYNPPHE
MUC5AC  QCSILHGPTFAACHAHVEPARYYEACVNDACACDSGGDCECFCTAVARYAQACHEVGTCVCLRTPSICPLFCDYYNPEGQ
MUC5B   QCSILHGPTFAACRSQVDSTKYYEACVNDACACDSGGDCECFCTAVAAYAQACHDAGLCVCWRTPDTCPLFCDFYNPHGG
MUC6    KCSVINSQTFATCHSKVYHLPYYEACVRDACGCDSGGDCECLCDAVAAYAQACLDKGVCVDWRTPAFCPIYCGFYNTHTQ

MUC2    -------------CEWHYEPCGNRSFETCRTINGIHSNISVSYLEGCYPRCPKDRPIYEEDLKKCVTADKCG-------C
MUC5AC  -------------CEWHYQPCGVPCLRTCRNPRG-DCLRDVRGLEGCYPKCPPEAPIFDEDKMQCVAT--CPTPPLPPRC
MUC5B   -------------CEWHYQPCGAPCLKTCRNPSG-HCLVDLPGLEGCYPKCPPSQPFFNEDQMKCVAQ--CG-------C
MUC6    DGHGEYQYTQEANCTWHYQPC----L--CPSQPQSVPGSNI---EGCYN-CSQDEYFDHEEGV-CVP---CMPPTTPQPP

MUC2    YVED-THYPPGASVPTEETCKSCVCTNSSQVVCRPEEG-K---------------ILNQTQDGAF-CYWEICGPNGTVEK
MUC5AC  HVHG-KSYRPGAVVPSDKNCQSCLCTERGVE-CTYKAE-ACVCTYNGQRFHPGDVIYHTT-DGTGGCISARCGANGTIER
MUC5B   YDKDGNYYDVGARVPTAENCQSCNCTPSGIQ-CAHSLE-ACTCTYEDRTYSYQDVIYNTT-DGLGACLIAICGTNGTIIR
MUC6    TTPQLPTTGSRPTQVWPMTGTSTTIGLLSSTGPSPSSNHTPASPTQTPLLPATLTSSKPTASSGEPPRPTTAVTPQATSG

MUC2    HFNICSITTR-PSTLTTFTTITLPTTPTSFTTTTT...
MUC5AC  RVYPCSPTTPVPPTTFSFSTPPLVVSTHTPSNGP...
MUC5B   KAVACPGTPATTPFTFTTAWVPHSTTSPALPVST...
MUC6    LPPTATLRSTATKPTVTQATTRATASTASPATTSTAQSTTRTTMTLPTPATSGTSPTLP..
```

Fig. 3. Alignment of protein sequence of MUC2, MUC5B, MUC5AC, and MUC6. The triangles represent the exon/intron boundaries and cysteine residues are shown in *grey*. The dashes represent spaces inserted to fit the alignment. (**a**) alignment of the sequences upstream of the tandem repeat domain, (**b**) alignment of the sequences downstream of the tandem repeat domain.

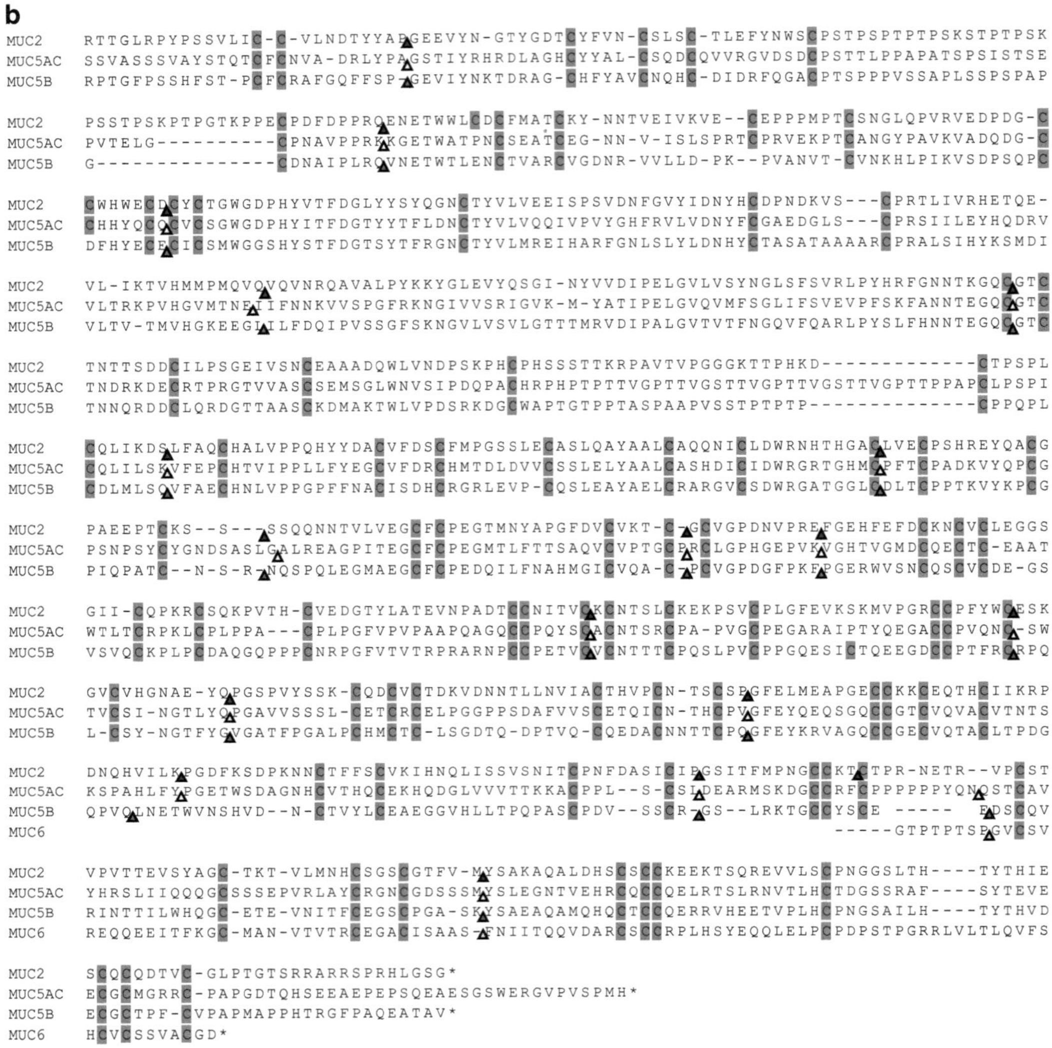

Fig. 3. (continued)

```
Human       MGLPLARLAAVCLALSLAGGSELQTEGRTRYHGR -NVCSTWGNFHYKTFDGDVFRFPGLCDYNFASDCRGSYKEFAVHLK
Mouse       MGLPLARLVAACLVLALAKGSELQKEARSRNH-----VCSTWGDFHYKTFDGDVYRFPGLCDYNFASDCRDSYKEFAVHLK
Chimp       MGLPLARLAAVCLALSLAGGSELQTEGRTRNHGH--NVCSTWGDFHYKTFDGDVFRFPGLCDYNFASDCRGSYKEFAVHLK
Orangutan   MGLPLARLAAVCLALSLAGGSELQTEGRTRNHGH--NVCSTWGDFHYKTFDGDVFRFPGLCDYNFASDCRGSYKEFAVHLK
Rhesus      MGLPLAPLAAVCLALSVAGGLELQTEGRARNHGH--NVCSTWGDFHYKTFDGDIFRFPGLCDYNFASDCRGSYKEFAVHLK
Marmoset    MGLPLARLAAVCLALSLARGSELQTEGRTRDHGH--NVCSAWGDFHYKTFDGDVFRFPGLCDYNFASDCRGSYKEFAVHLK
Guinea pig  MGLPLARLVATCLALASAGGLELQRV------GH--NVCSTWGNFHYKTFDGDIFQFPGLCNYNFASDCRDSYKEFAVHLK
Panda       MGLPLARLVAVCLALSSAGGAELQREGRSRNHGH--SVCSTWGDFHYKTFDGDVFRFPGLCDYNFASDCRDSYKEFAVHLK
Dog         MGLPLARLVAICLVLSSARGAELQREGRTRNHGH--SVCSTWGDFHYKTFDGDVFRFPGLCDYNFASDCRDSYKEFAVHLK
Elephant    MGLPLARLVAICLALSSAGGTQLQREKKGRTQNHGHTVCSTWGDGHYKTFDGDIYQFRGLCDYNFASDCRDSYKEFAVHLK

Human       RGPGQAEAPAGVESILLTIKDDTIYLTRHLAVLNGAVVSTPHYSPGLLIEKSDAYTKVYSRAGLTLMWNREDALMLELDTK
Mouse       RGLGEAGGHSQIESILITIKDDTIYLTHKLAVVNGAMVSTPHYSSGLLIEKNDAYTKVYSRAGLSLMWNREDALMVELDSR
Chimp       RGPGQAEAPAGVESILLTIKDDTIYLTRHLAVLNGAMVSTPHYSPGLLIEKSDAYTKVYSRAGLTLMWNREDALMLELDTK
Orangutan   RGPGQAEAPAGVESILLTIKDDTIYLTRHLAVLNGAMVSTPHYSPGLLIEKSDAYTKVYSRAGLTLMWNREDALMLELDSK
Rhesus      RGPGQAGAPAGVKSILLTIKDDTIYLTRHLAVLNGAMVSTPHYSPGLLIEKSDAYTKVYSRAGLTLLWNREDALMLELDSK
Marmoset    RGLDQAGATPQVESILLTIKDDTIYLTRHLAVLNGAVVSTPHYSPGLLIEKSDAYTKVYSRAGLALLWNREDAIMXXXXXX
Guinea pig  RGLGQSGSHPQVESVLLTIKDDIIYLTRQLTVVNGAMVSSPYYSSGLSIIKSEAYTKVYSRAGLLLLWNREDALMLELDSR
Panda       RSSGSDGGPPQLEYILLNIKDDTVYLTRQLAVVNGATVSTPHYSSGLLIEKSDAYTKVYSRAGLTLMWNREDSLMLELDSK
Dog         RGAGRDGKHPQLEYILLSIKDDTIYLTPQLVVVNGALVSTPHYSSGLLIEKNDAYTKVYSRAGLTLMWNREDSLMLELDSR
Elephant    RSLGHSRGQPELEYILLTIGDDTVYLTKQLTVVNGALVSTPHYSSGLLIEKNDVYTKVYSRAGLSLIWNREDSLMLELDSK

Human       FRNHTCGLCGDYNGLQSYSEFLS-DGVLFSPLEFGNMQKINQPDVVCEDPEEEVAPASCSEHRAECERLLTAEAFADCQDL
Mouse       FQNHTCGLCGDFNGMQTNYEFLSEEGIQFSAIEFGNMQKINKPEVQCEDPEAVQEPESCSEHRAECERLLTSAAFEDCQTR
Chimp       FRNHTCGLCGDYNGLQSYSEFLS-DGVLFSPLEFGNMQKINQPDVVCEDPEEEVAPASCSEHRAECERLLTAEAFADCQDL
Orangutan   FRNHTCGLCGDYNGLQSYSEFLS-DGVLFSPLEFGNMQKINQPDVVCEDPEEEAVPASCSEHRAECERLLTAEAFEDCQDL
Rhesus      FRNHTCGLCGDYNGLQSYSEFLS-DGVFFSPLEFGNMQKINEPDVVCEDPEETVALASCSEHRAECERLLTAEAFVDCQDL
Marmoset    XXXXXXXXXXXXXXXXXXXXXXXXXXXXXXXXXXXXXXXXXXXXXXXXXXXXXXXXXAASCAEHRAECERLLTAEAFADCPDR
Guinea pig  FQNHTCGLCGDYNGLQTYSEFLS-EGVTFSAIEFGNMQKINKPETECEDPQEAEEPESCSEHRAECESLLTVAAFEDCRDR
Panda       FQNHTCGLCGDYNGLQTYSEFLS-DGVLFSALEFGNMQKIDKPEVVCNDPEEVPALESCSEHRAECEKLLTAAAFADCLGL
Dog         FQNHTCGLCGDYNGLQTYSEFLS-DGVLFSAIEFGTMQKINKPGVVCDDPEEEPALESCSEHRAECERLLTASAFKDCLGL
Elephant    FRNHTCGLCGDFNGLQTYSEFLT-DGILFSPTQFGNMQKVNKPEVVCEDPEEEVVPATCDQHRAECERLLTAEAFSDCLER

Human       VPLEPYLRACQQDRCRCPGGD--TCVCSTVAEFSRQCSHAGGRPGNWRTATLCPKTCPGNLVYLESGSPCMDTCSHLEVSS
Mouse       VPVESYVRACMHDRCQCPKGG--ACECSTLAEFSRQCSHAGGRPENWRTASLCPKKCPNNMVYLESSSPCVDTCSHLEVSS
Chimp       VPLEPYLRACQQDRCRCPGGD--TCVCSTVAEFSRQCSHAGGRPGNWRTATLCPKTCPRNLVYLESGSPCMDTCSHLEVSS
Orangutan   VPLEPYLRACQQDRCRCPGGD--TCICSTVAEFSRQCSHAGGRPGNWRTATLCPKTCPGNLVYLESGSPCMDTCSHLEVSS
Rhesus      VPLEPYLRACQQDRCRCPGGD--TCICSTVAEFSRQCSHAGGRPGNWRTATFCPKTCPGNLVYLESGSPCMDTCSHLEVSS
Marmoset    VPLEPYLQACQQDRCRCSGGD--ACVCSTVAEFSRQCSHAGGRPRNWRTASLCPKTCPGNMVYLESGSPCMDTCSHLQVSS
Guinea pig  IPLELYVDACVRDRCQCTAAD--SCVCNSVAEYSRQCSHAGGRPGNWRTASFCAKSCPGNMVYLESSSPCMDTCSHLGVSH
Panda       VPLELYVQACARDRCQCPSGT--SCVCSTLAEFSRQCSHAGGRPGNWRTSELCPKSCPGNMVYLESGSPCMDTCSHLEVSS
Dog         LPLELYVQACAQDRCRCPVGT--SCACSTLAEFSRQCSHAGGQPGNWRTAELCPKSCPGNMVYLESGSPCMDTCSHLEVSS
Elephant    VPLELYVQACMEDRCRCAEGNGTACLCSTVAEFSRQCSHAGGRPRSWRSASLCSKSCPGNMVYLESGSPCVDTCSHLELSS

Human       LCEEHRMDGCFCPEGTVYDDIGDSGCVPVSQCHCRLHGHLYTPGQEITNDCEQCVCNAGRWVCKDLPCPGTCALEGGSHIT
Mouse       LCEEHYMDGCFCPEGTVYDDITGSGCIPVSQCHCKLHGHLYMPGQEFTNDCEQCVCNAGRWVCKDLPCPETCALEGGSHIT
Chimp       LCEEHRMDGCFCPEGTVYDDIGDSGCVPVSQCHCRLHGHLYTPGQEITNDCEQCVCNAGRWVCKDLPCPGTCALEGGSHIT
Orangutan   LCEEHRMDGCFCPEGTVYDDIGDSGCVPVSQCHCRLHGHLYTPGQEITNDCEQCVCNAGRWVCKDLPCPGTCALEGGSHIT
Rhesus      LCEEHRMDGCFCPEGTVYDDIGGSGCIPVSQCHCRLHGHLYTPGQEITNGCEQCVCNAGRWVCKDLPCPGTCALEGGSHIT
Marmoset    LCEEHRMDGCFCPEGTVYDDIAGGGCVPVSQCHCRLHGRLYTPGQGITSDCEQCICNDGRWVCKDLPCPGTCTLEGGSHIT
Guinea pig  LCEEHYMDGCFCPEGTVYDDITSSGCVPVSQCHCKLHGHLYAPGQEVTNECEQCVCNAGRWECQDLPCPGACALEGGAHIT
Panda       LCEEHRMDGCFCPEGTVYDDIAGSSCIPVSQCHCKLHGSLYSPGQQISSDCEECVCTAGRWVCKDLVCPGTCALEGGSHIT
Dog         LCEEHHMDGCFCPEGTVYDDITGSGCIPVSQCHCRLHGHLYSPGQQITNEC-EDVCTAGRWVCKDLPCPGTCSLEGGSHIT
Elephant    LCEEHSMDGCFCPEGTVYDDIAGSGCIPVSQCHCKLHGHLYSPGQQVTNQCEQCTCNSSRWVCKDLPCPGTCALEGGSHIT

Human       TFDGKTYTFHGDCYYVLAKGDHNDSYALLGELAPCGSTDKQTCLKTVVLLADKKKNAVVFKSDGSVLLNQLQVNLPHVTAS
Mouse       TFDGKKFTFHGDCYYVLTKSEHNDSYALLGELASCGSTDKQTCLKTVVLLTDDKKNVVAFKSGGSVLLNEMEVTLPHVAAS
Chimp       TFDGKTYTFHGDCYYVLAKGDHNDSYALLGELAPCGSTDKQTCLKTVVLLTDKKKNVVVFKSDGSVLLNELQVNLPHVTAS
Orangutan   TFDGKTYTFHGDCYYVLAKGDH-DSYALLGELAPCGSTDKQTCLKTVVLLADKKKNVVVFKSDGSVLLNELQVNLPHVTAS
Rhesus      TFDGKTYTFHGDCYYVLAKGDHNDSYALLGELAPCGSTDKQTCLKTVVLLADKKKNVVVFKSDGSVLLNELQVNLPHVTAS
Marmoset    TFDGKKFTFHGDCYYVLAKXXXXXXXXXXXXXXXXXXXXXXXXXXXXXXXXXXXXXXVVFKSDGSVLLNELQVNLPHVTAS
Guinea pig  TFDGKRFTFHGDCYYVLTKGDHNNSYSLLGELGPCGATDKQTCLKTVVLQMDNRKNVVAFKSDGSVLLNEMQVTLPHVAAS
Panda       TFDGKKYTFHGDCYYVLMQGDHNDSYALLGELAPCGSTDKQTCLKTVVLLADNKKNVVAFKSDGSVLLNELQVNLPHVTAS
Dog         TFDGKKYTFHGDCYYVLIKGDHNDSYALLGELAPCGSTDKQTCLKTVVLLADNRKNVLAFKSDGSVLLNELEVNLPHVTAS
Elephant    TFDGKKFTFHGDCYYILAKSDHNDSYTLLGELAPCGSTDKQTCLKTVVLLTDKKKNVVAFKSDGSVFLNELEVNLPHVTAS

Human       FSVFRPSSYHIMVSMAIGVRLQVQLAPVMQLFVTLDQASQGQVQGLCGNFNGLEGDDFKTASGLVEATGAGFANTWKAQST
Mouse       FSIFQPSSYHIVVNTKFGLRLQIQLLPVMQLFVTLDQAAQGQVQGLCGNFNGLESDDFMTSGGMVEATGAGFANTWKAQSS
Chimp       FSVFRPSSYHIMVSMAIGVRLQVQLAPVMQLFVTLDQASQGQVQGLCGNFNGLEGDDFKTASGLVEATGAGFANTWKAQSS
Orangutan   FSVFRPSSYHIVVSMAIGVRLQVQLAPVMQLFVTLDQASQGQVQGLCGNFNGLEGDDFKTASGLVEATGAGFANTWKAQSS
Rhesus      FSVFRPSSYHIIVSMASGLRLQVQLAPVMQLFLTLDQASQGQVQGLCGNFNGLEGDDFKTASGLVEATGAGFANTWKAQSS
Marmoset    FSIFRPSSYHIVVSVAVGLRLQVQLAPVMQLFLTLDQAAQGQVQGLCGNFNGLEGDDFKTASGLVEATGASFANSWKAQSS
Guinea pig  FSIFRPSSYHIVVSTVLGLRLQVQLEPIMQLFVTLDPSAQGQVQGLCGNFNGLETDDFETSSGLVEATGASFANSWKAQDS
Panda       FSIFQPSSYHLLVSTAFGLSLQVQLVPVMQLFVTLGQAAQGHVQGLCGNFNGLEGDDFKTAGGLVEATGAGFANTWKAQSS
Dog         FSVFQPSSYHLMVNTAFGLRLQVQLLPVMQLFLTLDQVAQGHVQGLCGNFNGQEGDDFKTPGGLVEATGASFANSWKAQSN
Elephant    FSIFRPSSYHIIVSTDLGLRLQVQLDPAMQLFATLDQGAQGQVQGLCGNFNGLEGDDFKTASGLVEATGAGFANTWKAQSS
```

Fig. 4. Partial alignment of MUC2 sequences from different species. The *dashes* represent spaces inserted to fit the alignment, crosses represent region of missing sequence in the genome. The *triangles* represent the location of the exon/intron boundaries. Alignment is shown for the first 562 amino acid residues of human MUC2.

4. Notes

1. The 18 MUC genes are *MUC1*, *MUC2*, *MUC3A*, *MUC3B*, *MUC4*, *MUC5AC*, *MUC5B*, *MUC6*, *MUC7*, *MUC8*, *MUC12*, *MUC13*, *MUC15*, *MUC16*, *MUC17*, *MUC19*, *MUC20*, and *MUC21*.
2. This kind of variant has been misinterpreted in dbSNP, since inter-repeat differences are confused with inter-allelic differences, e.g. rs72842456 in *MUC2*.
3. Two distinct mutations in the mouse *Muc2* gene, obtained by *N*-ethyl-*N*-nitrosourea mutagenesis, were found to have a profound effect on the MUC2 mucin (40). The mutations are located in the 5′ and 3′ cysteine-rich domains of *Muc2* leading to improper processing of Muc2 which failed to be secreted and resulted in the accumulation of Muc2 precursor (the non-glycosylated form) in the ER, causing ER stress. Mice with this mutation had a phenotype similar to ulcerative colitis in humans.
4. *MUC2* shows two TR domains, the larger one containing conserved 69-bp repeats and upstream from that a smaller one with poorly conserved 48-bp repeats. Although polymorphism in *MUC2* can be detected with a large number of restriction enzymes (8, 41), electrophoresis of *HinfI-digested* DNA (Fig. 1) reveals more than 12 distinct alleles (size range: 3.3–11.4 kb in the UK population tested; heterozygosity: 0.59).
5. *MUC5AC* is also highly polymorphic and polymorphism can be readily detected with a variety of enzymes (42, 43). Evidence of VNTR variation comes from the correspondence of patterns observed with several restriction enzymes. With *HinfI* and *PstI*, band sizes largely fall into two major classes [(a): *HinfI* 6.6 kb and *PstI* 8.4 kb; (b): *HinfI* 7.4 kb and *PstI* 9.0 kb]. The allele frequencies found in the UK population are a = 0.77 and b = 0.22, and rare allele 0.01 within 2,703 individuals.
6. Digestion of genomic DNA with *PvuII* reveals a very clear length polymorphism with the *MUC6* probe (43). A simple pattern of bands is observed with this enzyme, composed of one or two large bands in each individual, owing to 11 or more distinct alleles, ranging in size from 8 to 13.5 kb. The frequency distribution of these alleles is approximately unimodal, with a peak at about 10 kb. A heterozygosity of 0.70 was obtained in our previous studies for the unrelated chromosomes from the Centre d'Etude du Polymorphisme Humain (CEPH) families (43).
7. *MUC5B* contrasts with the other mucins in showing little variation (43). Multiple bands are detected in DNA digested with

several enzymes (e.g. *Mspl, Pstl,* and *Taql).* Relatively infrequent variant patterns involving the presence or absence of one or more small bands were detected with *PstI* and *TaqI.* A single large band is detected with *EcoRI* (27 kb) and with *HindIII* (25 kb). In most individuals (52/54), a single large band (16.5 kb) is detected in DNA digested with *BglII,* which cuts immediately outside the TR domain, but two individuals showed an additional band (19.5 or 15.5 kb). With *EcoRI,* these two individuals both showed the common phenotype of a single 27-kb band, suggesting that the variant phenotypes are owing to nucleotide changes within *BglII* sites rather than numbers of TR.

8. 10× TBE stock comes out of solution when cold.
9. Alternatives to ethidium bromide, which is said to be mutagenic, such as Safewhite (NBS Biologicals), are advertised as safer, but are less sensitive and more expensive. See also http://rrresearch.blogspot.com/2006/10/heresy-about-ethidium-bromide.html.
10. SSC for hybridization should be autoclaved to prevent bacterial growth.
11. Use blood or cultured cells for high-quality, long, fragment DNA.
12. This also provides a good indication of the concentration of the DNA.
13. Ethidium bromide should not be included in the gel because it distorts the electrophoretic separation and mobilities, particularly of *MUC2.*
14. Steps 8–10 are only for passive blotting. For vacuum blotting, the same solutions are used, but as recommended by the manufacturer of the vacuum blotter.
15. Several filters can be probed together, with a blank filter layered on top.
16. The molecular size markers are often visible after exposure since they are revealed with the *MUC* probes which conveniently bind non-specifically.
17. Slight gel-to-gel variations means that it is not easy to size the bands more accurately despite the use of size markers. Analysis of individual gels makes it apparent that several alleles exist within each size range, with the exception of *MUC6* (with a repeat unit of 507 bp), but it is not practicable to rerun large numbers of samples in different combinations to assign the alleles more precisely, and, in any case, alleles that differ by a single repeat unit are unlikely to be separated on these gels.
18. For long PCR to cover the entire TR region: Sets of primers that have been successfully used are MUC1 (12): forward

5′-AAGGAGACTTCGGCTACCCAG-3′, reverse 5′-TGTGCACCAGAGTAGAAGCTGA-3′; MUC2 (8): forward 5′-TGCCTCAACTACGAGATCAAC-3′, reverse 5′-ATTGGATGTGGTCAACTCAGC-3′; MUC5AC (R. Burgess and D Swallow, unpublished): forward, 5′-CGGTGACTTCGACACACTGGAGAAC-3′, reverse: 5′-GCAGAAGCAGGTTTGGGTGGAGTAAG-3′.

19. The annealing temperature is estimated to be a few degrees below the melting point of the oligonucleotide primers.
20. Check all sequencing files on receipt and make clear arrangements with the company if you wish it to retain samples of your DNA for future experiments.
21. Do not centrifuge the plate faster than 20 × *g* as this may result in the loss of the PCR product.
22. Because of the repetitive nature of the VNTR domain, if sequencing is to be attempted in or in the vicinity of this region, care has to be taken that the primer is not repeated several times in the PCR product to avoid annealing in several places.
23. Less DNA is required for small fragments.
24. If possible, the inclusion of a second non-polymorphic restriction site for the same enzyme provides an internal positive control.
25. It is recommended to use no more than 2.5 μL of PCR product per 10 μL digest reaction because of the risk of incompatibility of the Taq polymerase buffer with the conditions required for the restriction enzyme.
26. Many commercial companies recommend whole-genome amplification of DNA; however, our experience is that this technique, even with the most up-to-date kit, results in some allelic loss and this is likely to be more of a problem with the *MUC* genes than other genes because of the presence of the repetitive sequences.
27. For polymorphisms with a rare allele frequency, it is recommended to use as large a sample set as possible and also 384-well plates since genotype calling is more robust when there are bigger clusters of heterozygotes and if there are also homozygotes on the same plate.
28. The ARMS PCR method takes advantage of the observation that mismatch mutation at the 3′ end of oligonucleotide primer leads to failure of amplification, making it possible to design allele-specific primers.
29. *MUC6* is the most telomeric of the four mucin genes on chromosome 11, being located at chr11:1012824-1036706 on the reverse strand. The Refseq ID for *MUC6* is NM_005961. The central region of *MUC6* consists of 4.5 poorly conserved repeats in the sequence of the AC139749.4 clone; however, it

is likely that this represents a condensed version of the real *MUC6* TR sequence since Southern blot analyses have shown that in the individuals tested this region contains at least 15 repeat units of 507 bp each (44).

30. *MUC2* is located at chr11:1,074,875-1,104,416 on the forward strand and the Refseq ID is NM_002457.2. Sequencing of intestinal *MUC2* mRNA and inference from Southern blots showed that the central exon contains a region coding for a cysteine-rich domain followed by a region coding a non-variable domain rich in threonine and serine, another region coding for a cysteine-rich domain followed by a large region coding for a second threonine and serine-rich domain (45). The latter domain is polymorphic in size and alleles have been shown to range from 5 to 15 kb (43). This central region of the *MUC2* gene is condensed in the sequence of the AC139749.4 clone. The sequence from the first repetitive region is merged with the end of the second such that the region coding for the second cysteine-rich domain is deleted and the main tandem repeat region consists of only 4 repeats while the smallest allele is has been reported to contain at least 40 repeats (43).
31. *MUC5AC* is located just downstream of *MUC2*. Although most of the *MUC5AC* gene has been sequenced and deposited in databases, the human genome sequencing project still contains a large gap in the region of this gene. The genomic sequence encompasses the first 15 exons of *MUC5AC*, but this is followed by a gap, which should contain the genomic sequence for the next 15 exons and most of the large central exon. The sequence directly downstream of this gap corresponds to the end of the tandem repeat region of *MUC5AC* and is continuous until the end of the gene. The presence of this gap in the genomic sequence and the similarity of *MUC5AC* and *MUC5B* results in errors in annotation and confusion of the Refseq and Uniprot sequences. Importantly, the protein sequences A7Y9J9 and NIEHS 4586 are erroneous composites of MUC5AC and MUC5B.
32. *MUC5B* is the fourth gene of the *MUC* gene complex located at chromosome11p15.5. *MUC5B* is located at chr 11:1,244,296-1,283,406 and is contained in clone AC061979.17. Several groups have determined various partial sequences of the *MUC5B* gene. The Refseq sequence (NM_002458.2) represents several sequences combined and is the closest to full-length *MUC5B* sequence.
33. MUC19 is the fifth of the gel-forming secreted mucins. One mRNA sequence has been reported for this gene AY236870 (4), and recently the same research group characterized the *MUC19* gene by additional sequencing (46). The longest sequence put together has been submitted to Genbank as

Table 3
cDNA and protein accession numbers for the five gel-forming mucins

Gene	Refseq sequence	UniprotK ID
MUC6	NM_005961	Q6W4X9
MUC2	NM_002457.2	Q02817
MUC5AC	None	P98088*
MUC5B	NM_002458.2	Q9HC84

*At the time of writing, P98088 is the most complete correct MUC5AC sequence (10), though it is not complete because these authors did not sequence through the entire TR. No complete MUC19 Refseq sequence is available yet (unlike stated, sequence Q7Z5P9 is not complete and does not include the latest reported sequence from Zhu et al. (46))

HM801863. Interestingly, they have also found that approximately 7.5 kb of genomic DNA, coding for 11 exons (753bp), were missing in the human genome assembly (46).

34. The accession numbers for the most complete and accurate sequences of *MUC2, MUC6, MUC5B, MUC5AC*, and *MUC19* are shown in Table 3. The protein sequence A7Y9J9, if this remains on the databases at the time of publication of this article, should not be used since it corresponds to a mix of MUC5B and MUC5AC with the first 549 amino acid identical to MUC5AC while the rest seems to be mostly from the MUC5B sequence with regions identical to MUC5AC.

35. Whichever approach is selected for a disease association study, the ancestry of the individuals should be uniform, and in case–control studies age, sex, and environmental factors should be matched as far as possible to avoid confounding effects. TDTs suffer less from these confounders because of shared family background. A good account of these points can be found in 38. Consideration of power is also of paramount importance. This depends on the allele frequency and effect size. A case–control study of 50 cases and 50 controls would in an allele count test have >80% power to detect a twofold difference in allele frequency if the overall minor allele frequency is about 0.2, but if a greater significance level is required to take into account the problem of multiple testing many more samples are required. Most of positive associations in Table 2 can be criticized in this way, and may be false positives (type I error), or may fail to replicate because of small sample size in the original study.

36. HapMap (http://hapmap.ncbi.nlm.nih.gov/) is a good source of SNP information since alleles are reported consistently for the top strand with respect to the numbering along the chromosome, whereas the strand of DNA reported is arbitrary in the dbSNP database. The scrambled MUC5AC sequence described above (A7Y9J9) is used in the NIEH database, which leads to confusion regarding SNP assignment between *MUC5AC* and *MUC5B*. In addition, several databases report SNPs with no allele frequency information, and these should be treated with caution because they are not verified and could represent sequencing errors or possible confusion with differences between mucin genes or inter-tandem repeat domains.
37. Most studies indicate that there is linkage disequilibrium across the TR domains (18), but it is wrong to assume that all the variability, especially that of the TR domains, will be captured by SNP tagging. In our own studies, only in the case of *MUC7* have we found near 100% association with any SNP.
38. In the case of *Muc5b* (and the other *Muc* genes of the chromosome 11 complex), the AF086604 sequence is likely to align to the equine *Muc5b* gene but also to the other genes encoding secreted gel-forming mucins. The alignment with the highest identity and largest size of aligned sequence is likely to be the correct alignment.

Acknowledgements

The authors would like to thank Lynne Vinall (61), Lauren Johnson and Ralph Burgess whose work (Johnson PhD thesis UCL 2010; Burgess summer project, 2006) helped in the assembly of the information described in this chapter. KR was funded by the Horserace Betting Levy Board and the Medical Research Council.

References

1. Pigny, P., Guyonnet-Duperat, V., Hill, A. S., Pratt, W. S., Galiegue-Zouitina, S., d'Hooge, M. C., Laine, A., Van-Seuningen, I., Degand, P., Gum, J. R., Kim, Y. S., Swallow, D. M., Aubert, J. P., and Porchet, N. (1996) Human mucin genes assigned to 11p15.5: identification and organization of a cluster of genes. *Genomics 38*, 340–352.
2. Chen, Y., Zhao, Y. H., Kalaslavadi, T. B., Hamati, E., Nehrke, K., Le, A. D., Ann, D. K., and Wu, R. (2004) Genome-wide search and identification of a novel gel-forming mucin MUC19/Muc19 in glandular tissues. *Am J Respir Cell Mol Biol 30*, 155–165.
3. Thornton, D. J., Rousseau, K., and McGuckin, M. A. (2008) Structure and function of the polymeric mucins in airways mucus. *Annu Rev Physiol 70*, 459–486.
4. Swallow, D. M., Gendler, S., Griffiths, B., Corney, G., Taylor-Papadimitriou, J., and Bramwell, M. E. (1987) The human tumour-associated epithelial mucins are coded by an expressed hypervariable gene locus PUM. *Nature 328*, 82–84.
5. Vinall, L. E., King, M., Novelli, M., Green, C. A., Daniels, G., Hilkens, J., Sarner, M., and Swallow, D. M. (2002) Altered expression and

allelic association of the hypervariable membrane mucin MUC1 in Helicobacter pylori gastritis. *Gastroenterology 123*, 41–49.

6. Silva, F., Carvalho, F., Peixoto, A., Seixas, M., Almeida, R., Carneiro, F., Mesquita, P., Figueiredo, C., Nogueira, C., Swallow, D. M., Amorim, A., and David, L. (2001) MUC1 gene polymorphism in the gastric carcinogenesis pathway. *Eur J Hum Genet 9*, 548–552.
7. Silva, F., Carvalho, F., Peixoto, A., Teixeira, A., Almeida, R., Reis, C., Bravo, L. E., Realpe, L., Correa, P., and David, L. (2003) MUC1 polymorphism confers increased risk for intestinal metaplasia in a Colombian population with chronic gastritis. *Eur J Hum Genet 11*, 380–384.
8. Toribara, N. W., Gum, J. R., Jr., Culhane, P. J., Lagace, R. E., Hicks, J. W., Petersen, G. M., and Kim, Y. S. (1991) MUC-2 human small intestinal mucin gene structure. Repeated arrays and polymorphism. *J Clin Invest 88*, 1005–1013.
9. van de Bovenkamp, J. H., Hau, C. M., Strous, G. J., Buller, H. A., Dekker, J., and Einerhand, A. W. (1998) Molecular cloning of human gastric mucin MUC5AC reveals conserved cysteine-rich D-domains and a putative leucine zipper motif. *Biochem Biophys Res Commun 245*, 853–859.
10. Escande, F., Aubert, J. P., Porchet, N., and Buisine, M. P. (2001) Human mucin gene MUC5AC: organization of its 5'-region and central repetitive region. *Biochem J 358*, 763–772.
11. Engelmann, K., Baldus, S. E., and Hanisch, F. G. (2001) Identification and topology of variant sequences within individual repeat domains of the human epithelial tumor mucin MUC1. *J Biol Chem 276*, 27764–27769.
12. Fowler, J. C., Teixeira, A. S., Vinall, L. E., and Swallow, D. M. (2003) Hypervariability of the membrane-associated mucin and cancer marker MUC1. *Hum Genet 113*, 473–479.
13. von Mensdorff-Pouilly, S., Kinarsky, L., Engelmann, K., Baldus, S. E., Verheijen, R. H., Hollingsworth, M. A., Pisarev, V., Sherman, S., and Hanisch, F. G. (2005) Sequence-variant repeats of MUC1 show higher conformational flexibility, are less densely O-glycosylated and induce differential B lymphocyte responses. *Glycobiology 15*, 735–746.
14. Ng, W., Loh, A. X., Teixeira, A. S., Pereira, S. P., and Swallow, D. M. (2008) Genetic regulation of MUC1 alternative splicing in human tissues. *Br J Cancer*.
15. Kamio, K., Matsushita, I., Hijikata, M., Kobashi, Y., Tanaka, G., Nakata, K., Ishida, T., Tokunaga, K., Taguchi, Y., Homma, S., Nakata, K., Azuma, A., Kudoh, S., and Keicho, N. (2005) Promoter analysis and aberrant expression of the MUC5B gene in diffuse panbronchiolitis. *Am J Respir Crit Care Med 171*, 949–957.
16. Loh, A. X., Johnson, L., Ng, W., and Swallow, D. M. Cis-acting allelic variation in MUC5B mRNA expression is associated with different promoter haplotypes. *Ann Hum Genet 74*, 498–505.
17. Rousseau, K., Vinall, L. E., Butterworth, S. L., Hardy, R. J., Holloway, J., Wadsworth, M. E., and Swallow, D. M. (2006) MUC7 haplotype analysis: results from a longitudinal birth cohort support protective effect of the MUC7*5 allele on respiratory function. *Ann Hum Genet 70*, 417–427.
18. Rousseau, K., Byrne, C., Griesinger, G., Leung, A., Chung, A., Hill, A. S., and Swallow, D. M. (2007) Allelic association and recombination hotspots in the mucin gene (MUC) complex on chromosome 11p15.5. *Ann Hum Genet 71*, 561–569.
19. Pigman, W., Moschera, J., Weiss, M., and Tettamanti, G. (1973) The occurrence of repetitive glycopeptide sequences in bovine submaxillary glycoprotein. *Eur J Biochem 32*, 148–154.
20. Tettamanti, G., and Pigman, W. (1968) Purification and characterization of bovine and ovine submaxillary mucins. *Arch Biochem Biophys 124*, 41–50.
21. Blair, G. W., Folley, S. J., Malpress, F. H., and Coppen, F. M. (1941) Variations in certain properties of bovine cervical mucus during the oestrous cycle. *Biochem J 35*, 1039–1049.
22. Escande, F., Porchet, N., Aubert, J. P., and Buisine, M. P. (2002) The mouse Muc5b mucin gene: cDNA and genomic structures, chromosomal localization and expression. *Biochem J 363*, 589–598.
23. Escande, F., Porchet, N., Bernigaud, A., Petitprez, D., Aubert, J. P., and Buisine, M. P. (2004) The mouse secreted gel-forming mucin gene cluster. *Biochim Biophys Acta 1676*, 240–250.
24. Chen, Y., Zhao, Y. H., and Wu, R. In silico cloning of mouse Muc5b gene and upregulation of its expression in mouse asthma model. *Am J Respir Crit Care Med 164*, 1059–1066.
25. Bhargava, A. K., Woitach, J. T., Davidson, E. A., and Bhavanandan, V. P. (1990) Cloning and cDNA sequence of a bovine submaxillary gland mucin-like protein containing two distinct domains. *Proc Natl Acad Sci USA 87*, 6798–6802.
26. Jiang, W., Gupta, D., Gallagher, D., Davis, S., and Bhavanandan, V. P. (2000) The central domain of bovine submaxillary mucin consists of over 50 tandem repeats of 329 amino acids. Chromosomal localization of the BSM1 gene

and relations to ovine and porcine counterparts. *Eur J Biochem 267*, 2208–2217.

27. Jiang, W., Woitach, J. T., Gupta, D., and Bhavanandan, V. P. (1998) Sequence of a second gene encoding bovine submaxillary mucin: implication for mucin heterogeneity and cloning. *Biochem Biophys Res Commun 251*, 550–556.
28. Jiang, W., Woitach, J. T., Keil, R. L., and Bhavanandan, V. P. (1998) Bovine submaxillary mucin contains multiple domains and tandemly repeated non-identical sequences. *Biochem J 331 (Pt 1)*, 193–199.
29. Lang, T., Hansson, G. C., and Samuelsson, T. (2007) Gel-forming mucins appeared early in metazoan evolution. *Proc Natl Acad Sci USA 104*, 16209–16214.
30. Lang, T., Hansson, G. C., and Samuelsson, T. (2006) An inventory of mucin genes in the chicken genome shows that the mucin domain of Muc13 is encoded by multiple exons and that ovomucin is part of a locus of related gel-forming mucins. *BMC Genomics 7*, 197.
31. Gum, J. R., Byrd, J. C., Hicks, J. W., Toribara, N. W., Lamport, D. T., and Kim, Y. S. (1989) Molecular cloning of human intestinal mucin cDNAs. Sequence analysis and evidence for genetic polymorphism. *J Biol Chem 264*, 6480–6487.
32. Guyonnet Duperat, V., Audie, J. P., Debailleul, V., Laine, A., Buisine, M. P., Galiegue-Zouitina, S., Pigny, P., Degand, P., Aubert, J. P., and Porchet, N. (1995) Characterization of the human mucin gene MUC5AC: a consensus cysteine-rich domain for 11p15 mucin genes? *Biochem J 305 (Pt 1)*, 211–219.
33. Toribara, N. W., Roberton, A. M., Ho, S. B., Kuo, W. L., Gum, E., Hicks, J. W., Gum, J. R., Jr., Byrd, J. C., Siddiki, B., and Kim, Y. S. (1993) Human gastric mucin. Identification of a unique species by expression cloning. *J Biol Chem 268*, 5879–5885.
34. Dufosse, J., Porchet, N., Audie, J. P., Guyonnet Duperat, V., Laine, A., Van-Seuningen, I., Marrakchi, S., Degand, P., and Aubert, J. P. (1993) Degenerate 87-base-pair tandem repeats create hydrophilic/hydrophobic alternating domains in human mucin peptides mapped to 11p15. *Biochem J 293 (Pt 2)*, 329–337.
35. Newton, C. R., Graham, A., Heptinstall, L. E., Powell, S. J., Summers, C., Kalsheker, N., Smith, J. C., and Markham, A. F. (1989) Analysis of any point mutation in DNA. The amplification refractory mutation system (ARMS). *Nucleic Acids Res 17*, 2503–2516.
36. Ye, S., Humphries, S., and Green, F. (1992) Allele specific amplification by tetra-primer PCR. *Nucleic Acids Res 20*, 1152.
37. Strachan, T., Read, A. (2010) Human Molecular Genetics, 4th ed., Garland Science.
38. Spielman, R. S., McGinnis, R. E., and Ewens, W. J. (1993) Transmission test for linkage disequilibrium: the insulin gene region and insulin-dependent diabetes mellitus (IDDM). *Am J Hum Genet 52*, 506–516.
39. Offner, G. D., Nunes, D. P., Keates, A. C., Afdhal, N. H., and Troxler, R. F. (1998) The amino-terminal sequence of MUC5B contains conserved multifunctional D domains: implications for tissue-specific mucin functions. *Biochem Biophys Res Commun 251*, 350–355.
40. Heazlewood, C. K., Cook, M. C., Eri, R., Price, G. R., Tauro, S. B., Taupin, D., Thornton, D. J., Png, C. W., Crockford, T. L., Cornall, R. J., Adams, R., Kato, M., Nelms, K. A., Hong, N. A., Florin, T. H., Goodnow, C. C., and McGuckin, M. A. (2008) Aberrant mucin assembly in mice causes endoplasmic reticulum stress and spontaneous inflammation resembling ulcerative colitis. *PLoS Med 5*, e54.
41. Griffiths, B., Matthews, D. J., West, L., Attwood, J., Povey, S., Swallow, D. M., Gum, J. R., and Kim, Y. S. (1990) Assignment of the polymorphic intestinal mucin gene (MUC2) to chromosome 11p15. *Ann Hum Genet 54*, 277–285.
42. Pigny, P., Pratt, W. S., Laine, A., Leclercq, A., Swallow, D. M., Nguyen, V. C., Aubert, J. P., and Porchet, N. (1995) The MUC5AC gene: RFLP analysis with the Jer58 probe. *Hum Genet 96*, 367–368.
43. Vinall, L. E., Hill, A. S., Pigny, P., Pratt, W. S., Toribara, N., Gum, J. R., Kim, Y. S., Porchet, N., Aubert, J. P., and Swallow, D. M. (1998) Variable number tandem repeat polymorphism of the mucin genes located in the complex on 11p15.5. *Hum Genet 102*, 357–366.
44. Rousseau, K., Byrne, C., Kim, Y. S., Gum, J. R., Swallow, D. M., and Toribara, N. W. (2004) The complete genomic organization of the human MUC6 and MUC2 mucin genes. *Genomics 83*, 936–939.
45. Gum, J. R., Jr., Hicks, J. W., Toribara, N. W., Rothe, E. M., Lagace, R. E., and Kim, Y. S. The human MUC2 intestinal mucin has cysteine-rich subdomains located both upstream and downstream of its central repetitive region. *J Biol Chem 267*, 21375–21383.
46. Zhu, L., Lee, P., Yu, D., Tao, S., and Chen, Y. (2010) Cloning and Characterization of Human MUC19 Gene. *Am J Respir Cell Mol Biol In Press.*
47. Jia, Y., Persson, C., Hou, L., Zheng, Z., Yeager, M., Lissowska, J., Chanock, S. J., Chow, W. H., and Ye, W. (2010) A comprehensive analysis of common genetic variation in MUC1, MUC5AC, MUC6 genes and risk of stomach cancer. *Cancer Causes Control 21*, 313–321.

48. Ubell, M. L., Khampang, P., and Kerschner, J. E. (2009) Mucin gene polymorphisms in otitis media patients. *Laryngoscope 120*, 132–138.
49. Kwon, J. A., Lee, S. Y., Ahn, E. K., Seol, S. Y., Kim, M. C., Kim, S. J., Kim, S. I., Chu, I. S., and Leem, S. H. (2010) Short rare MUC6 minisatellites-5 alleles influence susceptibility to gastric carcinoma by regulating gene. *Hum Mutat 31*, 942–949.
50. Garcia, E., Carvalho, F., Amorim, A., and David, L. (1997) MUC6 gene polymorphism in healthy individuals and in gastric cancer patients from northern Portugal. *Cancer Epidemiol Biomarkers Prev 6*, 1071–1074.
51. Nguyen, T. V., Janssen, M., Jr., Gritters, P., te Morsche, R. H., Drenth, J. P., van Asten, H., Laheij, R. J., and Jansen, J. B. (2006) Short mucin 6 alleles are associated with H pylori infection. *World J Gastroenterol 12*, 6021–6025.
52. Ahn, E. K., Kim, W. J., Kwon, J. A., Choi, P. J., Kim, W. J., Sunwoo, Y., Heo, J., and Leem, S. H. (2009) Variants of MUC5B minisatellites and the susceptibility of bladder cancer. *DNA Cell Biol 28*, 169–176.
53. Vinall, L. E., Fowler, J. C., Jones, A. L., Kirkbride, H. J., de Bolos, C., Laine, A., Porchet, N., Gum, J. R., Kim, Y. S., Moss, F. M., Mitchell, D. M., and Swallow, D. M. (2000) Polymorphism of human mucin genes in chest disease: possible significance of MUC2. *Am J Respir Cell Mol Biol 23*, 678–686.
54. Jeong, Y. H., Kim, M. C., Ahn, E. K., Seol, S. Y., Do, E. J., Choi, H. J., Chu, I. S., Kim, W. J., Kim, W. J., Sunwoo, Y., and Leem, S. H. (2007) Rare exonic minisatellite alleles in MUC2 influence susceptibility to gastric carcinoma. *PLoS ONE 2*, e1163.
55. Moehle, C., Ackermann, N., Langmann, T., Aslanidis, C., Kel, A., Kel-Margoulis, O., Schmitz-Madry, A., Zahn, A., Stremmel, W., and Schmitz, G. (2006) Aberrant intestinal expression and allelic variants of mucin genes associated with inflammatory bowel disease. *J Mol Med 84*, 1055–1066.
56. Swallow, D. M., Vinall, L. E., Gum, J. R., Kim, Y. S., Yang, H., Rotter, J. I., Mirza, M., Lee, J. C., and Lennard-Jones, J. E. (1999) Ulcerative colitis is not associated with differences in MUC2 mucin allele length. *J Med Genet 36*, 859–860.
57. Chuang, S. C., Juo, S. H., Hsi, E., Wang, S. N., Tsai, P. C., Yu, M. L., and Lee, K. T. (2011) Multiple mucin genes polymorphisms are associated with gallstone disease in Chinese men. *Clin Chim Acta 412*, 599–603.
58. Barrett, J. C., Hansoul, S., Nicolae, D. L., Cho, J. H., Duerr, R. H., Rioux, J. D., Brant, S. R., Silverberg, M. S., Taylor, K. D., Barmada, M. M., Bitton, A., Dassopoulos, T., Datta, L. W., Green, T., Griffiths, A. M., Kistner, E. O., Murtha, M. T., Regueiro, M. D., Rotter, J. I., Schumm, L. P., Steinhart, A. H., Targan, S. R., Xavier, R. J., Libioulle, C., Sandor, C., Lathrop, M., Belaiche, J., Dewit, O., Gut, I., Heath, S., Laukens, D., Mni, M., Rutgeerts, P., Van Gossum, A., Zelenika, D., Franchimont, D., Hugot, J. P., de Vos, M., Vermeire, S., Louis, E., Cardon, L. R., Anderson, C. A., Drummond, H., Nimmo, E., Ahmad, T., Prescott, N. J., Onnie, C. M., Fisher, S. A., Marchini, J., Ghori, J., Bumpstead, S., Gwilliam, R., Tremelling, M., Deloukas, P., Mansfield, J., Jewell, D., Satsangi, J., Mathew, C. G., Parkes, M., Georges, M., and Daly, M. J. (2008) Genome-wide association defines more than 30 distinct susceptibility loci for Crohn's disease. *Nat Genet 40*, 955–962.
59. Guo, X., Pace, R. G., Stonebraker, J. R., Commander, C. W., Dang, A. T., Drumm, M. L., Harris, A., Zou, F., Swallow, D. M., Wright, F. A., O'Neal, W. K., and Knowles, M. R. (2011) Mucin variable number tandem repeat polymorphisms and severity of cystic fibrosis lung disease: significant association with MUC5AC. *PLoS One 6*(10), e25452.
60. Seibold, M. A., Wise, A. L., Speer, M. C., Steele, M. P., Brown, K. K., Loyd, J. E., Fingerlin, T. E., Zhang, W., Gudmundsson, G., Groshong, S. D., Evans, C. M., Garantziotis, S., Adler, K. B., Dickey, B. F., du Bois, R. M., Yang, I. V., Herron, A., Kervitsky, D., Talbert, J. L., Markin, C., Park. J., Crews, A. L., Slifer, S. H., Auerbach, S., Roy, M. G., Lin, J., Hennessy, C. E., Schwarz, M. I., and Schwartz, D. A. (2011) A common MUC5B promoter polymorphism and pulmonary fibrosis. *Engl J Med 364*(16), 1503–1512.
61. Vinall LE, Pratt WS, Swallow DM. Detection of mucin gene polymorphism. *Methods Mol Biol.* 2000;125:337–50(16), 1503–1512.
62. Guo X, Pace RG, Stonebraker JR, Commander CW, Dang AT, Drumm ML, Harris A, Zou F, Swallow DM, Wright FA, O'Neal WK, Knowles MR. Mucin variable number tandem repeat polymorphisms and severity of cystic fibrosis lung disease: significant association with MUC5AC. *PLoS One. 2011;6*(10):e25452. Epub 2011 Oct 6.
63. Seibold MA, Wise AL, Speer MC, Steele MP, Brown KK, Loyd JE, Fingerlin TE, Zhang W, Gudmundsson G, Groshong SD, Evans CM, Garantziotis S, Adler KB, Dickey BF, du Bois RM, Yang IV, Herron A, Kervitsky D, Talbert JL, Markin C, Park J, Crews AL, Slifer SH, Auerbach S, Roy MG, Lin J, Hennessy CE, Schwarz MI, Schwartz DA.A common MUC5B promoter polymorphism and pulmonary fibrosis. *N Engl J Med*, 2011 Apr 21;364(16):1503–12.

Chapter 2

Gel-Forming and Cell-Associated Mucins: Preparation for Structural and Functional Studies

Julia R. Davies, Claes Wickström, and David J. Thornton

Abstract

Secreted and transmembrane mucins are important components of innate defence at the body's mucosal surfaces. The secreted mucins are large, polymeric glycoproteins, which are largely responsible for the gel-like properties of mucus secretions. The cell-tethered mucins, however, are monomeric but are typically composed of two subunits, a larger extracellular subunit which is heavily glycosylated while the smaller more sparsely glycosylated subunit has a short extracellular region, a single-pass transmembrane domain, and a cytoplasmic tail. These two families of mucins represent high-molecular-weight glycoproteins containing serine and threonine-rich domains that are the attachment sites for large numbers of O-glycans. The high-M_r and high sugar content have been exploited for the separation of mucins from the majority of components in mucus secretions. In this chapter, we describe current and well-established methods (caesium chloride density-gradient centrifugation, gel-filtration and anion-exchange chromatography, and agarose gel electrophoresis) for the extraction and purification of gel-forming and cell-surface mucins which can subsequently be used for a variety of structural and functional studies.

Key words: Density-gradient centrifugation, Mucin extraction, Mucin purification, Guanidinium chloride, Gel-filtration chromatography, Anion-exchange chromatography, Agarose gel electrophoresis, Mucin

1. Introduction

Mucins are important components of innate defence at the body's mucosal surfaces (1). This family of multifunctional glycoproteins comprises transmembrane (cell surface) and secreted mucins. The transmembrane mucins are localised at the apical surface of epithelial cells and are part of the carbohydrate-rich glycocalyx within which they function to protect and "sense" the immediate environment at the epithelial cell surface. By providing the structural framework of mucus, the secreted gel-forming mucins form a dynamic protective barrier that lies above the glycocalyx. In normal physiology,

Michael A. McGuckin and David J. Thornton (eds.), *Mucins: Methods and Protocols*, Methods in Molecular Biology, vol. 842, DOI 10.1007/978-1-61779-513-8_2, © Springer Science+Business Media, LLC 2012

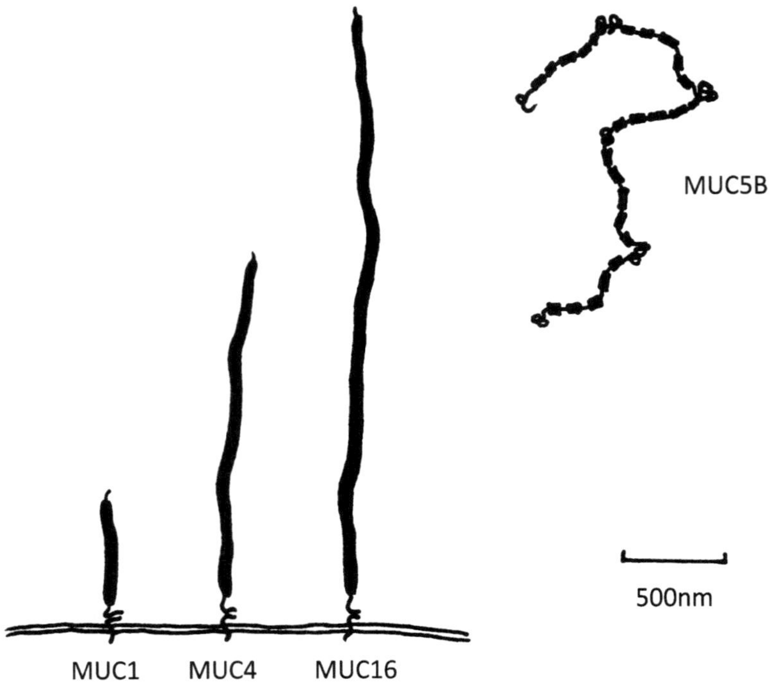

Fig. 1. Schematic drawing of MUC1, MUC4, MUC16, and MUC5B mucins. MUC1, MUC4, and MUC16 are depicted as cell-associated mucins with a putative cleavage site close to the cell membrane. The longest reported allele is shown for each mucin species. The polymeric gel-forming mucin comprises five MUC5B subunits. The four mucins are drawn to scale relative to each other with the lengths of the glycosylated domains (*thick lines*) estimated using the assumption that the length/amino acid is 0.25 nm in this part of the molecule.

these two mucin-rich environments act in concert to protect the mucosal surfaces from chemical, physical, and biological challenge.

The secreted mucins (MUC2, MUC5AC, MUC5B, MUC6, MUC7, and MUC19), with the exception of MUC7, are large, polymeric glycoproteins that are largely responsible for the gel-like properties of mucus secretions (2). The polymeric gel-forming mucins are assembled via disulphide linkages at the C- and N-termini of their constituent monomers (subunits), each monomer comprising a heavily glycosylated central region that is interspersed with cysteine-rich domains and flanked by cysteine-rich terminal domains (Fig. 1). In contrast, the cell-surface mucins (e.g. MUC1, MUC4, and MUC16) are monomeric but are typically composed of two subunits, the larger subunit is wholly extracellular and heavily glycosylated while the smaller more sparsely glycosylated subunit consists of a short extracellular region, a single-pass transmembrane domain, and a cytoplasmic tail (3) (Fig. 1). While these two families of mucins have quite different structural organisation, they do share common features; both are high-molecular-weight glycoproteins (0.25–50 MDa) with large, often tandemly repetitive sections of their polypeptides, rich in serine and threonine that are the attachment sites for O-glycans. This arrangement results in mucins having a high proportion of their mass as carbohydrate (30–80% of weight). These two features (high-M_r and high sugar content) have been exploited for the separation of mucins from the majority of proteins and other glycoproteins.

In this chapter, we describe current and well-established methods for the extraction and purification of secreted, gel-forming and cell-surface mucins. These protocols can be used to isolate mucins as the starting point for subsequent structural and functional studies, for example: analysis of mucin size distribution and morphology by electron microscopy, mass spectrometry-based analysis of the mucin polypeptide (identification of gene products or proteolytic processing) and mucin glycans, reagents for studying bacterial binding and drug interactions, or as a means of identifying mucin-binding proteins to better understand how the mucus layer is organised. Of course, the goals of the subsequent studies dictate the solvent conditions of extraction and purification (i.e. whether chaotropic denaturing agents can be used or if it is necessary for the mucins to be in a "native" state) (see Note 1).

2. Materials

2.1. Extraction of Mucins with Chaotropic Agents

1. Guanidinium chloride solution

   During each step in the mucin preparation, it is worth ensuring that the quality of the chemicals used is of sufficiently high quality not to interfere with subsequent analyses. Guanidinium chloride can either be purchased as ultrapure grade, which needs no further treatment, or as practical grade in which case a purification step is required. The use of the latter is considerably less expensive and is worth the extra work if large amounts are to be used. We prepare a *stock* solution with a molarity of approx. 7.5 M and use it to make all the necessary guanidinium chloride solutions.

   (a) Dissolve 765 g of guanidinium chloride in 1 L of distilled water and stir constantly.

   (b) Add 10 g of activated charcoal and stir overnight.

   (c) Filter solution through double-filter paper to remove the bulk of the charcoal.

   (d) To remove remaining charcoal, filter solution through an Amicon PM10 filter (Millipore, Billerica, MA) or equivalent using an ultrafiltration cell. The speed and capacity of filtration can be increased by using a Diaflow system.

   (e) Measure the density of the solution by weighing a known volume and calculating the molarity of the guanidinium chloride stock solution (see Note 2).

2. Solutions for mucin extraction
   (a) 6 M guanidinium chloride extraction buffer: 6 M guanidinium chloride (from stock), 5 mM EDTA, 10 mM sodium phosphate buffer, pH 6.5 (adjusted with NaOH). This solution can be stored at room temperature but should be cooled to 4°C for use. Immediately prior to extraction, add *N*-ethylmaleimide (NEM) and diisopropyl phosphofluoridate (DPF) or phenylmethylsulphonylfluoridate (PMSF) (see Note 3) to the solution to give concentrations of 5 mM NEM and 1 mM DFP or 0.1 mM PMSF.
   (b) Phosphate-buffered saline (PBS) containing protease inhibitors: 0.2 M sodium chloride, 10 mM EDTA, 10 mM NEM, 2 mM DFP, 10 mM sodium phosphate buffer, pH 7.4 (adjusted with NaOH).
   (c) 6 M guanidinium chloride reduction buffer: 6 M guanidinium chloride (from stock), 5 mM EDTA, 0.1 M Tris–HCl buffer, pH 8.0 (adjusted with HCl). This solution can be stored at room temperature (see Note 4).

2.2. Extraction of Mucins with Non-chaotrophic Agents

For some purposes, it may be necessary to prepare "native" mucins under non-denaturing or associative conditions. Under these conditions, the mucus gel is not completely solubilised and only those mucins that are easily released from the complex are present in solution. Thus, the risk that some mucin populations may be selectively lost in subsequent purification and centrifugation steps should be borne in mind.

1. 0.2 M sodium chloride solution (pH 7).
2. 10 mM Tris buffer, pH 8.0, containing detergent (0.5% (v/v) Triton X-100 or 0.5% (w/v) sodium deoxycholate).

2.3. Isopycnic Density-Gradient Centrifugation

Purification of mucins from complex secretions and tissues often requires a two-step isopycnic density-gradient centrifugation procedure using CsCl to remove proteins (first step) and nucleic acids (second step). Caesium chloride can be purchased as ultrapure grade, but we order small samples of less pure caesium chloride from several companies and test them to ensure that they give clear solutions with low absorbance at 280 nm and do not give spurious reactions in our assays before purchasing larger batches (see Note 5).

1. Beckman Quick seal polyallomer centrifuge tubes (Beckman Instruments, Palo Alto, CA) or equivalent.
2. 6 M guanidinium chloride extraction buffer, pH 6.5.
3. Sodium phosphate buffer: 10 mM sodium di-hydrogen phosphate, pH 6.5 (adjusted with NaOH).
4. 0.5 M guanidinium chloride buffer: 0.5 M guanidinium chloride (from stock), 5 mM EDTA, 10 mM sodium phosphate buffer, pH 6.5 (adjusted with NaOH).

2.4. Gel Chromatography

After CsCl density-gradient centrifugation, gel-filtration chromatography can be employed to separate high-M_r mucins from smaller proteins and glycoproteins. We use a variety of separation media, including Sephacryl S-1000, Sephacryl S-500HR Sepharose CL-2B, and Superose 6 gels (GE Healthcare Life Sciences) (see Notes 6 and 7).

1. Denaturing conditions
   (a) 4 M guanidinium chloride buffer: 4 M guanidinium chloride (from stock), 10 mM sodium phosphate buffer, pH 7.0 (adjusted with NaOH). This solution can be stored at room temperature.
2. Non-denaturing conditions
   (b) 0.15 M NaCl solution.

2.5. Anion-Exchange Chromatography

In our laboratories, we use Mono Q HR 5/5 or Resource Q columns (GE Healthcare Life Sciences) and eluants based on a piperazine buffer system with lithium perchlorate as the elution salt (see Note 8).

1. Buffer A: 10 mM piperazine/perchlorate buffer, pH 5.0 (adjusted with perchloric acid).
2. Buffer B: 0.25–0.4 M $LiClO_4$, 10 mM piperazine/perchlorate buffer, pH 5.0 (adjusted with perchloric acid).
3. Buffer C: 0.1% (w/v) 3-((3-cholamidopropyl)dimethylammonio)-1-propane sulfonate (CHAPS) in 6 M urea, 10 mM piperazine/perchlorate buffer, pH 5.0 (adjusted with perchloric acid).
4. Buffer D: 0.1% (w/v) CHAPS in 6 M urea, 0.25–0.4 M $LiClO_4$, 10 mM piperazine/perchlorate buffer, pH 5.0 (adjusted with perchloric acid).

2.6. Agarose Gel Electrophoresis

In addition to the preparative techniques above, analytical separation can be achieved by agarose gel electrophoresis and the separation monitored by lectin, immunochemical, or histochemical staining after transfer of the protein to membranes (for detection methods, see Chapter 3). This technique can be applied to "raw" mucus samples, although care must be taken to ensure that samples are protected from degradation. This separation, which works on the basis of inherent negative charge on the sialic acid and sulphate groups on the mucin glycans (4), can be used to separate different mucins (5) as well as glycosylated variants of the same mucin (6). Furthermore, this technique allows separation of mature and precursor forms of the mucins.

1. Horizontal gel electrophoresis apparatus for 25 × 15-cm or 15 × 15-cm gels (we are using Bio-RAD Subcell apparatus).
2. Ultrapure agarose.

3. Running buffer: 40 mM Tris–acetate, 1 mM sodium EDTA, pH 8.0, containing 0.1% (w/v) sodium dodecyl sulphate (SDS).
4. Loading buffer: 10× running buffer containing 50% (v/v) glycerol and 0.002% (w/v) bromophenol blue.
5. Transfer buffer: 0.6 M NaCl, 60 mM sodium citrate.

3. Methods

3.1. Mucin Extraction

3.1.1. Extraction of Gel-Forming Mucins from Mucus Secretions

Denaturing Conditions

For some purposes, it may be necessary to solubilise a viscoelastic mucus gel in order to extract the mucins. In this case, the protocol below, for extraction under denaturing conditions, should be followed.

1. Thaw secretions on ice, if necessary, mix with an equal volume of ice-cold PBS containing protease inhibitors and stir gently at 4°C for 1 h.
2. Centrifuge secretions (23,000×*g*, 4°C, 45 min).
3. Pour off the supernatant which represents the sol phase.
4. Add at least five volumes of 6 M guanidinium chloride extraction buffer to the pellet (which represents the gel phase) and stir gently overnight at 4°C. If the samples contain material that is difficult to disperse, they can be suspended using two or three strokes in a Dounce homogenizer (Kontes Glass Co., Vineland, NJ) with a loose pestle.
5. Centrifuge secretions (23,000×*g*, 4°C, 45 min).
6. Pour off the supernatant. This corresponds to the "guanidinium chloride soluble" gel-phase mucins.
7. If necessary, repeat steps 4–6 two more times and pool the extracts.
8. Add 6 M guanidinium chloride reduction buffer containing 10 mM dithiothreitol (DTT) to the pellet (corresponding to the "guanidinium chloride insoluble" mucins) (see Note 9) and incubate for 5 h at 37°C.
9. Add iodoacetamide (IAA) to give a 25 mM solution, and incubate overnight in the dark at room temperature. This sample corresponds to the reduced/alkylated guanidinium chloride "insoluble mucin complex".

Non-denaturing (Associative) Conditions

In cases where preparation of "native" mucins under non-denaturing (associative) conditions is required, the following protocol should be used (see Note 10).

1. Thaw secretions on ice, if necessary, mix with an equal volume of ice-cold PBS containing protease inhibitors, and stir gently at 4°C for 1 h.

2. Centrifuge secretions (4,400 × *g*, 4°C, 30 min).
3. Pour off and retain the supernatant for further purification.

3.1.2. Extraction of Gel-Forming Mucins from Tissue Samples

Since tissue extracts contain high levels of cell debris and extracellular matrix fragments, as well as proteolyic enzymes, we always prepare mucins from this source under denaturing conditions. If mucins are to be prepared from the surface epithelium and submucosa separately, begin at step 1, whereas if they are to be extracted from the whole tissue piece, begin at step 4.

1. Drench tissue pieces in 10 mM sodium phosphate buffer, pH 6.5, containing 1 mM DFP or 0.1 mM PMSF and allow to thaw if frozen.
2. Scrape the surface epithelium away from the underlying mucosa using a blunt instrument or a glass microscope slide.
3. Place the epithelial scrapings in at least five volumes of ice-cold 6 M guanidinium chloride extraction buffer and disperse with two or three strokes in a Dounce homogenizer with a loose pestle.
4. Cut the submucosal tissue (or whole tissue piece) into small pieces using scissors or a scalpel and submerge in liquid nitrogen. Pulverize or grind the tissue (we use a Retch tissue pulverizer, Retsch, Haan, Germany) for this purpose.
5. Mix the powdered tissue with at least five volumes of ice-cold 6 M guanidinium chloride extraction buffer and disperse with two or three strokes in a Dounce homogenizer with a loose pestle.
6. Gently stir samples overnight at 4°C.
7. Centrifuge secretions (23,000 × *g*, 4°C, 45 min).
8. Pour off the supernatant corresponding to the "guanidinium chloride soluble" mucins.
9. Repeat steps 5–8 two more times if necessary and pool the extracts.
10. Add 6 M guanidinium chloride reduction buffer containing 10 mM DTT to the extraction residue and incubate for 5 h at 37°C.
11. Add IAA to give a 25 mM solution, and incubate overnight in the dark at room temperature.
12. Pour off the supernatant corresponding to the reduced/alkylated guanidinium chloride "insoluble mucin complex".

3.1.3. Extraction of Gel-Forming Mucins from Mucin-Producing Cell Cultures

Mucin-producing cell lines and primary cell cultures represent a valuable complement to tissues and secretions for studies of the structure and function of these molecules. Many tumour cell lines overexpress cell-associated mucins making them good tools for investigation of, for instance, the biosynthesis and receptor functions of these molecules. Cells derived from intestinal and airway tumours or normal tissue have proved very useful for studies of biosynthesis and assembly of the large, gel-forming polymeric mucin species as well as providing a ready source of relatively "clean" secretions.

1. Cell culture medium
   Mucins may be extracted from aspirated cell culture medium under denaturing or non-denaturing conditions using the appropriate protocol given for extraction from mucus secretions (Subheading 3.1.1 or 3.1.2).
2. Cell layers
   Extraction of gel-forming mucins from the cell layer of a culture can be carried out as follows.
   (a) After removing the culture medium by aspiration, wash the cell layer with PBS.
   (b) Separate the cell layer from the substrate using, for instance, a cell scraper, and add a 0.5–1 mL of ice-cold 6 M guanidinium chloride extraction buffer per cm^2. Disperse the cells using a Pasteur pipette.
   (c) Gently stir the sample overnight at 4°C.
   (d) Centrifuge secretions (23,000 × *g*, 4°C, 45 min).
   (e) Pour off the supernatant corresponding to the "guanidinium chloride soluble" mucins.
   (f) Repeat steps 2–5 two more times if necessary and pool the extracts.
   (g) Add 6 M guanidinium chloride reduction buffer containing 10 mM DTT to the extraction residue and incubate for 5 h at 37°C.
   (h) Add IAA to give a 25 mM solution, and incubate overnight in the dark at room temperature.
   (i) Pour off the supernatant corresponding to the reduced/alkylated guanidinium chloride "insoluble mucin complex".

3.1.4. Extraction of Cell-Associated Mucins from Mucus Secretions

The extracellular domains of cell-associated mucins, such as MUC1, MUC4, and MUC16, may be present in mucus secretions as a result of shedding or secretion.

1. Denaturing conditions
   To isolate them under denaturing conditions, secretions are extracted with 6 M guanidinium chloride extraction buffer in the same way as for the gel-forming species (Subheading 3.1.1, steps 1–7). Cell-associated mucins are found together with the gel-forming ones in the sol phase and/or the "guanidinium chloride soluble fraction".

2. Associative conditions
   To isolate cell-associated mucins under non-denaturing conditions, secretions should be treated as shown in Subheading 3.1.2.

3.1.5. Extraction of Cell-Associated Mucins from Mucin-Producing Cell Lines

1. Cell culture medium
   Cell-associated mucins may be extracted from aspirated cell culture medium under denaturing or non-denaturing conditions using the appropriate protocol given for extraction from secretions (Subheading 3.1.1 or 3.1.2).
2. Cell membranes
   After preparation of a membrane fraction from the cells, the extraction method used depends upon whether the whole mucin complex (transmembrane and extracellular domains) or just the extracellular domains are required. In 0.5% Triton X-100 or 0.5% sodium deoxycholate, interaction between the subunits is maintained and the whole complex can be solubilised. However, in 6 M guanidinium chloride, the transmembrane and extracellular domains dissociate and the extracellular domain is extracted leaving the transmembrane domain in the membrane (7, 8).
   (a) Wash the cells with serum-free medium, harvest using a cell scraper if necessary, and suspend in 10 mM Tris, pH 8.0. Allow to stand on ice for 2 min prior to centrifugation (600 × *g*, 2 min).
   (b) Suspend the cell pellet in ten volumes of 10 mM Tris, pH 8.0, and homogenise using four to five strokes of a Dounce homogenizer with a tight pestle.
   (c) Bring the suspension to 3 mM Mg^{2+} by adding 30 mM $MgCl_2$ and 100 mM NaCl and centrifuge (1,000 × *g*, 1 min).
   (d) Collect supernatant and centrifuge (10,000 × *g*, 10 min).
   (e) Collect supernatant and centrifuge (100,000 × *g*, 90 min). Retain pellet containing membrane fragments.
   (f) Resuspend pellet by vigorous vortexing in (1) 10 mM Tris buffer, pH 8.0, containing 0.5% Triton X-100, (2) 10 mM Tris buffer, pH 8.0, containing 0.5% sodium deoxycholate or (3) 6 M guanidinium chloride extraction buffer and incubate on ice for 30 min.
   (g) Centrifuge sample (35,000 rpm, 70.1 Ti rotor, 1 h) and retain the supernatant (see Note 11).

3.2. Mucin Purification

Although some mucus gels contain only one polymeric mucin species, e.g. MUC2 in the intestine, in general, mucus secretions contain more than one large gel-forming species (e.g. airways and cervix—MUC5AC and MUC5B; stomach—MUC5AC and MUC6). In addition, the extracellular domain of some cell-associated mucins may also be present in such secretions (e.g. airways—MUC4 and

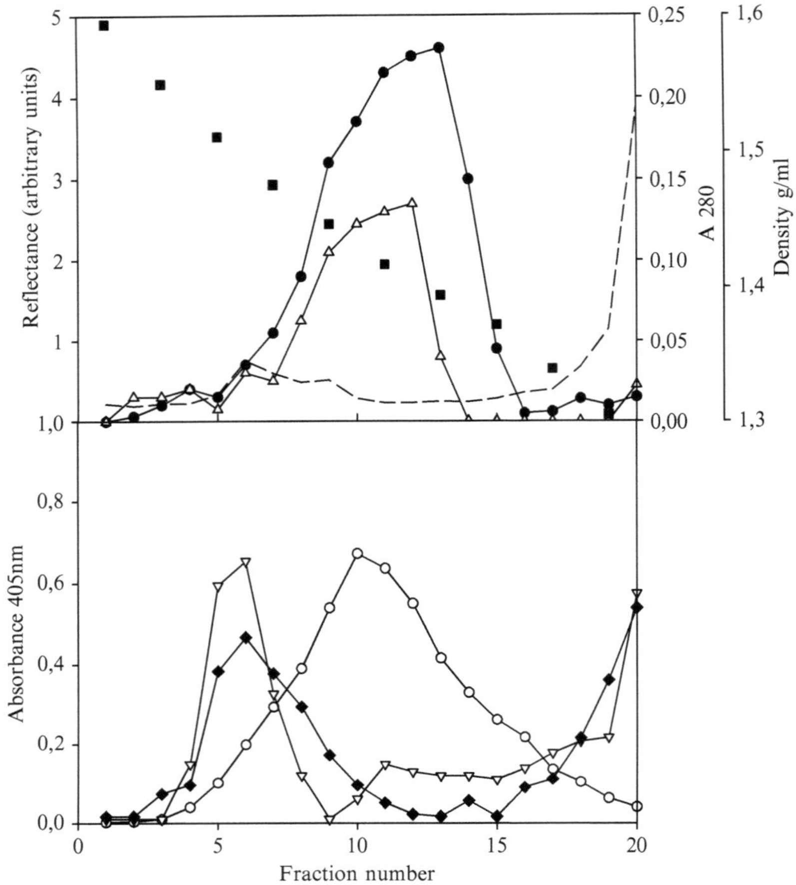

Fig. 2. Isopycnic density-gradient centrifugation in 4 M guanidinium chloride/CsCl of secretions from normal human bronchial epithelial (NHBE) cells in culture. The gel phase of apical washings from cultured NHBE cells was subjected to density-gradient centrifugation in a Beckman Optima centrifuge (36,000 rpm, 15°C, approximately 80 h, Beckman 50.2 Ti rotor) in CsCl, 4 M guanidinium chloride (starting density 1.39 g/mL). Fractions were analysed for density (*filled square*), A280 (*dashed line*), and the MAN5AC-I antibody recognizing MUC5AC (*filled circle*), LUM5B antibody recognizing MUC5B (*open triangle*), LUM16-2 antibody recognizing MUC16 (*inverted triangle*), LUM4-18 antibody recognizing MUC4 (*filled diamond*), and 214D4 antibody recognizing MUC1 (*open circle*).

MUC16). While there are some differences between the different mucins, overall they share many common structural features, in particular their high molecular weights and high sugar content. While these features allow their purification from the majority of proteins and other glycoproteins in mucus secretions, they make it difficult to separate different mucin species at the level of the intact polymer (Fig. 2). Routinely, after solubilisation using guanidinium chloride (as described above) or non-denaturing solvents such as NaCl, CsCl density-gradient centrifugation is employed to purify mucins from nucleic acids and proteins present in secretions and tissue or cell extracts. Further separation of mucins from other smaller glycoproteins can then be achieved by gel-filtration chromatography. However, due to their similar properties (size and buoyant density), it may be difficult to cleanly separate different populations using this combined procedure. It may be possible to separate mucins by anion-exchange chromatography, but this is only achievable using mucins that have been depolymerised into their constituent monomers after treatment with reducing agents.

3.2.1. Isopycnic Density-Gradient Centrifugation Under Denaturing Conditions

This procedure can be applied to guanidinium chloride extracts of mucus secretions, tissues, cell culture medium, and cell layers (see Note 12).

1. Isopycnic density-gradient centrifugation in CsCl/4 M guanidinium chloride
   - (a) Dialyse samples against ten volumes of 6 M guanidinium chloride extraction buffer. The volume of the sample that can be run in each tube is two-thirds of the total volume held by the tube.
   - (b) For practical purposes, the preparation of gradient is carried out by weighing rather than measuring volumes. Check the volume by weighing (the density of 6 M guanidinium chloride is 1.144 g/mL; see Note 2). If the sample volume is less than two-thirds of the total, fill up to the required volume with 6 M guanidinium chloride.
   - (c) Weigh the required amount of CsCl to give the correct density into a beaker (see Note 13).
   - (d) Add the sample to the CsCl stirring gently.
   - (e) The final weight of the sample is calculated from the volume of the tube and the final density of the solution. Add sodium phosphate buffer to give the final weight and stir the sample gently.
   - (f) Measure the density of the sample prior to loading with a syringe and cannula into the tubes. Balance the tubes carefully and seal according to the manufacturer's instructions.
   - (g) Centrifuge the samples. We use a Beckman Optima LE-80 K centrifuge and a 50.2 Ti rotor (tube capacity 40 mL) with a starting density 1.39 g/mL, a 70.1 Ti rotor (tube capacity 13 mL) with a starting density of 1.40 g/mL, or a Ti 45 rotor with a starting density of 1.45 g/mL. Samples are centrifuged at 36,000 rpm (50.2 Ti rotor) or 40,000 rpm (70.1 Ti and Ti 45 rotor) at 15°C for 72–96 h. These conditions give gradients of approx. 1.25–1.60 g/mL but vary according to the rotor geometry, starting density, and speed used (see Note 11). Care should be taken to ensure that the starting concentration of CsCl at a given rotor speed and temperature does not exceed that

recommended so that CsCl does not precipitate at the bottom of the tubes during the centrifugation run. This information should be available in the manufacturer's rotor handbook.

(h) After centrifugation, recover 20–40 fractions from the gradients by piercing the bottom of the tubes and collecting fractions with a fraction collector equipped with a drop counter. Analyse the fractions for density (by weighing a known volume) and absorbance at 280 nm, as well as the appropriate carbohydrate (see Chapter 3 in this book for details) and antibody analyses (see Note 14).

(i) Large amounts of proteins/lipids in the samples, due to the use of insufficient volumes of extraction buffer, may lead to a poor separation between these molecules and mucins. In this case, mucin-containing fractions may be pooled and subjected a second time to density-gradient centrifugation in CsCl/4 M guanidinium chloride. SDS-polyacrylamide gel electrophoresis of the mucin-containing fractions may be used to determine whether all proteins have been removed.

2. Isopycnic density-gradient centrifugation in CsCl/0.5 M guanidinium chloride

Density-gradient centrifugation in CsCl/4 M guanidinium chloride may be followed by subjecting the mucin-containing fractions to a second density-gradient step in CsCl/0.5 M guanidinium chloride, which gives a better separation between mucins and DNA (see Note 15).

(a) Dialyse samples against ten volumes of 0.5 M guanidinium chloride buffer.

(b) Measure the volume of the sample by weighing (the density of 0.5 M guanidinium chloride is 1.015 g/mL; see Note 2).

(c) Weigh caesium chloride to give the required density into a beaker (see Note 13).

(d) Add the sample (volume must not exceed three-quarters of the total volume held by the tube).

(e) If required, add 1% CHAPS solution to give a final concentration of 0.01% (i.e. 1% of the total volume) (see Note 15).

(f) The concentration of guanidinium chloride in the final volume must be 0.5 M, and the volume of the CsCl and CHAPS must therefore be compensated by the addition of a small volume of 8 M guanidinium chloride.

(g) The final weight of the sample is calculated from the volume of the tube and the final density of the solution.

Add sodium phosphate buffer to give final weight and stir the sample gently.

(h) Measure the density of the sample and load into tubes with a syringe and cannula. Seal the tubes according to the manufacturer's instructions.

(i) Centrifuge the samples at 36,000 rpm (50.2 Ti rotor, starting density 1.50 g/mL) or 40,000 rpm (70.1 Ti rotor, starting density 1.52 g/mL, or Ti 45 rotor, starting density 1.52 g/mL) at 15°C for 72–96 h. These starting conditions give gradients of approximately 1.35–1.67 g/mL, but vary according to the rotor geometry, starting density, and speed used (see Note 11). Care should be taken to ensure that the starting concentration of CsCl at a given rotor speed and temperature does not exceed that recommended so that CsCl does not precipitate at the bottom of the tubes during the centrifugation run. This information is available in the manufacturer's rotor handbook.

(j) After centrifugation, recover 24–40 fractions from the gradients by piercing a hole in the bottom of the tubes and collecting fractions with a fraction collector equipped with a drop counter or by manually counting drops. Analyse the fractions for density (by weighing a known volume) and absorbance at 280 nm, as well as the appropriate carbohydrate (see Chapter 3 in this book for details) and antibody analyses (see Note 14).

3.2.2. Isopycnic Density-Gradient Centrifugation Under Associative Conditions

This procedure can be applied to extracts of mucus secretions, cell culture medium, and membranes from cultured cells obtained without the use of guanidinium chloride (see Note 16).

Isopycnic Density-Gradient Centrifugation in CsCl/0.1 M Sodium Chloride

1. Check the volume of the sample by weighing (it should not exceed 4/5 of the total volume held by the tube).
2. Weigh the required amount of CsCl to give the correct starting density into a beaker (see Note 13).
3. Add the sample to the CsCl stirring gently.
4. The final weight of the sample is calculated from the volume of the tube and the final density of the solution. Add 0.1 M sodium chloride if necessary to give the final weight and stir gently.
5. Measure the density of the sample prior to loading with a syringe and cannula into the tubes. Balance the tubes carefully and seal according to the manufacturer's instructions.
6. Centrifuge the samples. We use a Beckman Optima LE-80 K centrifuge and a 50.2Ti rotor (tube capacity 40 mL) with a starting density of 1.45 g/mL, a 70.1 Ti rotor (tube capacity 13 mL)

with a starting density of 1.40 g/mL, or a Ti 45 rotor with a starting density of 1.45 g/mL. Samples are centrifuged at 36,000 rpm (50.2Ti rotor) or 40,000 rpm (70.1 Ti and Ti 45 rotor) at 15°C for 72–96 h. These conditions give gradients of approx 1.30–1.60 g/mL but vary according to the rotor geometry, starting density, and speed used (see Note 11). Care should be taken to ensure that the starting concentration of CsCl at a given rotor speed and temperature does not exceed that recommended so that CsCl does not precipitate at the bottom of the tubes during the centrifugation run. This information should be available in the manufacturer's rotor handbook.

7. After centrifugation, recover 20–40 fractions from the gradients by piercing the bottom of the tubes and collecting fractions, either using a fraction collector equipped with a drop counter or by measuring volume. Analyse the fractions for density (by weighing a known volume) and absorbance at 280 nm, as well as the appropriate carbohydrate and antibody analyses (see Note 14).

3.2.3. Gel Chromatography

1. Sephacryl S-1000, Sepharose CL-2B, Sepahacryl S-500, or Superose 6
   (a) Elute columns with 4 M guanidinium chloride buffer at flow rates well below the maximum recommended for each column.
   (b) Dialyse samples against 4 M guanidinium chloride and apply them to the column through a loading loop.
   (c) Monitor the eluate online with an ultraviolet (UV) monitor and collect fractions using a fraction collector and subject them to the appropriate carbohydrate (see Chapter 3 in this book for details) and antibody analyses (see Note 14).

3.2.4. Ion-Exchange Chromatography

Run Mono Q or Resource Q columns on a system comprising an HPLC system with an automated injector. We routinely use an Akta Purifier at a flow rate of 0.5 mL/min. All connections are made using Teflon or Peak tubing.

1. Equilibrate the column with ten volumes of buffer A or buffer C.
2. Dialyse the sample exhaustively or dissolve sample in buffer A or buffer C and apply the sample to the column.
3. Run the column with 5 volumes of buffer A or buffer C and then with a linear gradient up to 100% buffer B or buffer D (gradient equivalent to 30-column volumes).
4. Monitor the eluate online with a UV monitor and collect fractions (we collect 0.5 mL) using a fraction collector and subject to the appropriate carbohydrate (see Chapter 3 in this book for details) and antibody analyses (see Note 14).

3.2.5. Agarose Gel Electrophoresis and Transfer to Membrane Support

1. Dissolve agarose in running buffer in a microwave. We routinely use 0.7% gels for intact mucin polymers and 0.7–1.0% gels for reduced mucins.
2. Allow to cool to around 40°C and pour the gel into the casting trough. Insert well-forming comb (see Note 17).
3. Leave to set for at least 1 h prior to use.
4. Load sample in situ in the gel tank and then submerge gel with running buffer. Run samples for 16 h at 30 V at room temperature. Alternatively, electrophoresis can be performed for 3 h at 30 V.
5. Transfer the gel to transfer buffer for 5 min. If samples were unreduced, the mucins can be reduced by incubating the gel with 10 mM DTT in transfer buffer for 30 min at room temperature prior to transfer (see Note 18).
6. Transfer mucins to membrane support (nitrocellulose or PVDF) by vacuum blotting (we use a Vacu-Gene XL); suction pressure 45 mBar for 2 h.
7. The membrane can be stained with PAS reagent (see Chapter 3 in this book for details) or probed with mucin-specific antisera (see Note 14).

4. Notes

1. While commercially available preparations such as freeze-dried porcine gastric and bovine submaxillary mucins can be used for some purposes, they contain mixtures of gel-forming mucins, which, in our experience, are often highly degraded. They are, thus, of limited value for use in most of the applications presented here and this should be taken into account when interpreting data obtained using such preparations.
2. The molarity of a guanidinium chloride solution can be calculated from the density according to the following formula:

$$M = \frac{\rho - 1.003}{0.02359},$$

   where M is the molarity and ρ is the density in grams per millilitre.
3. Inhibitors are added to the 6 M guanidinium chloride extraction buffer in order to block the activity of the three classes of proteolytic enzymes: metalloproteases, serine proteases, and thiol proteases. The action of metalloproteases is inhibited by the addition of 5 mM sodium EDTA to the buffer. This can be added during the initial preparation of the buffer since it is

stable at room temperature. Thiol proteases are inactivated by the addition of 5 mM NEM. NEM has the added advantage that it blocks exchange reactions between free thiol groups and disulphide bonds. DFP is a potent inhibitor of serine proteases and esterases, including acetylcholinesterase, and should therefore be handled with extreme care. DFP is supplied in 1-g vials with a septum, and prior to dilution vials should be cooled on ice to reduce the vapour pressure. Working in a fume cupboard, the septum should be pierced with a needle to equilibrate the pressure, and the DFP should be transferred using a syringe and needle. The contents of the vial should be placed directly into the correct volume of ice-cold dry propan-1-ol to give a 100 mM solution. DFP is unstable in water but can be stored at −20°C in propan-1-ol. After dilution, the vial as well as the needles and syringes should be rinsed with 1 M NaOH to inactivate the DFP. PMSF can be used at a concentration of 0.1 mM in place of DFP. However, the low solubility of PMSF in water makes it a less attractive option, although it is possible to prepare a stock solution of PMSF in an organic solvent which is miscible with water.

4. Reduced subunits (mucin monomers) can be prepared from polymeric gel-forming mucins by treatment with reducing agents. For this purpose, the sample should be dialysed into, or dissolved in, 6 M guanidinium chloride reduction buffer. To break the disulphide bonds between the monomers, the reducing agent DTT should then be added to give a final concentration of 10 mM and the sample incubated for 5 h at 37°C. IAA is then used to alkylate the free thiol groups and prevent their re-association. IAA should be added to give a 25 mM solution and the sample maintained overnight at room temperature in the dark.
5. In our laboratories, we generally use CsCl, rather than CsBr or Cs_2SO_4, as the density-gradient forming salt since our gradients are often run in the presence of guanidinium chloride and the use of CsBr or $CsSO_4$ would give rise to mixed caesium salts. Gel-forming MUC5B mucins and nucleic acids have, however, been shown to be well-resolved in density gradients of Cs_2SO_4/10 mM sodium phosphate buffer, pH 6.5, whereas the separation obtained in CsBr/10 mM sodium phosphate buffer, pH 6.5, is poor (9).
6. In our experience, mucins show a tendency to adhere to Sephacryl columns, which is seen to a lesser extent with the Sepharose CL-2B or Superose 6 gels.
7. Sephacryl S-500HR and Superose 6 columns can be purchased pre-packed or packed in the laboratory into custom-sized columns according to manufacturer's instructions. Sepharose

CL-2B and Sephacryl S-1000 columns cannot be purchased pre-packed and must be prepared in the laboratory. The columns should be sufficiently large so that sample volumes do not exceed 5% of the total column volume. Sephacryl S-1000 gel is sufficiently porous for some polymeric gel-forming mucin species, e.g. MUC5B from saliva, to be separated from each other. On Sepharose CL-2B and Sephacryl S-500 columns, the whole polymeric gel-forming mucins, as well as the cell-associated MUC16 mucin, elute in the void volume and can thus be resolved from reduced subunits or cell-associated species, such as MUC1, which are included on the gels. Superose 6 columns are of value for the separation of reduced mucin subunits that elute in the void volume from mucin glycopeptides which elute in a more included volume.

8. We use lithium perchlorate as the eluting salt because many of the assays we originally employed were compatible with this salt; however, sodium chloride could be used instead. The technique we employ works well for cell-associated mucins and reduced mucin subunits from the gel-forming species but not for intact polymeric mucins. For the reduced subunits, 6 M urea and CHAPS are included in the buffer (buffers C and D) to minimise non-specific hydrophobic interactions with the column matrix that can occur as a result of opening up hydrophobic regions of the polypeptide by reduction.

9. For mucus gels from most sources, the degree to which the large polymeric mucins are solubilised by treatment with guanidinium chloride depends largely on the extraction conditions. Our operational definition of the "guanidinium chloride insoluble" fraction is the material that appears in the centrifugation pellet after three rounds of extraction with 6 M guanidinium chloride extraction buffer. However, for MUC2 mucus gels from the intestine, most of the material is resistant to extraction with 6 M guanidinium chloride extraction buffer and is only brought into solution after reduction and alkylation. The presence of an additional cross-linking bond that stabilises the MUC2 gel network has been postulated to account for this observation (10).

10. Preparation of mucins from mucus secretions without the addition of guanidinium chloride as a solubilising agent can lead to enrichment of non-gel-forming species or degraded molecules since complexes of gel-forming mucins may be pelleted during the centrifugation step.

11. The rotor type, speed, and starting densities rather than the g-force are given for the density-gradient runs. In our experience, for fixed angle rotors such as the 70.1Ti, 50.2Ti, or Ti45, the rotor geometry (the tube angle within the rotor), starting

density, and speed at which the run is conducted are the most important factors determining the gradient formed and the conditions cannot necessarily be reproduced using the same g-force in another rotor.

12. For cell-associated mucin complexes (e.g. MUC4α and MUC4β or ASGP-1 and ASGP-2), isolated by detergent extraction from cell membranes, isopycnic density-gradient centrifugation in the presence of guanidinium chloride leads to dissociation of the two subunits in the gradient.
13. The amount of caesium chloride needed to give a required density in 4 M or 0.5 M guanidinium chloride can be calculated according to the following formula:

$$x = v\,(1.347\rho - 0.0318M - 1.347),$$

where x is CsCl (grams), v is the final volume, M is the molarity of the guanidinium chloride (4 or 0.5), and ρ is the starting density (grams per milliliter). If no guanidinium chloride is used, the amount of caesium chloride needed (x) is equal to the final volume (v) multiplied by the starting density (ρ).
14. A range of different antibodies is available for the detection of gel-forming and cell-associated mucins. Table 1 shows antibodies that have been used successfully in our laboratories for this purpose.
15. Some mucins show a tendency to precipitate in the presence of CsCl at low concentrations of guanidinium chloride, and 0.1% CHAPS is sometimes added to the gradient to counteract this effect.
16. For complexes of cell-associated mucins isolated from cell membranes using Triton X-100 extraction, Triton X-100 or sodium deoxycholate should be added to the gradient to give a final concentration of 0.5% in the solution. Under associative conditions (i.e. in the absence of guanidinium chloride), no dissociation between the subunits should occur in the gradient.
17. The comb size (width relative to thickness) is an important factor in the quality of the data. Thin combs yield better quality data resulting in sharper bands.
18. Because of the large size of the intact polymers, their transfer from the gel onto the membrane support (needed for mucin detection) is problematic. Therefore, prior to vacuum transfer, the mucin polymers are converted into their constituent monomers.

Table 1
Antibodies used for the detection of mucin proteins

Mucin	Antibody name	Recognition site	Supplier/references
MUC1	214D4 (mouse monoclonal)	Extracellular domain	Millipore
MUC2	LUM2-3 (rabbit polyclonal)	C-terminal side of large tandem repeat	(11)
MUC4	LUM4-18 (rabbit polyclonal)	Extracellular domain	Not published
MUC5AC	LUM5-1 (rabbit polyclonal)	Unique sequences flanking tandem repeats	(11, 12)
	MAN5AC-I (rabbit polyclonal)	Unique sequences flanking tandem repeats	(13)
	45 M1 (mouse monoclonal)	Unique sequences flanking tandem repeats	Abcam, Santa Cruz
	Mab2011	Tandem repeat (precursor)	Chemicon Int. CA
MUC5B	LUM5B-2 (rabbit polyclonal)	Cysteine-rich domains of central region	(14)
	LUM5B-3 (rabbit polyclonal)	Cysteine-rich domains of central region	(14)
	LUM5B-4 (rabbit polyclonal)	C-terminal domain	(14)
	MAN5B-I (rabbit polyclonal)	Cysteine-rich domains of central region	(15)
	MAN5B-III (rabbit polyclonal)	Unique sequence flanking tandem repeats	(16)
	EUMUC5B-2a (mouse monoclonal)	Cysteine-rich domains of central region	(17)
	EUMUC5B-2b (mouse monoclonal)	Cysteine-rich domains of central region	(17)
	LUM5B-13 (rabbit polyclonal)	Tandem repeat (precursor)	Not published
MUC6	LUM6-1 (rabbit polyclonal)	Tandem repeat	(18)
MUC7	LUM7-1 (rabbit polyclonal)	N-terminal of tandem repeat domain	(19)
	LUM7-2 (rabbit polyclonal)	N-terminal of tandem repeat domain	(19)
	EUMUC7-1a (mouse monoclonal)	N-terminal of tandem repeat domain	(17)
MUC16	LUM16-2 (rabbit polyclonal)	Extracellular domain	(20)
	OC125 (mouse monoclonal)	Extracellular domain	NeoMarkers, CA, USA
	M11 (mouse monoclonal)	Extracellular domain	NeoMarkers, CA, USA
	LUM16-4 (rabbit polyclonal)	Cell-associated domain	(20)

Acknowledgements

The authors acknowledge support from The Crafoord Foundation, Sweden; The Knowledge Foundation (Biofilms—Research Centre for Biointerfaces), Malmö University, Sweden; The Swedish Patent Revenue Fund for Research in Preventive Odontology; The Swedish Dental Association; BBSRC; The Wellcome Trust; MRC; and the Horserace Betting Levy Board, UK.

References

1. Lindén SK, Sutton P, Karlsson NG, Korolik V, McGuckin MA. (2008) Mucins in the mucosal barrier to infection. *Mucosal. Immunol.* **1**, 183–197.
2. Thornton DJ, Rousseau K, McGuckin MA. (2008) Structure and function of the polymeric mucins in airways mucus. *Annu. Rev. Physiol.* **70**, 459–486.
3. Hattrup CL, Gendler SJ. (2008) Structure and function of the cell surface (tethered) mucins. *Annu. Rev. Physiol.* **70**, 431–457.
4. Holmén JM, Karlsson NG, Abdullah LH, Randall SH, Sheehan JK, Hansson GC, Davis, CW. (2004) Mucins and their O-Glycans from human bronchial epithelial cell cultures. *Am J Physiol Lung Cell Mol Physiol.* **287**, 824–834.
5. Kirkham S, Sheehan JK, Knight D, Richardson PS, Thornton DJ. (2002) Heterogeneity of airways mucus: variations in the amounts and glycoforms of the major oligomeric mucins MUC5AC and MUC5B. *Biochem. J.* **361**, 537–546.
6. Thornton DJ, Khan N, Mehrotra R, Howard M, Veerman E, Packer NH, Sheehan JK. (1999) Salivary mucin MGl is comprised almost entirely of different glycosylated forms of the MUC5B gene product. *Glycobiology* **9**, 293–302.
7. Sherblom A, Carraway KL. (1980) A complex of two cell surface glycoproteins from ascites mammary adenocarcinoma cells. *J. Biol. Chem.* **255**, 12051–12059.
8. Hull SR, Sheng Z, Vanderpuye O, David C, Carraway KL. (1990) Isolation and partial characterization of ascites sialoglycoprotein-2 of the cell surface sialomucin complex of 13762 rat mammary adenocarcinoma cells. *Biochem. J.* **265**, 121–129.
9. Davies JR, Carlstedt I. (1999) Isolation of large gel-forming mucins. In 'Methods in Molecular Biology. Glycoprotein methods and protocols.' pp 3–13. Humana Press NJ, USA.
10. Herrmann A, Davies JR, Lindell G, Mårtensson S, Packer N, Swallow D, Carlstedt I. (1999) Studies on the "insoluble" glycoprotein complex from human colon. Identification of reduction-insensitive MUC2 oligomers and C-terminal cleavage. *J. Biol. Chem.* **274**, 15828–15836.
11. Hovenberg HW, Davies JR, Herrmann A, Lindén CJ, Carlstedt I. (1996) MUC5AC, but not MUC2, is a prominent mucin in respiratory secretions. *Glycoconj. J.* **13**, 839–847.
12. Hovenberg HW, Davies JR, Carlstedt I. (1996) Different mucins are produced by the surface epithelium and the submucosa in human trachea: identification of MUC5AC as a major mucin from the goblet cells. *Biochem. J.* **318**, 319–324.
13. Thornton DJ, Carlstedt I, Howard M, Devine PL, Price M, Sheehan JK (1996) Respiratory mucins: identification of core proteins and glycoforms. *Biochem. J.* **316**, 967–975.
14. Wickström C, Davies JR, Ericsen G, Veermann E, Carlstedt I. (1998) MUC5B is a major gel-forming, oligomeric mucin from human salivary gland, respiratory tract and endocervix: identification of glycoforms and C-terminal cleavage. *Biochem. J.* **334**, 685–693.
15. Thornton DJ, Howard M, Khan N, Sheehan JK. (1997) Identification of two glycoforms of the MUC5B mucin in human respiratory mucus. Evidence for a cysteine-rich sequence repeated within the molecule. *J. Biol. Chem.* **272**, 9561–9566.
16. Thornton DJ, Gray T, Nettesheim P, Howard M, Koo JS, Sheehan JK. (2000) Characterisation of the MUC5AC and MUC5B mucins synthesised by cultured normal human tracheobronchial epithelial cells. *Am. J. Physiol. Lung Cell. Mol. Physiol.* 278, L1118–1128.
17. Rousseau K, Wickström C, Whitehouse DB, Carlstedt I, Swallow DS (2003) New monoclonal antibodies to non-glycosylated domains of the secreted mucins MUC5B and MUC7. *Hybrid. Hybridom.* 22, 293–299.

18. Sylvester PA, Myerscough N, Warren BF, Carlstedt I, Corfield AP, Durdey P, Thomas MG. (2001) Differential expression of the chromosome 11 mucin genes in colorectal cancer. *J. Pathol.* **195**, 327–335.
19. Wickström C, Christersson C, Davies JR, Carlstedt I. (2000) Macromolecular organization of saliva: identification of 'insoluble' MUC5B assemblies and non-mucin proteins in the gel phase. *Biochem. J.* **351**, 421–428.
20. Davies JR, Kirkham S, Svitacheva N, Thornton DJ, Carlstedt I. (2007) MUC16 is produced in tracheal surface epithelium and submucosal glands and is present in secretions from normal human airway and cultured bronchial epithelial cells. *Int. J. Biochem. Cell Biol.* **39**, 1943–1954.

Chapter 3

Detecting, Visualising, and Quantifying Mucins

Ceri A. Harrop, David J. Thornton, and Michael A. McGuckin

Abstract

The extreme size, extensive glycosylation, and gel-forming nature of mucins make them a challenge to work with, and methodologies for the detection of mucins must take into consideration these features to ensure that one obtains both accurate and meaningful results. In understanding and appreciating the nature of mucins, this affords the researcher a valuable toolkit which can be used to full advantage in detecting, quantifying, and visualising mucins. The employment of a combinatorial approach to mucin detection, using antibody, chemical, and lectin detection methods, allows important information to be gleaned regarding the size, extent of glycosylation, specific mucin species, and distribution of mucins within a given sample. In this chapter, the researcher is guided through considerations into the structure of mucins and how this both affects the detection of mucins and can be used to full advantage. Techniques including ELISA, dot/slot blotting, and Western blotting, use of lectins and antibodies in mucin detection on membranes as well as immunohistochemistry and immunofluorescence on both tissues and cells grown on Transwell™ inserts are described. Notes along with each section advice the researcher on best practice and describe any associated limitations of a particular technique from which the researcher can further develop a particular protocol.

Key words: Mucin, Detection, ELISA, Western blotting, Immunohistochemistry, Antibody, Lectin, Bronchial epithelial cell culture

1. Introduction

The extreme size, extensive glycosylation, and gel-forming nature of mucins make them a challenge to work with, and caution must be taken when considering methodologies for detection, visualisation, and quantification of these molecules to ensure that one obtains both accurate and meaningful results. In order to accurately detect and quantify mucins, it is vital to understand the nature of these molecules: mucins are a family of high-molecular-weight glycoproteins (0.25–20 MDa) containing one or more domains that are heavily O-glycosylated (up to 90% of the weight of the molecule is made up from carbohydrate). Mucins can be

Michael A. McGuckin and David J. Thornton (eds.), *Mucins: Methods and Protocols*, Methods in Molecular Biology, vol. 842, DOI 10.1007/978-1-61779-513-8_3, © Springer Science+Business Media, LLC 2012

divided into two major subcategories, one being monomeric and the other polymeric. The monomeric mucins are mostly membrane bound while the polymeric mucins are secreted from the cell.

Membrane-bound mucins are typically made up of two subunits: the larger of the two subunits is the extracellular mucin domain, which is heavily glycosylated, and the smaller subunit is made up of a short extracellular region, the transmembrane region, and a cytoplasmic tail region (1). Between the mucin domain and the transmembrane domain is the sea urchin sperm protein, enterokinase, and agrin (SEA) domain which contains a proteolytic cleavage site.

In common with the membrane-bound mucins, the central domain of the secreted mucins is dominated by a series of tandem repeats, rich in serine, threonine, and proline. Serine and threonine are the sites of attachment of a wide variety of O-glycans via the linkage sugar N-acetylgalactosamine (GalNAc). These O-glycans are often sialylated or sulphated, making mucins negatively charged and providing a means of chemical detection. The cysteine-rich domains at the N- and C-termini enable mucin multimerisation, and speculation in a recent review discussed the idea that mutations in the N- and C-termini of mucins might lead to multimerisation defects (detectable by agarose gel electrophoresis, western blotting, and antibody detection) (2). The central cysteine domains are dispersed throughout the extensive O-glycosylated central region (3–6). In comparison to the heavily glycosylated serine and threonine residues, the cysteine domains (or "cys" domains) have previously been termed the "naked" domains due to their apparent lack of glycosylation (7, 8). The functions of the central cys domains are largely unknown, although antibodies raised specifically against sequences within these domains provide useful tools for detecting mucins regardless of their state of glycosylation.

To consider detecting mucins in a given sample, it is crucial that the researcher understands how mucins behave in solution, paying particular attention to their relative insolubility in conventional physiological buffers. In order to prepare secreted mucins for analysis, for example from clinical samples, mucins must first be extracted in chaotropic agents, such as 6 M urea or 6 M guanidinium chloride. Further purification using caesium chloride density gradients, both with and without 4 M guanidinium chloride, is employed to purify mucins from other contaminating proteins and nucleic acids, and mucin detection methodologies must be suitable for use in the presence of such agents. Researchers are referred to Chapter 2 of this volume for methods and discussion of the preparation and purification of secreted and membrane-associated mucins.

In light of points addressed regarding the polymeric and glycosylated nature of mucins, when choosing methods for their detection, the researcher must consider both the core protein sequence and the extent and nature of glycosylation. From a practical perspective, the number of samples to analyse, the degree of quantitation required, and, moreover, the availability of reagents and methodologies need to be considered when deciding on the most suitable approach. Broadly, there are two main methods for the detection of mucins: using chemical stains or lectins to utilise the highly glycosylated nature of the mucins or the use of peptide sequence-specific antibodies for detection.

Several commonly used chemical stains are employed to detect mucins, including Periodic Acid/Schiff's (PAS) reagent (9), high iron diamine (HID), and alcian blue (AB). PAS works by the acid oxidising vicinal diols on the sugar residues and then reacting with the Schiff's reagent to yield a distinctive pink–purple colour. HID colours sulphated residues on mucins black while alcian blue staining can be used at pH 1 to stain sulphate residues and at pH 2.5 to stain both sulphate and sialic acid groups. Assays directed towards detecting mucins via their glycosylated domains are probably most useful as a fast detection of mucins during intermediate purification steps (for example, to identify in which fractions mucins are present following caesium chloride density gradient centrifugation, Chapter 2). However, in unspecified samples, staining of non-mucin glycoproteins is an obvious but significant disadvantage of chemical staining techniques. Lectin staining goes some way to address specificity, by allowing staining of specific carbohydrate groups, although both chemical and lectin staining methods fail to distinguish between specific mucin gene products. In order to detect specific mucins, antibodies raised against specific mucin peptide sequences become important. It is worth noting here that, perhaps rather obviously, the domain of the mucin to which the antibody has been raised alters what the antibody is most useful for: antibodies raised to sequences in the STP-rich domains of mucins detect mucin precursors, for example from cell lysates, but are of little use in detecting the secreted glycosylated whole mucin, for example in cell culture media, washes, and secretions. Although deglycosylation methods can be used to increase the effectiveness of these probes in the detection of mature mucins in a range of samples (10–13), it is probably more accurate and efficient to use antibodies raised against sequences that are exposed in both the precursor and mature mucin (N- and C-termini and the central "cys" domains). While antibodies provide a more specific detection toolkit than chemical stains, it should be noted that different *MUC* gene products can share regions of homology and therefore a certain amount of cross-reactivity (14). For this reason, use of multiple detection methods with differing principles is recommended. By way of a worked example, if one were to study the production of two specific mucins by a cell culture model, the use of one or more antibodies specific for each mucin would be the best technique to use. In contrast, using a single specific mucin antibody would not be suitable to gain a global picture of mucin composition within a clinical sample of mucus as this would likely result in one or more mucin species being missed. In this case, chemical and/or lectin stains may be more appropriate to detect mucins throughout the mucin purification strategy. Combining both antibody and chemical stains can be a powerful tool, for example in immunohistochemical staining of tissue sections, as can dual-antibody labelling approaches. Sections through cultures of primary bronchial epithelial cells, which produce MUC5AC and MUC5B, can be first stained with a mucin-specific antibody and then counterstained with, for example, alcian blue, allowing the visualisation of both mucins within one population of cells. Such dual staining also goes someway to identify if cells are likely to be producing both mucin species or if populations are distinct. In summary, the extent and nature of mucin glycosylation influence detection techniques, whether they are carbohydrate or peptide specific, and both should be considered when planning which mucin detection methodologies to be utilised.

2. Materials

1. Immunoassays are most conveniently performed in 96-well plates using 50–100 μL incubation volumes; plates with a range of protein binding properties are commercially available.
2. Immunoassay buffers: CB = 0.1 M carbonate buffer, pH 9.6; PBS = 0.05 M phosphate, 0.9% (w/v) NaCl, 0.02% (w/v) KCl, pH 7.2; TBS = 0.01 M Tris–HCl, 0.9% (w/v) NaCl, pH 7.5.
3. Blocking solutions for ELISA and immunoblotting: 10% (w/v) skim milk powder, 1–5% (w/v) BSA, 1–5% (w/v) casein, or 10% (v/v) serum (of a different species type to detection antibodies) in PBS; non-ionic detergents: 0.05% (v/v) NP40 or Tween-20.
4. Soluble ELISA enzyme substrates: 2-2′-azinobis(3-ethylbenzothiazoline-6-sulfonic acid)-diammonium salt (ABTS, 1 mg/mL, A_{405nm}), O-phenyldiamine (OPD, 1 mg/mL, A_{492nm}), tetramethylbenzidine (TMB, 0.01 mg/mL, A_{450nm}) in sodium acetate with 0.01% H_2O_2 (pH 6.0), *p*-nitrophenyl phosphate (PNPP, 1 mg/mL, A_{405nm}) in 10 mM diethanolamine with 0.5 mM $MgCl_2$ (pH 9.5).
5. Secondary antibodies (see Note 1): Peroxidase-labelled secondary antibodies are used with soluble ELISA substrates (see Subheading 2, item 4) or with enhanced chemiluminescence (ECL) western blot detection reagent for Western blots. Alkaline phosphatase (AP)-conjugated secondary antibodies are used with nitroblue tetrozolium (NBT) and 5-bromo-4-

chloroindol-3yl-phosphate (BCIP) as the substrate (both at 50 mg/mL). Biotinylated secondary antibodies for immunohistochemistry (Santa Cruz) are used with a streptavidin complex horseradish peroxidase (HRP) kit (e.g. Dako) and with diaminobenzadine (DAB) detection solution. Fluorescent secondary antibodies are used for fluorescence microscopy.

6. Reagents for immunohistochemistry of tissues and cultured cells:
   (a) Fixing and processing cells on transwells: 10% neutral buffered formalin, pH 7.4, aqueous industrial-grade methylated spirit (IMS) at 50% (v/v), 70% (v/v), 90% (v/v), and 100% (v/v), 50% aqueous IMS and 50% chloroform (v/v), VA5 RalWax (Raymond A. Lamb).
   (b) Fixing tissues with Carnoy's fixative: 60% (v/v) dry methanol, 30% (v/v) chloroform, 10% (v/v) glacial acetic acid.
   (c) Non-fading mounting media with or without 4′,6-diamidino-2-phenylindole (DAPI), e.g. Vectashield (Vector laboratories) or Mowiol (Polysciences).

3. Methods

3.1. Immunoassay in Solution: ELISA and Radioisotope Assays (see Note 1)

3.1.1. Detection of Mucins in Solution Using Double-Determinant Immunoassays (see Notes 2–4)

1. Coating the capture antibody: Antibodies need to be purified to optimise coating; the concentration should be optimised for each antibody, buffer (CB or PBS), and plate type (range 0.1–2 μg/well). Incubate overnight at RT. Wash 3× for 1 min in PBS (if using alkaline phosphatase, avoid phosphate buffers, e.g. use TBS).
2. Blocking: Block non-specific binding on coated plates with protein blocking solution and/or non-ionic detergent. Block for 1–24 h at RT or 4°C. Wash 3× for 1 min in PBS. Blocked plates can be used immediately, stored in PBS for several days at 4°C or dried thoroughly, vacuum sealed in a bag with silica gel, and stored at 4°C (storage time can be over 6 months, addition of 5% (w/v) sucrose to the blocking buffer can substantially increase the shelf-life of dried plates).
3. Sample incubation (see Note 5): Incubate in humidified environment for 1–24 h at 4–37°C. Wash as per Note 5.
4. Detection antibody incubation: The required concentration of the detection antibody needs to be determined for each application (usual range 0.1–10 μg/mL). Buffers as above (do not use sodium azide if the antibody is HRP conjugated); incubation time 1–24 h at 4–37°C. Wash as above.

5. Secondary labelled antibody: Only required if detection antibody is not labelled. Optimisation and conditions as per step 4. Wash as above.
6. Detection: For enzyme assays, the choice of substrate and buffer depends on the enzyme: ABTS, OPD, or TMB for HRP; PNPP for AP. Incubate at RT or 37°C for 20 min, 1 h. Reactions can be stopped with an equal volume of 2.5% NaF or 1 M H_2SO_4 (HRP) or 0.1 M EDTA (AP) and plates read at the appropriate wavelength. For radioisotope detection, gel-forming scintillant should be added to the wells after step 6 and the radioactivity determined using a microplate isotope counter.
7. Quantitation: Quantitation is best achieved using a standard curve fitted using an appropriate line of best fit; programs are available to interface with microplate readers and isotope counters that store data and compute standard curves.

3.1.2. Detection of Mucins in Solution Using Antibody Capture Competitive Binding Immunoassays (see Notes 6–8)

1. Assay optimisation: Serially dilute the mucin down one or more 96-well plates and incubate overnight at RT, and leave one column with buffer only to control for non-specific binding (see Subheading 3.1.1 for plates and coating buffers). Wash 3× for 1 min in PBS. Block plate as in Subheading 3.1.1, step 2. Repeat wash. Prepare antibody at 10 μg/mL in selected assay buffer (see Note 5) and serially dilute across the plates. Incubate for 1–24 h at RT. Wash as in Subheading 3.1.1, step 3, and detect as in steps 5 and 6.
2. Assay: Select a dilution of antibody and antigen that gives an absorbance of about 1.5 (or about 75% of maximal radioactivity for isotope detection) and uses the least amount of coating mucin or peptide. Coat and block plates as above; coated plates can be dried and stored in vacuum-sealed bags for at least several weeks at 4°C. Prepare duplicate or triplicate samples and standards (serial dilution of mucin in sample buffer) in assay buffer containing the detection antibody at the final dilution selected above. The sample/antibody mix can be pre-incubated (1–24 h at 4–37°C) prior to transfer to the mucin-coated plate. Incubate, wash, and detect as in step 1.
3. Quantitation: Absorbance values, or radioactivity, are normally expressed as a percentage of the non-inhibited (sample blank) controls and appropriate standard curves fitted as in Subheading 3.1.1, step 7.

3.2. Dot/Slot and Western Blotting

3.2.1. Preparation of Dot/Slot Blots for Detection of Mucins (see Note 9)

1. Application of samples: Samples can be either directly applied to membranes (see Note 10) in volumes of 0.5–2.5 μL or added using a commercially available vacuum manifold device (these are preferable due to more even sample distribution, greater sample volume, and superior washing). For quantitation and comparison across blots, a standard in the same buffer

as samples should be titrated for use as a standard curve, and samples should be included on all blots to determine inter-assay variation. Equivalent amounts of a non-mucin protein should also be titrated to act as a measure of non-specific binding.

2. Washing and storage: Wash the wells in distilled water (for manifold devices) and then the entire membrane in three changes of PBS or TBS. Either proceed directly to PAS, antibody, or lectin detection (see Subheadings 3.2.2 and 3.2.3) or store the membrane either sealed in a bag in buffer at 4°C or dry thoroughly and store sealed at −20°C.

3.2.2. Detection of Mucins on Membranes Using PAS (see Notes 11–13)

1. Wash the dot/slot or Western blots in three changes of water (1 mL/cm^2) and transfer to a freshly prepared solution of 1% (v/v) periodic acid in 3% (v/v) acetic acid (1 mL/cm^2) for 30 min at RT.
2. Rinse twice (2 min, 1 mL/cm^2) in freshly prepared 0.1% (w/v) sodium metabisulfite in 1 mM HCl (SM). Transfer to Schiff's reagent (commercially available) for 15 min (0.5 mL/cm^2). PAS-reactive glycoproteins stain a pink–red colour. Wash 3× for 2 min in SM and dry the membrane in a warm air stream.

3.2.3. Detection of Mucins on Membranes Using Antibodies or Lectins (see Notes 11–14)

1. Blocking: Membranes need to be blocked with protein and/or non-ionic detergents (as for Subheading 3.1.1). The optimal blocking protein and buffer, wash buffers, and antibody buffers (to prevent non-specific binding) vary with different antibodies and lectins but can readily be optimised using 1× 1-cm^2 pieces of membrane incubated within 24-well plate wells taken through the entire staining detection procedure. In open trays with agitation, incubations and washes should use at least 0.25 mL/cm^2 of membrane. Block for 1–24 h at RT or 4°C. Wash 3× for 1 min in TBS.
2. Incubation with antibody or lectin: The concentration of antibody or lectin needs to be optimised to give the best signal-to-background ratio. As a guideline, optimal antibody and lectin concentrations usually are in the range 0.1–10 μg/mL. Antibodies can be used either unconjugated (to be followed with a conjugated antibody against the primary antibody species and class) or conjugated directly with an enzyme (e.g. HRP, AP), a ligand for a secondary enzyme conjugate (e.g. biotin, digoxigenin), or be radioactively labelled (e.g. ^{125}I). Lectins need to be conjugated usually with biotin either through amino or carbohydrate groups; biotinylated lectins are commercially available. It is recommended that replicate blots are probed with same species/isotype-irrelevant antibody to control for non-specific binding. Incubation buffers need to contain either protein (50% of blocking concentration) and/or

non-ionic detergent. Incubate for 1–24 h at 4–37°C with agitation. To save valuable reagents, this step can be performed in sealed bags with 0.125 mL/cm^2 of membrane.

3. Washing: Thorough washing with agitation is critical in immunoblotting; a good starting point is 3× for 2 min in TBS, 3× for 2 min in TBS plus 0.05% Tween-20, 3× for 2 min in TBS. Less-stringent washing may suffice for some antibodies. If non-specific binding is a problem, try washing in 1% NP40, 0.05% sodium deoxycholate, 0.1% SDS in TBS, or increase the NaCl concentration until non-specific binding is reduced and specific binding retained.
4. Secondary antibodies: The concentration of secondary antibody (or streptavidin peroxidase for biotin) also needs to be optimised to give the best signal-to-background ratio. Affinity-purified antibodies with cross-reactivity with other species antibodies deleted are best. A recommended dilution for blotting (usually in the range 1/500 to 1/20,000) is often provided with commercial conjugates. Incubation details as for primary antibody. Wash as in step 3.
5. Detection: Detection of bound enzyme conjugates can be achieved using insoluble chromogens which leave a coloured precipitate on the blot (e.g. 3,3′-diaminobenzidine, 4-chloronapthol mix for HRP (15); 5-bromo-4-chloro-3′-indoly phosphate toluidine salt, nitro blue tetrazolium chloride mix for AP). Alternatively, chemiluminescent substrates [e.g. ECL (Amersham) for HRP] can be utilised which have the major advantages of high sensitivity, allowing for several different exposures to be recorded on X-ray film, and compatibility with stripping and re-probing blots. However, beware possible non-specific results with chemiluminescent substrates on membranes distorted by vacuum manifold devices. Detection of ^{125}I-labelled antibodies is achieved by direct autoradiography with X-ray film.
6. Quantitation: Densitometry can be used to quantitate results provided that the samples have not been overloaded (exceeded the membrane binding capacity or detection system capacity). Titration of samples may aid in quantitation. Inter-assay control samples need to be included on each membrane (gel for Western blotting) if quantitation between membranes, and particularly between different electrophoresis/transfer/immunodetection runs, is required.
7. Stripping: Antibody-probed chemiluminescent-detected blots can be stripped after thorough washing in TBS by incubation at 50°C for 30 min in 2% SDS, 100 mM 2-mercaptoethanol in 62.5 mM Tris–Hcl pH 6.8. Thoroughly washed stripped blots can be stored in TBS at 4°C before re-blocking and probing.

Up to four probings are often possible and, although sensitivity gradually diminishes with each cycle, signal-to-noise ratio often increases concurrently allowing longer development times.

3.2.4. Visualising and Quantifying Mucins in Tissue Sections and Primary Bronchial Cell Cultures by Immunohistochemistry (see Notes 15–20)

Primary human bronchial epithelial cultures have been available for several years and are now used in laboratories worldwide for numerous applications, including the study of mucin production and regulation (see Note 15). Chapter 15 details establishment of these cultures, and Chapter 16 details techniques for measuring mucin secretion from these cultures.

Fixation of Tissue for Immunohistochemistry and Immunofluorescence

Standard formalin-based fixation and paraffin embedding can be used to detect mucins in animal and human tissues. However, to preserve mucus layers, fixation in Carnoy's fixative is required (specific procedures for preserving the mucus layer are detailed in Chapter 13).

Fixation of Cells on Inserts

Wash the apical cell surface 1× for 60 min with 250–500 μL PBS per insert at 37°C to remove secreted mucus. Wash the apical surface a further 3× for 1 min in PBS and the basolateral surface 3× for 1 min in PBS to remove cell culture media. Add 0.5 and 1 mL of 4% (v/v) neutral buffered formalin to apical and basolateral surfaces, respectively. Use a cork borer or similar to remove the inserts from the plastic surround. Bisect insert with scalpel blade and transfer both halves to plastic histology cassettes between sponges. Fix cells in 4% (v/v) neutral buffered formalin for 24–48 h before processing (see Note 16).

Processing and Sectioning of Inserts

It is not recommended to use standard processing and embedding but to use the following protocol (see Notes 16–18). Process samples (preferably under vacuum in automated processor) as follows: 1 h 10% (v/v) neutral buffered formalin, 1 h 50% (v/v) aqueous IMS, 1 h 70% (v/v) aqueous IMS, 1.5 h 90% (v/v) aqueous IMS, 1.5 h absolute IMS 1, 2 h absolute IMS 2, 1 h absolute IMS 3, 2 h 50% IMS 50% chloroform (v/v), 2 h chloroform 1, 2 h chloroform 2, 2 h wax 1, 2 h wax 2. Remove inserts from cassettes embedded on end (see Note 18). Cool wax blocks on ice before sectioning. Sectioned at 3-μm thickness, float sections on warm water bath (50°C) (see Note 17) and mount two duplicate sections onto each glass slide. Dry sections overnight at 37°C and place sections in a 60°C incubator for 30 min before staining.

Immunohistochemical Staining of Mucins

1. De-wax and rehydrate: Transfer slides to xylene for 30 min, then gradually from absolute IMS to 70% IMS, and block endogenous peroxidase in 3% H_2O_2 (v/v) in methanol (10 min).
2. Some antigens may require "antigen retrieval" to reveal epitopes following formalin fixation and paraffin embedding.

A range of commonly used antigen retrieval techniques are applicable to detection of mucins, including heating in citric acid buffer at pH 6 or alkaline solutions at pH 9, and use of enzymes, such as trypsin. However, due to extensive disulphide bonds in the N- and C-terminal domains of the secreted mucins, antibodies that react with epitopes in these domains often react much better after reduction of disulphide bonds and alkylation to prevent reformation of these bonds.

3. Wash and block: Wash slides in gently running tap water for 5 min. Wash 2× for 5 min in PBS. Block overnight in 10% (v/v) serum (see Note 19) and 1% (w/v) BSA in PBS.
4. Antibody detection: Rinse off blocking solution and incubate slides in primary antibody in PBS overnight at 4°C. Remove from cold and allow slides to reach RT (30 min). Wash slides 3× for 5 min in PBS. The appropriate biotinylated secondary antibody was applied in PBS for 30 min. Wash 3× for 5 min in PBS. Detection can be performed using Dako ABComplex Kit HRP (Dako, Glostrup, Denmark) or equivalent, diluted 1:200 in Tris–HCl buffer, and applied to sections for 30 min. Wash 3× for 5 min in PBS. Add DAB solution for 30 min. Rinse in distilled water and wash in running water for 5 min.
5. Counterstain: Treat slides with 1% (w/v) alcian blue in 2.5% acetic acid at pH 2.5 to stain goblet cells not identified by antibody staining. Wash slides in running water for 5 min and counterstain nuclei in Coles haematoxylin for 1 min. Wash in running water, rinse in distilled water, and transfer very briefly (4 s) to 0.5% acid alcohol to remove excess haematoxylin.
6. Clear and mount slides: Rinse in distilled water, wash in running water for 5 min, and dehydrate in graded IMS. Clear slides in xylene and apply coverslip with mounting medium.
7. Imaging: Photograph sections along their length and antibody staining can be quantified by measuring the epithelium length and the area of staining per length or volume of epithelium quantified using morphometry software, such as Imaging Associates (Carl Zeiss Ltd.) (see Note 20).

3.2.5. Visualising and Quantifying Mucins in Cell Cultures by Immunofluorescence (see Notes 21–22)

In order to visualise intracellular mucins produced by adherent mucin-producing non-primary cell lines, cells need to first be grown on sterile glass coverslips (see Note 21).

Immunofluorescence Microscopy of Cells on Coverslips

1. Fixing and permeabilising cells: Wash cells 2× for 1 min in PBS and fix in 4% (w/v) paraformaldehyde in PBS for 15 min (see Note 22). Reactive groups need to be quenched in 100 mM glycine and PBS for 5 min. Permeabilise cells in 0.1% (v/v) Triton X-100 and PBS for 5 min at RT.

2. Antibody detection: Incubate cells in the relevant primary antibody for 30 min, wash slides in PBS for 10 min, and then incubate with the relevant fluorophore-conjugated secondary antibody for 15 min in the dark. Wash slides 3× for 5 min in PBS and mount with non-fading mounting media containing DAPI to detect cell nuclei. Alternatively, cells can be pretreated with DAPI and mounted with standard non-fading mounting media onto glass slides (see Note 23).

Immunofluorescence Microscopy of Transwell™ Insert-Grown Cells

This method refers specifically to fluorescent staining on intracellular mucins in human bronchial epithelial cells cultured at air–liquid interface on Transwell™ inserts (see Note 15) (16–18), but also applicable to other mucin-producing cells in culture.

1. Fixing cells: Once fully confluent and established at air–liquid interface, fix cells in 4% (w/v) PFA for 1 h at room temperature by adding 500 μL to the apical surface of cells and then adding 1 mL to the basolateral surface. As per Subheading "Immunofluorescence Microscopy of Cells on Coverslips", reactive groups need to be quenched in 100 mM glycine and PBS for 5 min. Alternative fixatives include coagulative fixatives (e.g. 100% methanol at −20°C for 10 min before 10-min PBS wash to thoroughly rehydrate the samples can be used if issues with antibody reactivity and/or autofluorescence prove problematic). We advise to try the PFA fixation first and troubleshoot should problems arise. However, the subsequent labelling is the same for both fixation methods.
2. Remove from inserts: Using a scalpel blade and tweezers, carefully cut around the edge of the Transwell™ insert to remove from the hanging plastic frame and lower the insert into the well beneath (see Note 22).
3. Permeablise the cells in 0.1% (w/v) Triton in PBS for 15 min at RT before incubation with relevant primary antibodies. Primary antibody incubation time varies between 1 and 24 h, and then wash insert 3× for 1 min in PBS prior to incubation with fluorophore-conjugated secondary antibody plus 0.5 μM DAPI for 1 h. Wash inserts again 3× for 1 min in PBS to remove excess secondary antibody and mount insert between a glass slide and a coverslip using non-fading mounting media containing 0.25% (v/v) DABCO™. Mounted samples are left overnight in the dark at 4°C before visualising using microscope with argon laser. Use the 488-nm line to visualise Alexa-Fluor 488-labelled proteins and the 568-nm line to visualise the Alexa-Fluor 594-labelled mucins. Quantitation of mucin staining can be performed using ImageJ or similar software by quantifying the number of red/green cells compared to the number of DAPI-stained nuclei.

4. Notes

1. Secondary antibodies are available commercially from many sources but should be selected for specificity, with antibodies depleted of cross-reactivity for other species of immunoglobulin required when multiple primary antibodies are to be used.
2. Immunoassay techniques rely on reactions between antigens (in this case, mucins) and antibodies or lectins; the protocols refer to antibodies, but lectins are interchangeable. Chapter 2 in this volume details some available mucin antibodies. Immunoassays are suitable for quantitative and sensitive detection of mucins in large numbers of samples. A variety of different techniques can be devised with either both antigen and antibody being free in solution or with either being fixed to a solid phase, such as a tube, bead, or 96-well plate. All the variations require that the antigen, antibody, or a secondary antibody is labelled with an enzyme, a ligand (e.g. biotin, digoxigenin) for a labelled secondary conjugate, or a radioisotope. Optimisation of conditions for these assays is required and comprehensive texts concerning the theory and practical aspects of immunoassays are available (19). Double-determinant assays are especially useful for detection, quantitation, and characterisation of mucins due to their large size and the multivalent nature of many mucin peptide and carbohydrate epitopes. For example, a core protein epitope-specific antibody can be used for capture and then several antibodies reactive with different carbohydrate epitopes can be used for detection to both quantitate and characterise a particular mucin. However, it is extremely important that the previous warnings regarding the influence of mucin glycosylation on antibody reactivity are heeded (see Subheading 1). For example, almost all the commercially available MUC1 assays utilise a double-determinant enzyme or radioisotope format. We have shown that assays using antibodies with similar VNTR epitopes can have substantially different capture and detection characteristics. Subheading 3.1.1 describes a double determinant format and Subheading 3.1.2 a competitive binding assay using antibody capture.
3. In this protocol, purified mucin-reactive antibodies or lectins are coated onto microtitre plates and used to capture mucins in biological samples. Detection antibodies or lectins are then introduced to react with the captured mucins. If the detection antibody or lectin is not labelled, then secondary enzyme (HRP or alkaline phosphatase) or radioisotope-labelled antibody is used to quantify the amount of captured mucin.

4. The capture antibody/lectin is critical as it determines which mucin molecules are available for detection by the detection antibody/lectin. Choice is governed by knowledge of the mucin to be measured and availability of specific antibodies. If capture antibodies are to be detected by secondary antibody conjugates, then capture and detection antibodies need to be of differing species or isotypes. Although it is possible to use combinations of both capture and detection antibodies with different specificities, the interpretation of binding is problematical, and performing distinct assays, while more time consuming, is more informative.
5. Samples need to be added in a buffer compatible with antibody–antigen reactions. Avoid high concentrations of chaotropic agents, SDS, and reducing reagents (because immunoassays are sensitive, this can often easily be achieved through dilution); interference by specific factors can be tested easily by progressive addition. Some biological fluids can be assayed neat but often cause interference problems. Addition of protein (50% of blocking concentration) and/or 0.05% non-ionic detergent should be trialled. Serial dilution of antigen in antigen-free assay fluid needs to be performed to validate the assay; this should also be the form of the standard curve included on each plate along with a sample buffer blank. Each sample should be assayed at least in duplicate. Multiple aliquots of several samples at different levels of the standard curve should be prepared for inclusion on each plate as a measure of inter-assay variation. Thorough washing is important (e.g. 3× for 1 min in PBS–0.05% Tween-20, 3× for 1 min in PBS). More or less stringent washing may be needed for some antigens/antibodies; if non-specific binding is a problem, try different detergents and gradually increasing the NaCl concentration of the wash buffer. Enzymatic or chemical deglycosylation can be employed before, during, or following antigen capture; however, it must be ensured that the techniques are compatible with maintenance of the antibody–antigen reaction. For example, treatment of serum with neuraminidase (0.1 U/mL in 50 mM sodium acetate, 1 mM $CaCl_2$, 154 mM NaCl, pH 5.5) for 1 h at 37°C prior to the sample incubation resulted in substantially increased signal in an MUC1 immunoassay (20).
6. In this assay, non-labelled semi-purified mucin or synthetic mucin peptides or carbohydrates are coated to microtitre plates and used to capture specific anti-mucin antibodies or lectins. Samples are introduced to this reaction and those containing the epitopes recognized by the antibody or lectin compete for antibody binding to the solid-phase antigen. The amount of bound antibody or lectin is then determined using a secondary enzyme or radioisotope-labelled antibody.

7. The concentration of coating mucin antigen and detecting antibody needs to be determined using a checkerboard serial dilution. Higher binding is achieved if the mucin is purified, but crude preparations can work; synthetic peptides and fusion proteins work well in these assays. Selection of the starting dilution for the mucin is somewhat empirical; however, the protein binding capacity of the plate wells should not be exceeded.
8. A standard curve, sample blank, and appropriate inter-assay control samples should be included on each plate. These inhibition assays can be more sensitive than double-determinant assays but can also be subject to greater inter-assay variation unless rigid consistency in technique is employed.
9. Blotting techniques rely on binding of mucins onto a membrane filter support and subsequent detection using chemical, lectin, or antibody detection. Dot/slot blotting is suitable for semi-quantitative detection of mucins in reasonably large numbers of samples. The advantages over in-solution assays include increased sensitivity due to the potential for concentration of sample on the membrane and reduced problems with interfering substances which can be filtered through the membrane. The main disadvantages of direct blotting compared with Western blotting is the potential for false-positive results due to non-specific antibody binding and the lack of separation and data regarding the molecular weight of the reactive proteins. False-negative results also occur if sample protein concentrations are very high. However, direct blotting is more amenable to inclusion of standards than Western blotting (due to restrictions on the number of lanes per gel). Therefore, dot/slot blotting is often the method of choice, especially for monitoring mucins during purification. However, it is highly recommended that representative samples are subjected to electrophoretic separation and Western blotting (see below) to confirm the specificity of dot/slot blot results by demonstrating that the reactivity is restricted to proteins of an expected molecular weight. The choice of chemical, lectin, polyclonal, or monoclonal antibodies differs with the application and the availability of reagents. Subheading 3.2.1 outlines the procedures for preparing the membranes, and Subheading 3.2.2 describes the use of the PAS reagent to detect mucins immobilised on membranes. It should be noted that other classical histological reagents (e.g. Alcian blue and HID) have also been employed to probe mucins immobilised on membranes (21). Subheading 3.2.3 describes detection of specific mucin epitopes using antibodies or lectins (these are also applicable to Western blotting).

10. Nitrocellulose is the most commonly used membrane, although both PVDF and nylon (inferior protein binding) can be used. PVDF is less brittle than nitrocellulose, and is therefore more likely to survive several rounds of stripping and re-probing and is also resistant to chemical deglycosylation with TFMSA (10). The total protein added should not exceed the protein binding capability of the membrane and samples should be titrated if relative quantitation is required. Some mucins and in particular small mucin glycopeptides (mucin fragments prepared by extensive proteolysis) may not bind well to nitrocellulose, and addition of poly-L-lysine (100 μg/mL) or a lectin (e.g. WGA) to the membrane prior to application of the samples can increase retention (21, 22).
11. Western blotting refers to detection of proteins first separated by gel electrophoresis and then transferred to membrane filter supports for subsequent detection using chemical, lectin, or antibody detection. This technique is suitable for specific, sensitive, semi-quantitative detection of mucins in moderate numbers of samples. The main advantages are the potential separation of different mucins and the provision of data regarding molecular weight. The most frequent mistake in published mucin Western blotting concerns not the immunodetection but the electrophoretic separation. Polyacrylamide gels that are of a percentage that does not allow migration of high-Mr mucins even into the stacking gels are often used. Even at the lower limit of polyacrylamide gel formation (3%), many known mucins are too large to penetrate the gel. Agarose gel electrophoresis is often required to achieve separation of these large mucins (10); appropriate electrophoretic techniques are described in Chapter 2 of this volume. Mucins can be detected by chemical methods within acrylamide or agarose gels (23); however, transfer to membranes is necessary for antibody or lectin detection. Transfer of mucins from polyacrylamide or agarose gels can be achieved by electrophoretic elution (wet or semi-dry) and vacuum or capillary transfer (PAS staining of gels can be used to evaluate the transfer (9)).
12. This protocol describes the detection of mucins on membranes following dot/slot blotting or transfer following electrophoretic separation. Mucin carbohydrate groups are reacted with periodic acid and then detected using the Schiff's reagent; for more details, see ref. 21.
13. PAS-stained dots/slots or bands on Westerns can be readily quantitated using densitometry equipment and the mucin content determined relative to standards included on the same blot.
14. This protocol describes the detection of mucins on membranes following dot/slot blotting or transfer following electrophoretic separation. Blots are incubated with antibodies or lectins

reactive with the mucins and these are in turn detected with secondary antibodies/ligands labelled with enzymes or isotopes. Chemical and/or enzymatic deglycosylation can be performed before starting these detection procedures (10).

15. Human bronchial epithelial cells can be grown and differentiated at air–liquid interface and are a valuable tool to study MUC5AC and MUC5B production and regulation (16–18).
16. It is important *not* to leave the cells in fixative for more than 48 h, as it can make the insert very brittle and later sectioning becomes extremely tricky. Chloroform clearing rather than xylene is also important as xylene makes the cell layer more brittle and hence more difficult to section.
17. VA5 RalWax (Raymond A. Lamb) is recommended as it is harder than conventional paraffin wax, so offers more support to the cell layer, making it easier to section. VA5 also has a higher melting temperature; hence, the water bath needs to be set at 50°C for sections.
18. Sections should be embedded in wax on end and in parallel so that sections are seen as two parallel images, making imaging and quantifying later on much more straightforward.
19. When blocking sections in serum, use serum that is not from the same species to which the secondary antibody is raised against.
20. The basis of the quantitative analysis is to calculate the length of epithelium and the area of staining per length of epithelium in each section. For each section, the full length of the epithelium was analysed, and for each condition at least three sections were analysed. Quantification of stained area per length or volume of epithelium is performed in preference to enumeration of goblet cells as individual goblet cell counts are often inaccurate and misrepresentative due to variations in thecal size affecting the likelihood of sectioning.
21. Cells are transferred to a 10-cm-diameter cell culture-treated Petri dish containing a number of sterile glass coverslips. Culturing cells for 2–3 days should be sufficient to allow cell attachment and some growth on the coverslips, although this is cell-line dependent and should be altered accordingly.
22. Fixing and washing cells grown on coverslips/Transwell™ inserts are most easily performed in 12- or 24-well plate format with a vacuum to easily change buffers. This helps keep coverslips/inserts upright and separate to allow multiple antibody combinations to be used. If epitopes are not preserved in 4% PFA, try 1–2% PFA, coagulative fixatives (e.g. methanol, acetone, or Carnoy's fixative which also preserves secreted mucus).
23. Area of fluorescent staining can be quantified relative to DAPI staining using Image J software or similar.

Acknowledgements

Michael McGuckin is supported by an NHMRC Senior Research Fellowship. Ceri Harrop is supported by a grant from the Dr. Hadwen Trust for Humane Research.

References

1. Hattrup, C. L., and Gendler, S. J. (2008) Structure and function of the cell surface (tethered) mucins. *Annu Rev Physiol* **70**, 431–457.
2. Thornton, D. J., Rousseau, K., and McGuckin, M. A. (2008) Structure and function of the polymeric mucins in airways mucus. *Annu Rev Physiol* **70**, 459–486.
3. Desseyn, J. L., Aubert, J. P., Van Seuningen, I., Porchet, N., and Laine, A. (1997) Genomic organization of the 3′ region of the human mucin gene MUC5B. *J Biol Chem* **272**, 16873–16883.
4. Desseyn, J. L., Buisine, M. P., Porchet, N., Aubert, J. P., and Laine, A. (1998) Genomic organization of the human mucin gene MUC5B. cDNA and genomic sequences upstream of the large central exon. *J Biol Chem* **273**, 30157–30164.
5. Buisine, M. P., Desseyn, J. L., Porchet, N., Degand, P., Laine, A., and Aubert, J. P. (1998) Genomic organization of the 3′-region of the human MUC5AC mucin gene: additional evidence for a common ancestral gene for the 11p15.5 mucin gene family. *Biochem J* **332 (Pt 3)**, 729–738.
6. Desseyn, J. L., Guyonnet-Duperat, V., Porchet, N., Aubert, J. P., and Laine, A. (1997) Human mucin gene MUC5B, the 10.7-kb large central exon encodes various alternate subdomains resulting in a super-repeat. Structural evidence for a 11p15.5 gene family. *J Biol Chem* **272**, 3168–3178.
7. Carlstedt, I., Lindgren, H., Sheehan, J. K., Ulmsten, U., and Wingerup, L. (1983) Isolation and characterization of human cervical-mucus glycoproteins. *Biochem J* **211**, 13–22.
8. Carlstedt, I., and Sheehan, J. K. (1984) Is the macromolecular architecture of cervical, respiratory and gastric mucins the same? *Biochem Soc Trans* **12**, 615–617.
9. Mantle, M., and Allen, A. (1978) A colorimetric assay for glycoproteins based on the periodic acid/Schiff stain [proceedings]. *Biochem Soc Trans* **6**, 607–609.
10. Thornton, D. J., Howard, M., Devine, P. L., and Sheehan, J. K. (1995) Methods for separation and deglycosylation of mucin subunits. *Anal Biochem* **227**, 162–167.
11. Gerken, T. A., Gupta, R., and Jentoft, N. (1992) A novel approach for chemically deglycosylating O-linked glycoproteins. The deglycosylation of submaxillary and respiratory mucins. *Biochemistry* **31**, 639–648.
12. Raju, T. S., and Davidson, E. A. (1994) New approach towards deglycosylation of sialoglycoproteins and mucins. *Biochem Mol Biol Int* **34**, 943–954.
13. Hong, J. C., and Kim, Y. S. (2000) Alkali-catalyzed beta-elimination of periodate-oxidized glycans: a novel method of chemical deglycosylation of mucin gene products in paraffin embedded sections. *Glycoconj J* **17**, 691–703.
14. Kim, Y. S., Gum, J., Jr., and Brockhausen, I. (1996) Mucin glycoproteins in neoplasia. *Glycoconj J* **13**, 693–707.
15. Young, P. R. (1989) Enhancement of immunoblot staining using a mixed chromogenic substrate. *J Immunol Methods* **121**, 295–296.
16. Randell, S. H., Walstad, L., Schwab, U. E., Grubb, B. R., and Yankaskas, J. R. (2001) Isolation and culture of airway epithelial cells from chronically infected human lungs. *In Vitro Cell Dev Biol Anim* **37**, 480–489.
17. Holmen, J. M., Karlsson, N. G., Abdullah, L. H., Randell, S. H., Sheehan, J. K., Hansson, G. C., and Davis, C. W. (2004) Mucins and their O-Glycans from human bronchial epithelial cell cultures. *Am J Physiol Lung Cell Mol Physiol* **287**, L824–L834.
18. Gray, T. E., Guzman, K., Davis, C. W., Abdullah, L. H., and Nettesheim, P. (1996) Mucociliary differentiation of serially passaged normal human tracheobronchial epithelial cells. *Am J Respir Cell Mol Biol* **14**, 104–112.
19. Diamandis, E. P., Christopoulos, T.K. (1989) *Immunoassay, Academic Press, San Diego.*
20. McGuckin, M. A., Devine, P. L., Ramm, L. E., and Ward, B. G. (1994) Factors effecting the measurement of tumor-associated MUC1 mucins in serum. *Tumour Biol* **15**, 33–44.
21. Thornton, D. J., Holmes, D. F., Sheehan, J. K., and Carlstedt, I. (1989) Quantitation of mucus glycoproteins blotted onto nitrocellulose membranes. *Anal Biochem* **182**, 160–164.

22. Ayre, D., Hutton, D. A., and Pearson, J. P. (1994) The use of wheat germ agglutinin to improve binding of heterogeneous mucin species to nitrocellulose membranes. *Anal Biochem* **219**, 373–375.

23. Jay, G. D., Culp, D. J., and Jahnke, M. R. (1990) Silver staining of extensively glycosylated proteins on sodium dodecyl sulfate-polyacrylamide gels: enhancement by carbohydrate-binding dyes. *Anal Biochem* **185**, 324–330.

Chapter 4

Mass Spectrometric Analysis of Mucin Core Proteins

Mehmet Kesimer and John K. Sheehan

Abstract

Mucins are difficult to handle for their identification and characterization via proteomic applications due to their heavily glycosylated nature (up to 90%), high molecular weight (200 kDa–200 MDa), and size (Rg 10–300 nm). Their core proteins are extremely large and highly substituted with oligosaccharides, which only allow access to a highly restricted portion of their protein. For this reason, conventional 1D or 2D polyacrylamide gel-based proteomic approaches are not effective for identification and characterization of mucin molecules. In this chapter, we present our current protocol employing a modified shotgun proteomic approach to identify these complex glycoproteins.

Key words: Mucins, Mucus, Glycoproteins, Proteomics, Mass spectrometry, UPLC, Shotgun proteomics

1. Introduction

Mucins are heavily glycosylated, large glycoproteins found mainly on mucosal surfaces, such as respiratory, gastrointestinal, and urogenital systems. They are produced by both mucosal surface goblet cells and submucosal gland mucous cells (1). Mucins are densely coated with O-linked oligosaccharides that make them large both in size and mass. As protein products of *MUC* genes, mucins show complex multidomain structures (2, 3). The large gel-forming mucins, such as MUC5AC and MUC5B, have very large, defined, heavily glycosylated regions that are encoded in a single large exon which contains a number of short cysteine-rich, unglycosylated regions. Their NH_3- and COOH-terminal regions are composed of sparsely glycosylated protein regions, which contain cysteine-rich von Willebrand factor (vWF)-like domains. Other mucins, like MUC1 and MUC4, are characterized by small transmembrane regions at one end and their glycosylated regions containing numbers of repeats termed VNTRs which are characterized by a variable number of tandem repeat

Michael A. McGuckin and David J. Thornton (eds.), *Mucins: Methods and Protocols*, Methods in Molecular Biology, vol. 842,
DOI 10.1007/978-1-61779-513-8_4,

regions found in different individuals. Some membrane-associated mucins contain only one glycosylated region and display complexity on their COOH-terminal ends by having N-glycosylated regions, cysteine-rich regions, SEA regions, transmembrane, and cytoplasmic domains. The so-called membrane-associated mucins can also be found as secreted soluble forms released from the membrane-tethered molecule probably after protein cleavage (4).

The presence of mucins and other large glycoconjugates, such as proteoglycans, glycoproteins, and glycolipids, and also the presence of large protein complexes prevent the use of conventional 1D or 2D gel approaches for complete proteome analysis of the mucous samples, such as saliva and cervical and tracheobronchial mucus gels (5, 6). In particular, mucins are difficult to handle for their identification and characterization via proteomic applications due to their heavily glycosylated nature (up to 90% carbohydrate by weight), high molecular weight ($0.2–200 \times 10^6$ Da), and size (Rg 10–300 nm). The high number of attached oligosaccharides only allows limited access to their protein regions (see Note 1).

Over the last decade, the shotgun proteomics approach has become the preferred method to analyze complex protein mixtures (7–9). With this method, the complex protein mixture after treatment with specific proteases is subsequently fractionated and analyzed by liquid chromatography coupled to tandem mass spectrometry (LC-MS/MS). It provides a powerful alternative to gel-based proteomics applications and has been developed to a high degree of sophistication such that now it is routine to employ a multidimensional, sequential cation exchange and reverse-phase chromatography approach to feed samples directly into the mass spectrometer (10, 11). The method has also been improved by introducing a nanoscale variable flow interface (called peak trapping or peak parking (PP)-LC/MS-MS) which enhances the analysis and peptide coverage of the proteome (12).

In this chapter, we describe the protocol for identification and characterization of the mucins and their interacting proteins (see Note 2) by shotgun mass spectrometric methods.

2. Materials

2.1. Mucin Extraction and Purification

See Chapter 2 for a more detailed protocol for extraction of mucins.

2.2. Preparation of Mucin Samples for Digestion (Reduction/Carboxymethylation and Desalting) and Trypsin Digestion

1. Guanidine hydrochloride (GuHCl, high purity grade, Sigma, USA).
2. Mucin reduction buffer: 6M GuHCl, 0.1M Tris–HCl, 5 mM EDTA, pH 8.0.
3. Reducing agent: 5 mM dithiothreitol (prepare fresh).

4. Cysteine blocking: 15 mM iodoacetamide (prepare fresh).
5. Digestion buffer: 50 mM ammonium hydrogen carbonate (NH_4HCO_3) (prepare fresh).
6. G25 HiTrap™ (GE Bio-sciences Uppsala, Sweden) or equivalent desalting column.
7. Proteomics grade modified trypsin (Promega, CA, USA, or Sigma, USA) (see Notes 3 and 4).
8. Trypsin resuspension buffer, 0.1M HCl.
9. Trypsin dilution/dissolving buffer, 100 mM ammonium hydrogen carbonate (NH_4HCO_3).

2.3. Separation and Concentration of Mucin Peptides from Glycopeptides

1. Conditioning and running buffer: 50 mM NH_4HCO_3.
2. Superdex 200 HR 10/30 (GE Bio-sciences, Uppsala, Sweden) (24 mL) or equivalent gel permeation chromatography column.
3. Ettan LC with Frac-950 fraction collector (GE Bio-sciences, Uppsala, Sweden) or equivalent HPLC or FPLC system with fraction collector.
4. Silicon-coated, low-binding, 1.5-mL microcentrifuge tubes (see Note 5).
5. Formic acid.
6. A Speedvac vacuum concentrator with a cooling trap (e.g., Heto vacuum centrifuge).

2.4. Mass Spectrometry

1. A hybrid Quadrupole/Time Of Flight (Q-TOF) mass spectrometry system (e.g., Q-TOF micro, Waters, Milford, MA).
2. A capillary HPLC or UPLC system, e.g., CapLC® system or nanoAcquity® (Waters, MA, USA) ultrahigh-pressure LC system, both comprise a nano flow capillary LC pump and an autosampler.
3. Mass spectrometry grade water (18 mΩ), e.g., Milli-Q Synthesis® 18 φ water purification system (Millipore, MA, USA).
4. Acetonitrile, LC/MS grade (Optima®, Fisher, USA).
5. Formic acid (Fluka Chemical, USA).
6. Mobile phase A for CapLC®: 5% acetonitrile in 0.1% (v/v) formic acid (200 mL). Mix 10 mL acetonitrile and 190 mL of 18 mΩ water and degas for 20 min (see Note 6). Add 200 μL formic acid and mix gently. Prepare once a week.
7. Mobile phase B for CapLC®: 95% (v/v) acetonitrile in 0.1% formic acid (v/v) (100 mL): Mix 95 mL acetonitrile and 5 mL of 18 mΩ water and degas for 20 min. Add 100 μL formic acid and mix gently. Prepare once a week.

8. Mobile phase A for nanoAcquity®: 0.1% (v/v) formic acid in water (200 mL). Degas 200 mL of 18 mΩ water for 20 min. Add 200 μL formic acid and mix gently. Prepare once a week.
9. Mobile phase B for nanoAcquity®: 100% acetonitrile, 0.1% formic acid (100 mL): Degas 100 mL of acetonitrile for 20 min. Add 100 μL formic acid and mix gently. Prepare once a week.
10. Trap column: PepMap™ C18, 300 μm ID × 5 mm (LC Packing, CA, USA), Symmetry® C18 180 μm × 20 mm, 5 μm (Waters, MA, USA) or equivalent preconcentration column.
11. Analytical column: Atlantis® (Waters) dC18 NanoEase™ (75 μm × 150 mm) nanoscale analytical column or 100 μm × 100 mm 1.7 μm, BEH130 nanoscale column (Waters. Milford, MA).
12. Spray tip: PicoTip™ (New Objectives, MA, USA) capillary with 10 μm diameter.
13. Calibration standard: Glu-Fibrinopeptide B (Sigma, MO, USA).
14. Quality control digest standard (see Note 7): Enolase (Sigma, MO, USA), alcohol dehydrogenase (ADH) (ICN Biomedicals, OH, USA) or MassPREP™ (Waters, Milford, MA, USA) enolase and/or ADH standards.
15. Masslynx 4.0 or equivalent data acquisition software with Proteinlynx or equivalent data processing module.
16. Mascot Search Engine v 2.0 or later (Matrix Science, London, UK) or an equivalent database search engine.

3. Methods

A schematic diagram of the strategy from mucin isolation to identification is given in Fig. 1 and an example mass spectra for MUC5B is shown in Fig. 2.

3.1. Isolation and Purification of Mucins

Although mucins are highly abundant in the mucus samples, there are over a hundred other proteins some of which are also very abundant (5). We have found that in order to maximize both numbers of proteins identified and mucin protein coverage the initial complexity of the mixture has to be reduced by prefractionation using CsCl/4MGuHCl density-gradient centrifugation. The advantage of separating the major mucins, especially gel-forming mucins MUC5B and MUCAC, from the other proteins using CsCl was apparent (5). Density-gradient centrifugation is a powerful approach for obtaining a group separation of mucins from proteins and lipids (13) on the basis of solute density. It has the key advantage of concentrating rather than diluting the molecules of interest

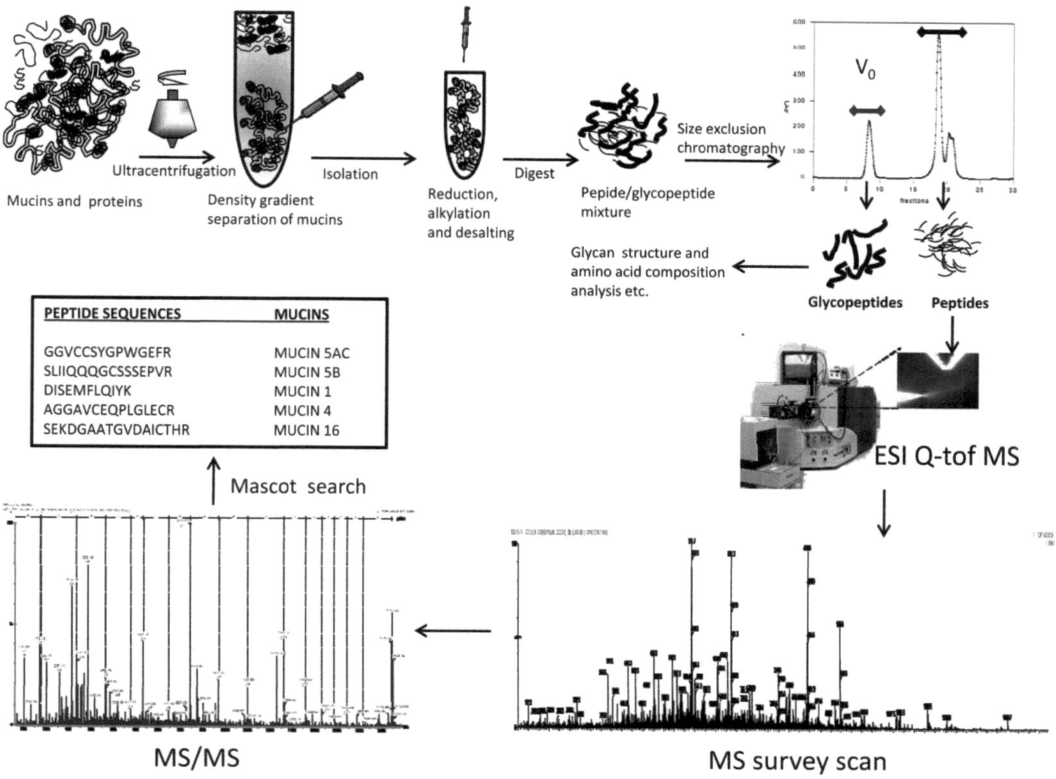

Fig. 1. Schematic summary of the approach used for mucin isolation and identification. V_0 Void volume of the column; *ESI Q-TOF MS* electron spray ionization hybrid quadrupole-time-of-flight mass spectrometry.

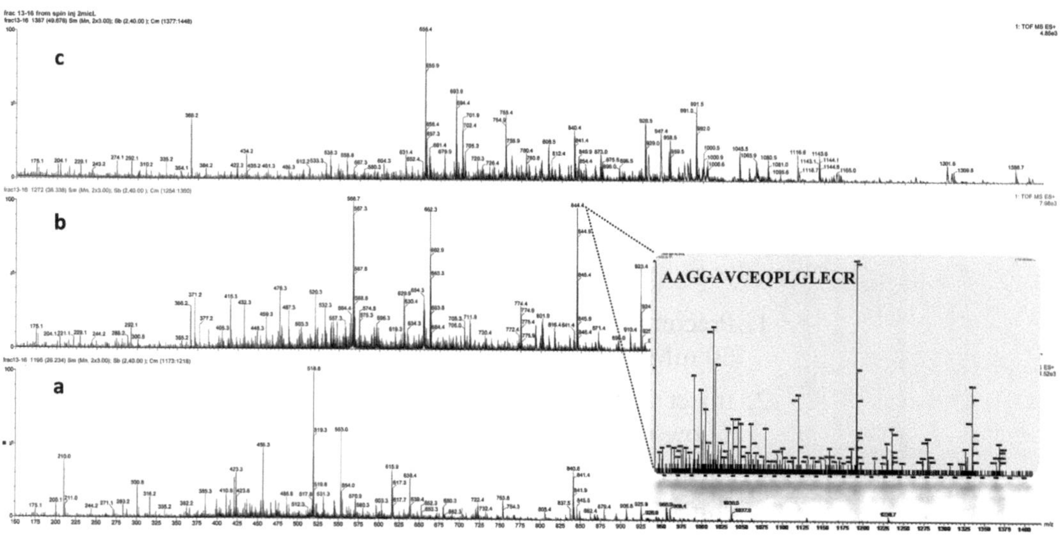

Fig. 2. Mass spectra of the MUC5B digest over different retention time scans during the capillary reverse-phase separation. Typical survey scan mass spectra of: (**a**) early (22–28 min), (**b**) middle (30–40 min), and (**c**) late (42–50 min) eluted peptides. *Inset spectra*: MS/MS spectra of the doubly charged precursor ion 844.41 Da, **AAGGAVCEQPLGLECR.**

that are being separated. Separations of the proteins with CsCl/4MGuHCl density-gradient centrifugation gave us another dimension of separation, lowered the complexity of the sample, and improved the detection of both mucin-rich and protein-rich fractions (5) (see Note 8).

3.2. Preparation of the Mucin Samples for Digestion (Reduction/Carboxymethylation and Desalting and Trypsin Digestion)

Breaking large mucin oligomeric structures and making mucin subunits (monomers) employing reduction and alkylation:

1. Dialyze mucins into reduction buffer (see Notes 9–12).
2. Add DTT to mucin preparation to make 10 mM to a final concentration and reduce mucins at 37°C for 2 h or 45 min at 65°C.
3. Block free thiol groups with the addition of iodoacetamide to a final concentration of 20 mM at room temperature in the dark for 0.5 h.
4. Prewash/equilibrate a Sephadex G25 (5 × 5 cm) desalting column with 10 mL of digestion buffer (50 mM NH_4HCO_3).
5. Take 1.5 mL of the reduced and carboxymethylated mucin preparation and load to the preconditioned Sephadex G25 column.
6. Elute the samples from the column with 50 mM NH_4HCO_3 and collect first 2 mL of eluant.
7. Add 10 μL of trypsin solution (100 ng/μL) and incubate overnight at 37°C.

3.3. Gel Permeation Chromatography for Peptide/Glycopeptide Separation

Digestion of mucins with proteases yields, in addition to small peptides, *very high molecular weight* glycopeptides (10^4–10^6 Da). These major fragments, on occasion, poisoned our capillary nanoscale analytical columns in the LC-MS/MS. Therefore, we introduced a gel permeation chromatography step, after digestion, to separate the heavily glycosylated mucin glycopeptides from the smaller peptides. The glycopeptides can be used for some other applications, such as sugar and amino acid analysis, for their further characterization. Peptides are pooled and subjected to mass spectrometry analysis.

1. Precondition the Superdex 200 HR 10/30 column with 50 mM NH_4HCO_3.
2. Inject 2 mL of the digested mucin preparation and elute with a flow rate of 0.3 mL/min. Collect 32 × 1-mL fractions.
3. In this particular size column at this flow rate, glycopeptides eluted in the V_0 volume (8–10 mL) while peptides with different sizes elute in broader region, between 17 and 22 mL (fractions 16–21). The V_t volume of this column is about 24 mL.

4. Wash column for additional 15 mL or until getting a steady baseline to be ready for next injection.
5. Add 10 μL of 1% formic acid to each 1-mL fraction (fractions 16–21) and place them to the vacuum concentrator. Concentrate the volume down to ~10–20 μL. This step takes about 3–4 h depending on the concentrator.
6. Pool all the fractions (50–100 μL) and concentrate again down to 10–20 μL.
7. Add 1–2 μL of 1% formic acid to the final concentrated peptides.
8. Centrifuge for 10 min at 5,000 × *g*.
9. Transfer peptides to "maximum recovery" (Waters, MA, USA) sample vial.

3.4. LC-ESI Mass Spectrometry

The shotgun proteomic method usually entails the delivery of a very complex peptide mixture into the mass spectrometer using an in-line or off-line 1D or 2D chromatographic separation (9, 10, 14). The success of this technique depends on the optimization of peptide preseparation and the delivery rate of the peptides to the sprayer tip, on the one hand, and the data-directed analysis (DDA) duty cycle of the MS on the other. Identification can be enhanced significantly in some cases by nanoscale chromatography in 1 or 2 more dimensions combined with variable flow (peak parking, see Note 13) (12).

3.4.1. CapLC/Nano Acquity Conditions

Nanoscale capillary liquid chromatography is one of the crucial pieces of the LC-MS/MS identification of complex protein/peptide mixtures. New cutting-edge LC technologies, e.g., ultrahigh (10,000 psi)-pressure LC systems which permit the use of smaller particle (1.7 μm)-size media, provide improvement in chromatographic performance, such as increased analytical speed, resolution, and sensitivity. Please see ref. 15 for more information, including troubleshooting tips, on nano-liquid chromatography.

We have experience in two different LC sample/solvent delivery system, CapLC® and nanoAcquity UPLC® (Waters, MA, USA). The nanoAcquity UPLC system was introduced relatively recently and has many advantages for end users, such as reliable, reproducible high-resolution chromatography with splitless nano flow. Importantly, the system is extremely robust and requires very little user time for the maintenance. The technical details and experimental running conditions of these sample delivery systems are given below.

1. Digested samples (1–2 μL) are introduced either via a Waters CapLC® system or Waters nanoAcquity® LC system; both comprise a nano flow capillary LC pump and an autosampler.

2. The CapLC analytical system is connected to a ten-port valve called a stream select valve.
3. The stream select module is attached directly to the Z spray source and is configured with a PepMap™ C18 (LC Packing, 300 μm ID × 5 mm) preconcentration column in series with an Atlantis® (Waters) dC18 NanoEase™ (75 m × 150 mm) nanoscale analytical column.
4. Program the pump to deliver 9 μL/min, and a splitter gave a resultant flow through the analytical column 200 nL/min with a split ratio of 45 at about 1,200 psi backpressure.
5. Elute the peptides from analytical column with a gradient of 5% acetonitrile in 0.1% formic acid to 60% acetonitrile in 0.1% formic acid over 65 min.
6. Nano acquity UPLC system consists of only one six-port valve and is capable of delivering mobile-phase splitless as low as 200 nL/min.
7. The tryptic digest is introduced via a nano acquity system configured with a Symmetry C18 180 μm × 20 mm, 5 μm (Waters) preconcentration/desalting column in series with a nanoAcquity UPLC® column, 100 μm × 100 mm 1.7 μm, BEH130 nanoscale analytical column. Typical flow rate is 400 nL/min at about 1,900 psi backpressure.
8. Peptides are trapped in the trapping column for 3 min and eluted with the following gradient: 1–50% solvent B in 45 min followed by 50–85% solvent B for 5 min.
9. Wash column with 85% solvent B for 10 min.

3.4.2. Mass Spectrometry Settings

1. Mass spectrometry analysis of purified and concentrated peptide mixtures is performed with a hybrid Q-TOF instrument (micro, Waters, MA, USA).
2. For such a machine, set capillary, sample cone, and extraction cone voltage to 3 kV, 45 V, and 2 V, respectively. Set source temperature to 80°C.
3. To calibrate the instrument: Inject 1 μL of 1 pmol/L Glufib standard and acquire MS/MS data.
4. Trap GluFib in the trapping column for 3 min with 5 μL/min flow rate. Elute GluFib 40% solvent B for 20 min.
5. Create an MSMS data acquisition file with the following parameters: Set mass 785.8, low mass 100, high mass 1,500, scan time 2.4 s, interscan time 0.1 s, and collision energy 30 V. You should get an intensity of at least 50 for the 684.35 Da fragment ion.
   (a) Perform a quality control test before injecting the mucin sample by injecting 2 μL of 500 fmol/mL enolase digest.

Perform a sample run and you should get at least 12 peptides identified 60% coverage from the enolase suggesting a good sensitivity for your Q-TOF micro (see Note 14).

(b) For sample analysis: Create a DDA to define MS and MS/MS method using the following variables:

- Mass survey range 150–2,000
- Survey scan time 0.9 s, interscan time 0.1 s
- MS/MS start mass 100, end mass 1,700
- Number of MS/MS component 3
- Exclude singly charged ions, and include doubly and triply charged ions
- MS/MS to MS switch criteria: Total ion count (TIC) rising above 3,000
- MS/MS switch after 4.5 s
- MS/MS scan time 1.4 s, interscan time 0.1 s
- MS/MS cone voltage 35 V
- Use charge state recognition for collision energy as recommended by the manufacturer

(c) Inject 2 μL of mucin peptide mixture. Repeat experiment two times; exclude peptides identified in previous runs with other samples.

3.5. Raw Data Analysis/Data Processing

1. Create *.pkl (peak list) files using Proteinlynx module of Masslynx 4.0 or 4.1.*.pkl files are suitable for the MS/MS ions database search via search engines, such as Mascot, Sequest, etc.

   Use the following parameters for Proteinlynx:

   (a) Set the peptide QA filter 30 to eliminate poor-quality spectra,

   (b) Subtract background with polynomial order 10 and below curve % 10.

   (c) Set smoothing parameters to: Smooth x2 with Savitzky Golay with smooth window 3.0.

   (d) Set the minimum peak width at half height to 4 to eliminate background noise peaks.

   (e) Take centroid spectra at 80% of the peak height.

3.6. Database Search

1. Submit your peak list file to online or in-house search engines for database search.
2. Search data against National Center for Biotechnology Information (NCBI) nonredundant (nr) and/or Swiss-Prot protein databases (use the newest version possible).

3. Use following search parameters:
   - (a) Taxonomy: human (or the species you are working on)
   - (b) Use 0.2 Da mass accuracy for parent ions and 0.3 Da accuracy for fragment ions
   - (c) Allow one missed cleavage
   - (d) Use carbamidomethyl-Cys and methionine oxidation as fixed and variable modifications, respectively
4. Accept the matching peptide/protein if MASCOT probability-based Mowse individual ion score is above 40 indicating identity or extensive homology ($p < 0.05$).
5. Examine MS/MS spectra which have scores between 20 and 40 individually with the acceptance criteria being the parent and fragment ion masses were within the calibrated tolerance limits and that the spectrum contained an extended series of consecutive y- or b-ions.

4. Notes

1. Mucins are very large, heavily glycosylated molecules. Electrophoretic separation and following gel-based proteomic applications of mucins or their subunits are not achievable by conventional 1D or 2D polyacrylamide gel electrophoresis since in their fully glycosylated and undegraded form they barely penetrate a 3% acrylamide gel. An agarose gel is the convenient electrophoretic methods to resolve and separate oligomeric mucins or their subunits. However, for technical reasons, recovering mucin from an agarose gel and following proteomic applications are also not feasible. This necessitated the shotgun approach to be adopted.
2. GuHCl is one of the most effective chaotropic agents for denaturing proteins and breaking noncovalent protein–protein interactions. In this work, employing 4M GuHCl, we deliberately disrupted noncovalent association between the mucins and other proteins. However, 200 mM NaCl or any other low-salt/nonchaotropic agent instead of GuHCl can be used to allow mucin–protein interactions to remain intact and subsequently allow us to identify mucin-interacting proteins.
3. Trypsin is the commonly used endoprotease for both gel- and shotgun-based peptide mapping and peptide-sequencing proteomic applications. It cleaves lysine and arginine residues at the C-terminal region that results in a charged amino acid in acidic environment.

4. Use modified trypsin (modification of lysine and arginine residues by reductive alkylation on the protein) to prevent autolysis of the enzyme.
5. Use silicon-coated microcentrifuge tubes to collect peptides to avoid peptide loss.
6. Proper degassing is crucial to get rid of dissolved air in the HPLC/UPLC solvents. Streaming helium bubbles into the buffers removes the dissolved air. Hook up a frit to a helium source and flow helium through the solvent with a fine stream of bubbles. Sonication is also an effective degassing method. However, sonication may heat up your mobile phase. In this case, evaporation might be an issue, especially for the mobile phase, which has a high percentage of ACN. If you are using sonication, take extra care against evaporation and cool down the mobile phases after the sonication before use.
7. Homemade MS/MS quality control standards: Prepare 1 nmol/L of either ADH or enolase in 50 mM NH_4HCO_3. Add 10 μL of trypsin (100 ng/μL) and incubate for 16–18 h. Prepare 100-μL aliquots and store at −20°C for months.
8. Lowering the complexity of the proteins in the samples is essential for shotgun proteomics approaches. Therefore, to maximize the detection and coverage of mucin core proteins, separation of mucins from the globular proteins is necessary. See Chapter 2 for mucin isolation and separation.
9. Reduction is essential for breaking oligomeric mucins to their monomeric form and opening the cysteine-rich regions. This is crucial for a proper digestion and maximum protein coverage. We typically use 6M GuHCl reduction buffer and DTT for denaturing the mucin and its subsequent reduction. Using other denaturants or acid-labile surfactants (such as Rapigest and PPS Silent Surfactant) is not as effective as 6M GuHCl for mucin denaturation.
10. Urea is another potent chaotropic agent for lysing cells and solubilizing and denaturing proteins. However, most urea solutions may degrade to isocyanic acid which reacts with protein amino groups. This causes N-terminal carbamylation of the resultant peptides. To avoid this, always use fresh urea solutions and pretreat urea with an ion exchange resin to remove isocynic acid. If urea is used during sample preparation, include N-terminal carbamylation as a variable modification in the database search.
11. Using protease inhibitors: When mucins are stored in at least 4M GuHCl, there is no need for protease inhibitors. However, when mucins are stored below 2M GuHCl or any other nondenaturing buffers, protease inhibitors (e.g., Roche Complete mini) should be added according to manufacturer's suggestion.

12. Keratin is one of the major contaminations in the lab environment. Use Clean Hood Benches, and wear gloves and head cap to avoid contamination.
13. In variable flow (peak parking) mode, when the DDA acquisition system detects an eligible precursor ion to do MS/MS, a signal is sent to the PP valve such that the mass spectrometry extends the analysis time on coeluting species by reducing the flow up to 20 nL/min, and thus prevented the elution of other peptides.
14. After injecting the 1 pmol of enolase or ADH digest, typically 35–45 total peptides should be identified with a total mascot score of 1,200–1,500. You should get at least 16–20 unique peptides identified and the coverage of the proteins around 60–65%. If the identified peptides are lower than 12 and coverage lower than 50%, troubleshoot HPLC/UPLC and/or MS system for potential source of sensitivity decrease.

Acknowledgment

This work was supported by National Heart, Lung, and Blood Institute/National Institutes of Health (NHLBI/NIH) grants HL103940 (MK) and HL084934 (JKS).

References

1. Hovenberg, H. W., Davies, J. R., and Carlstedt, I. (1996) Different mucins are produced by the surface epithelium and the submucosa in human trachea: identification of MUC5AC as a major mucin from the goblet cells, *Biochem J 318 (Pt 1)*, 319–324.
2. Buisine, M. P., Desseyn, J. L., Porchet, N., Degand, P., Laine, A., and Aubert, J. P. (1998) Genomic organization of the 3 -region of the human MUC5AC mucin gene: additional evidence for a common ancestral gene for the 11p15.5 mucin gene family, *Biochem J 332 (Pt 3)*, 729–738.
3. Desseyn, J. L., Buisine, M. P., Porchet, N., Aubert, J. P., and Laine, A. (1998) Genomic organization of the human mucin gene MUC5B. cDNA and genomic sequences upstream of the large central exon, *J Biol Chem 273*, 30157–30164.
4. Hattrup, C. L., and Gendler, S. J. (2008) Structure and function of the cell surface (tethered) mucins, *Annu Rev Physiol 70*, 431–457.
5. Kesimer, M., Kirkham, S., Pickles, R. J., Henderson, A. G., Alexis, N. E., Demaria, G., Knight, D., Thornton, D. J., and Sheehan, J. K. (2009) Tracheobronchial air-liquid interface cell culture: a model for innate mucosal defense of the upper airways?, *Am J Physiol Lung Cell Mol Physiol 296*, L92–L100.
6. Nicholas, B., Skipp, P., Mould, R., Rennard, S., Davies, D. E., O'Connor, C. D., and Djukanovic, R. (2006) Shotgun proteomic analysis of human-induced sputum, *Proteomics 6*, 4390–4401.
7. Utleg, A. G., Yi, E. C., Xie, T., Shannon, P., White, J. T., Goodlett, D. R., Hood, L., and Lin, B. (2003) Proteomic analysis of human prostasomes, *Prostate 56*, 150–161.
8. Wu, C. C., and MacCoss, M. J. (2002) Shotgun proteomics: tools for the analysis of complex biological systems, *Curr Opin Mol Ther 4*, 242–250.
9. Lu, B., Xu, T., Park, S. K., McClatchy, D. B., Liao, L., and Yates, J. R., 3 rd. (2009) Shotgun protein identification and quantification by mass spectrometry in neuroproteomics, *Methods Mol Biol 566*, 229–259.
10. Link, A. J. (2002) Multidimensional peptide separations in proteomics, *Trends Biotechnol 20*, S8–13.

11. Washburn, M. P., Ulaszek, R. R., and Yates, J. R., 3 rd. (2003) Reproducibility of quantitative proteomic analyses of complex biological mixtures by multidimensional protein identification technology, *Anal Chem 75*, 5054–5061.
12. Vissers, J. P., Blackburn, R. K., and Moseley, M. A. (2002) A novel interface for variable flow nanoscale LC/MS/MS for improved proteome coverage, *J Am Soc Mass Spectrom 13*, 760–771.
13. Thornton, D. J., Carlstedt, I., Howard, M., Devine, P. L., Price, M. R., and Sheehan, J. K. (1996) Respiratory mucins: identification of core proteins and glycoforms, *Biochem J 316 (Pt 3)*, 967–975.
14. Lohrig, K., and Wolters, D. (2009) Multidimensional protein identification technology, *Methods Mol Biol 564*, 143–153.
15. Frohlich, T., and Arnold, G. J. (2009) A newcomer's guide to nano-liquid-chromatography of peptides, *Methods Mol Biol 564*, 123–141.

Chapter 5

O-Glycoprotein Biosynthesis: Site Localization by Edman Degradation and Site Prediction Based on Random Peptide Substrates

Thomas A. Gerken

Abstract

The characterization of mucin-type O-glycosylation is fraught with extreme difficulty at almost every level of analysis: from difficulties in obtaining glycopeptides suitable for study, their structural heterogeneity, lack of broad acting glycosidase tools capable of simplifying the glycans, and finally the vast complexity of performing analysis on multiply glycosylated glycopeptides. This, along with a lack of known peptide sequence motif(s) for the transferases that initiate mucin-type O-glycosylation, significantly hinders our understanding of mucin-type O-glycosylation at almost every level from their biosynthesis to their biological and biophysical properties. In this chapter, the use of partial chemical deglycosylation coupled with Edman amino acid sequencing is described to quantify sites of O-glycosylation. In addition, the use of oriented random peptide substrates is described for providing the specificities of the polypeptide α-*N*-acetylgalactosaminyltransferases, which can be used to estimate transferase-specific sites of O-glycosylation.

Key words: Mucin-type O-glycosylation, ppGalNAc T, Edman amino acid sequencing, O-glycans, Mucin

1. Introduction

Mucin-type protein O-glycosylation is initiated in the Golgi compartment by a large family (~20) of UDP-GalNAc:polypeptide α-*N*-acetylgalactosaminyltransferases (ppGalNAc-Ts) that transfers α-GalNAc, from UDP-GalNAc, to Ser and Thr residues of polypeptide acceptors (for a review, see ref. 1). With subsequent action of a series of specific glycosyltransferases, the O-linked glycans are further elongated to produce a vast array of glycan structures; see refs. 2, 3. Possible O-glycan core structures are shown in Fig. 1. The characterization of mucin-type O-glycosylation is fraught with difficulty at many levels. Unlike N-linked glycans, there is no

Michael A. McGuckin and David J. Thornton (eds.), *Mucins: Methods and Protocols*, Methods in Molecular Biology, vol. 842, DOI 10.1007/978-1-61779-513-8_5, © Springer Science+Business Media, LLC 2012

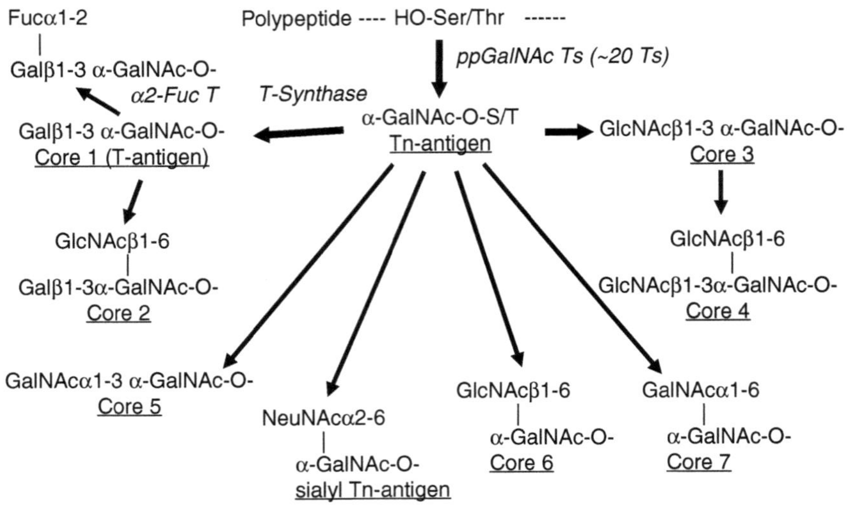

Fig. 1. Possible mucin-type O-glycan core structures. Each core structure may be further elongated producing a vast array of structures (2, 3). Transferases involved in the biosynthesis of the porcine submaxillary gland mucin (PSM) O-glycans and which have been partially characterized with respect to peptide sequence are also given (6, 32, 34).

specifically required sequence motif (e.g., Asn Xaa (not Pro) Ser/Thr) for O-glycosylation nor is there a universal glycosidase that can remove the majority of O-linked glycans (see Note 1). This, coupled with the protease resistance that most native O-glycans impart to mucin domains, makes the characterization of the glycosylation of these domains an analytical challenge. However, by exploiting a series of unique chemical and enzymatic approaches specific to α-GalNAc-linked glycans, which vastly simplify the O-glycan structures, followed by Edman amino acid sequencing, the detailed and quantitative glycosylation pattern of the porcine submaxillary mucin (PSM) O-glycosylated tandem repeat domain has been obtained (4–6). These and other studies have revealed that O-glycosylation (i.e., degree of substitution and glycan length) may indeed be modulated by peptide sequence and neighboring glycosylation in a relatively reproducible manner (6–8). Under favorable circumstances, the approach has been utilized for the characterization of the core glycosylation of other mucin tandem repeat domains, including the canine (CSM) (8) and the bovine (BSM) (unpublished) submaxillary gland mucins and the human MUC1 (9) tandem repeat. Edman sequencing has also been utilized for quantifying the glycosylation of peptide and glycopeptide ppGalNAc T substrates (for example: 7, 8, 10–12).

The database analyses of known mucin-type O-glycosylation sites have provided a number of algorithms (see Note 2) (13, 14) for the approximate prediction of mucin-type O-glycosylation sites. However, none of these approaches readily account for the wide range and remarkable reproducibility of O-glycan site-to-site occupancy observed to date (6, 8). Nor do these approaches take into account the specificities of the individual ppGalNAc T isoforms. This latter point is critical as several ppGalNAc T isoforms have been shown to be necessary for, or associated with, normal development, cellular processes, or specific disease states (15–30) by possessing specific protein targets that other coexpressed ppGalNAc T isoforms fail to recognize. To address transferase specificity directly, we recently developed the use of a series of oriented random peptide and glycopeptide substrate libraries for quantitatively determining ppGalNAc T transferase-specific amino acid residue preferences (31–34). To date, the catalytic domains (see Note 3) of ppGalNAc-T1, T2, and T10 and the fly orthologues of T1 and T2 have been characterized against these substrates utilizing Edman amino acid sequencing (31–33). Similarly, the peptide preferences of T-synthase, that adds α -Gal (1–3) to the peptide-linked GalNAc residue forming the T- antigen or Core 1 structure, were obtained using a random glycopeptide substrate (34). These studies are continuing in the Gerken laboratory with the goal of fully characterizing the additional ppGalNAc T isoforms as well as several of the Core-elongating transferases. It is anticipated that this data can be utilized for predicting transferase-specific sites of glycosylation and identifying transferase peptide targets. The goal of this chapter, therefore, is to provide detailed methods on the most critical analytical aspects for characterizing mucin O-glycosylation used in the Gerken laboratory, i.e., the mild trifluoromethanesulfonic acid (TFMSA) reaction, the Edman amino acid sequencing of TFMSA-treated mucin-type glycoproteins, and the use of lectin affinity chromatography in the isolation of α -GalNAc-O- containing random glycopeptides and the subsequent determination of ppGalNAc T isoform specificity.

2. Materials

2.1. Trimming O-Glycans by Mild Trifluoromethanesulfonic Acid

1. Anisole, anhydrous 99.7% septum capped.
2. TFMSA, reagent grade 98%, 10 g scored ampules.
3. Diethyl ether, anhydrous.
4. 50% (v/v) pyridine/water solution.
5. Mucin or glycoprotein (see Note 4): Salt free, exhaustively lyophylized from 30-mL teflon capped reaction tube.
6. Second dry teflon capped 30-mL reaction tube.
7. 125-mL Erlenmeyer flask.
8. Thin glass stirring rod and spatula long enough to reach to the bottom of reaction tube.

9. 50-mL separatory funnel.
10. Pasture pipettes and bulbs.
11. Dry ice/ethanol slurry.
12. Water ice bath.
13. Dry argon for purging tubes.
14. Standard dialysis tubing.
15. Procedure should be performed using two persons for transferring reagents and purging reaction vessels. Must use chemical-resistant nitrile gloves, full eye protection, and lab coat. Work should be performed in a vented chemical fume hood. Have in hood: Pasteur pipettes with bulbs, ethanol wash bottle (used to neutralize TFMSA), beakers for waste, and paper towels for spills and drying glassware.

2.2. Automated Edman Degradation

16. ABI Procise Edman sequencer (or equivalent), equipped with necessary standard peptide-sequencing reagents.
17. α -GalNAc-O-Ser, α -GalNAc-O-Thr, Thr and Ser standards, or equivalent glycopeptide standards obtainable from Sussex Research (Ottawa, ON Canada) or Bachem America (Torrance, CA)

2.3. Isolation of α-GalNAc-Glycopeptide on Lectin Chromatography

1. Random peptide substrate (see below): GAGAXnTXnAGAG, where $n = 3$–5 and X = randomized residues (obtain from Bachem America or Sussex Research) (see Note 5).
2. Transferase reaction buffer: 10 mM $MnCl_2$, 2 mM UDP-GalNAc (^{3}H-labeled 0.1 μCi/100 μL), and appropriate buffer. Protease inhibitors (Sigma P8340 and P8849 (see Note 6)) and 2-mercaptoethanol may be optionally included.
3. The source of transferases may be media supernatant, affinity-tagged transferase bound to affinity beads, or purified transferase.
4. Sephadex G-10 (0.7 × 113 cm column) (GE Healthcare Bio-Sciences Corp., Piscataway, NJ) equilibrated in 50 mM acetic acid made to pH 4.4 with NH_4OH, attached to a fraction collector.
5. Dowex 1-X8 anion exchange resin, 2–3 mL in Pasteur pipette (Bio Rad, Hercules, CA).
6. Peptide-O-α-GalNAc (Tn antigen)-specific, mixed-bed, immobilized lectin column (10 mL, 22 × 0.8 cm) containing: 2 mL each of the agarose bound lectins, Sophora japonica (SJA), SBA (Glycine max), and Helix pomatia (HPA) from EY Laboratories (San Mateo, CA) in addition to 4 mL of the Vicia villosa (VVA, predominately the B4 isoform) lectin from Vector Laboratories (Burlingame, CA) run at 4°C and attached to a fraction collector.
7. 1× lectin column buffer: 10 mM NaCl, 2.5 mM tris, pH 8.0, and 0.002% NaN_3.

8. 10× lectin column buffer: 100 mM NaCl, 25 mM tris, pH 8.0, and 0.02% NaN_3.
9. Suitable scintillation counter.

3. Methods

3.1. Site Localization by Edman Degradation

The key to this approach is the observation that mucin-type O-linked glycans could be quantitatively trimmed to the peptide-linked GalNAc residue by mild TFMSA (for a review of the use of TFMSA, see ref. 35) and that upon Edman amino acid sequencing the phenylthiohydantion (PTH) derivatives of the GalNAc-O Ser/Thr residues are identifiable in the PTH-amino acid chromatograms. This general approach has two highly favorable features: first, it eliminates O-glycan heterogeneity, thereby simplifying the analysis, and second it vastly eliminates the protease resistance that the full-length O-glycans impart onto the mucin or glycoprotein, thus allowing the use of specific proteases for peptide mapping approaches. Although all glycan side-chain structure is lost, the site of glycosylation and extent of substitution are preserved and readily quantified. Furthermore, under favorable conditions, one can selectively (chemically or enzymatically) remove/modify specific glycans prior to the TFMSA treatment, thus yielding additional glycan structural information (by difference), as was performed for the PSM tandem repeat (6). There are two key advantages of using the TFMSA/Edman sequencing approach over mass spectrometry approaches. The first being that site-specific core glycan occupancy can be easily and relatively accurately determined on a single sequencing run. Second, the very large (i.e., 20+ residues), highly glycosylated glycopeptides (and mucin tandem repeat domains), that would be too large for most MS analysis, can be readily sequenced and fully characterized by Edman sequencing, even up to 40–60 residues in our hands (4–6, 8).

To further simplify the isolation of the mucin glycosylated tandem repeated domains, we have typically performed an exhaustive trypsinolysis of the reduced and carboxymethylated mucin followed by the isolation of the high-molecular-weight tandem repeat glycosylated domain on gel filtration as described previously (4). This approach effectively removes the nonglycosylated N- and C-terminal domains from the mucin, leaving the protease-resistant glycosylated tandem repeat domain.

3.1.1. Trimming O-Glycans by Mild Trifluoromethanesulfonic Acid

The success of this reaction relies on keeping moisture out of the reaction vessels by prior, careful Ar purging and on the chilling (on dry ice) of the lyophilized glycoprotein, TFMSA/anisole reagent, and quenching reagents prior to their use. For convenience, the procedure below is designed for utilizing one 10 g sealed ampoule of TFMSA reagent which is sufficient to react with 100–200 mg of mucin glycoprotein. The reaction can be scaled accordingly for smaller or larger sample sizes. For mucins suspected to contain sialic acid residues 1–3 linked to the peptide GalNAc residue, prior treatment with neuraminidase improves the efficiency of the TFMSA treatment (36). The neuraminidase treatment can be performed as described (37) (see Note 7).

1. In a ~30-mL (~2 × 14 cm) teflon screw-top glass tube (make sure that the edges of the top of the tube are in good condition to maintain a tight seal), freeze and exhaustively lyophylize 100–200 mg of the trypsinized glycosylated tandem repeat mucin domain dissolved in 5–15 mL of water. Before use, the sample must be completely and exhaustively lyophylized. Once removed from the lyophylizer, the sample must be immediately and carefully purged with Ar gas and quickly capped to keep moisture out (see Note 8).
2. Prepare a crushed dry ice/ethanol bath in an ice bucket of sufficient volume to completely cover the sides of both 30-mL reaction vessels. Place the Ar purged and tightly sealed lyophylized mucin sample into the bath, completely covering the sides of the vessel to nearly the height of the cap. No condensate should appear on the inside of the tube.
3. Purge the second 30-mL dry reaction vessel at room temperature with Ar and cap—to this, you transfer 3 mL of anhydrous anisole. Draw into a dry glass 5-mL syringe, equipped with an 8-gauge needle ~4 mL of dry Ar gas. Next, inject this into the septum cap of the anisole bottle, turn upside down, and then pull out 3 mL of anisole into the syringe. Transfer the anisole into the second reaction vessel. After another Ar purge, tightly cap and freeze the anisole in the dry ice bath.
4. Remove 10 g TFMSA ampoule from its protective packaging and cool on water ice. Safety warning: Use extreme caution with TFMSA; spilled TFMSA can be neutralized with ethanol from wash bottle. Should TFMSA get onto your nitrile glove (typically noted by color change of glove), quickly neutralize with ethanol. If necessary, replace with fresh nitrile glove. Fully dry off the cold TFMSA ampoule, place firmly on hood surface, and carefully snap off ampoule top. With help from a second person, immediately transfer TFMSA into the reaction tube of frozen anisole (i.e., place inverted open TFMSA ampoule directly on top of tube), purge with Ar, and cap. Remove anisole/TFMSA-containing tube from dry ice and gently shake mixture until the anisole fully thaws and both reagents are well-mixed. Note: Reagents turn a light yellow color. Once mixed, return TFMSA/anisole reagent tube to the dry ice bath and cool until slightly viscous (dry ice solution should stop bubbling around tube).

5. With the help of a second person, open and Ar purge the reaction tube containing the cold lyophylized mucin and quickly transfer the TFMSA/anisole reagent to it (see Note 9). Carefully purge with Ar, tightly cap, and return tube to dry ice bath.
6. Begin the reaction by transferring the tube to an ice bucket containing an ice-water slurry. Be sure to have enough ice to fully cover sides of tube. After placing in ice for a short time and warming to 0°C, remove and vigorously shake the solution to begin to solubilize the mucin. Repeat until mucin is fully solubilized. This can take up to several hours with repeated vigorous shaking. Once the solution becomes clear, the reaction tube can stand fully covered in the ice bath with occasional shaking. Depending on the carbohydrate content, the solution turns from orange to dark purple over time. Typical reaction times at 0°C range from 6 to 8 h (see Note 10).
7. Reaction is stopped by placing the reaction tube in a dry ice/ethanol bath until equilibrated. Using a Pasteur pipette, transfer 1–2 volumes of dry ice-cold diethylether to the reaction mix (watch out for the cold diethyl ether squirting out of the warm pipette). Mix the ether and reagents thoroughly together while on dry ice using a glass stirring rod. Once mixed well, the mixture is added dropwise, using a Pasteur pipette, to 20 mL of a frozen slush of 50% pyridine in water. This mixture should be constantly stirred with a spatula and kept on the dry ice to maintain the slush. Once the contents are fully combined, the slush can be allowed to thaw to room temperature.
8. The reaction mixture is placed in a separatory funnel and extracted 3–4 times with an equal volume of diethyl ether. Vent the separatory funnel periodically after shaking and save the lower aqueous layer while discarding the upper ether layer as hazardous waste. Once fully extracted, the aqueous phase is returned to a clean Erlenmeyer flask which is nested in larger clean beaker. Using a clamped Pasteur pipette, gently bubble the solution with house air to further remove trace diethyl ether. Foam and bubbles appear when ether is fully removed.
9. In the cold, exhaustively dialyze the diethyl ether-free reaction mixture using an appropriate molecular weight cutoff (MWCO) dialysis tubing (typically, 12–15 kDa MWCO for mucins) (see Note 11). Once exhaustively dialyzed (5–6 changes of water, 2× a day), the sample may be lyophilized.
10. After lyopyhylization, the sample is brought up in 1–5 mL of water and if necessary spun at high speed to clarify. Typically, the pellet is discarded and the supernatant passed through a 0.45- or 0.22-μm low-protein-binding filter and relyophylized. Confirmation of the removal of all but the peptide-linked GalNAc residue can be performed by carbon-13 NMR

spectroscopy as described in 6, 36, 37. This material is ready for digestion with trypsin (or other appropriate protease) and subsequent peptide or tandem repeat isolation via gel filtration or other chromatographic approaches. The detailed description of these procedures is beyond the scope of this method as each mucin glycoprotein has its own unique requirements (however, see refs. 4–6 for PSM and 8 for CSM tandem repeat isolation).

3.1.2. Edman Amino Acid Sequencing of GalNAc-O-Ser/Thr Containing Glycopeptides

In our laboratory, the isolated, mild TFMSA trimmed, O-linked glycopeptides are sequenced by Edman degradation on an Applied Biosystems Procise Peptide Sequencer (Applied Biosystems, Foster City, CA) typically utilizing the standard manufacturer-recommended pulse liquid cycles. Since the α-GalNAc-O-Ser and α-GalNAc-O-Thr glycosidic linkages are highly stable to acid (see Note 12), they readily survive intact the post phenylisothiocyanate coupling, anhydrous TFA cleavage reaction, and the subsequent dilute TFA conversion cycles of the Edman sequencer. Furthermore, the α-GalNAc-O-Ser and α-GalNAc-O-Thr residues are sufficiently hydrophobic that their anilinothazolinone derivatives are efficiently transferred to the conversion cycle flask, unlike Ser or Thr residues containing O-glycans of longer length which do not efficiently transfer under standard conditions (see Note 13). Finally, the final phenylthiohydantion (PTH) derivatives of α-GalNAc-O-Ser and α-GalNAc-O-Thr elute in unique positions in the C18-PTH amino acid chromatograms.

1. General conditions: Sequencing is performed on an Applied Biosystems Procise 494 protein sequencer utilizing standard manufacturer-recommended pulse liquid cycles. Samples 0.05–5 nmol are dried on trifluoroacetic acid-washed glass fiber filters, previously spotted with 1.5 mg BioBrene Plus (ABI 400385). Amino acid PTH derivatives are chromatographed on a standard ABI 5-μm C18 PTH column, typically at 55°C, using the Fast Normal 1 gradient program and are monitored by absorbance at 269 nm (see Note 14).
2. For best results, α-GalNAc-O-Ser/Thr standards (as free glycosylated amino acids and/or glycopeptide(s)) should be run on the sequencer as a means of optimizing separation on the PTH analyzer and for determining the α-GalNAc-O-Thr-PTH diastereotopic ratios. This can be performed by running known mixtures of free Ser, Thr, α-GalNAC-O-Ser, and α-GalNAC-O-Thr and adjusting the column temperature to achieve optimal separation. An example of the chromatograms of the glycosylated Thr and Ser residues in the PSM glycopeptide GAT*GAS*IGQPETSR (38) is shown in Fig. 2a, b. As previously reported (38, 39), the α-GalNAc-O-Thr-PTH appears as two peaks, one eluting just after the position of the

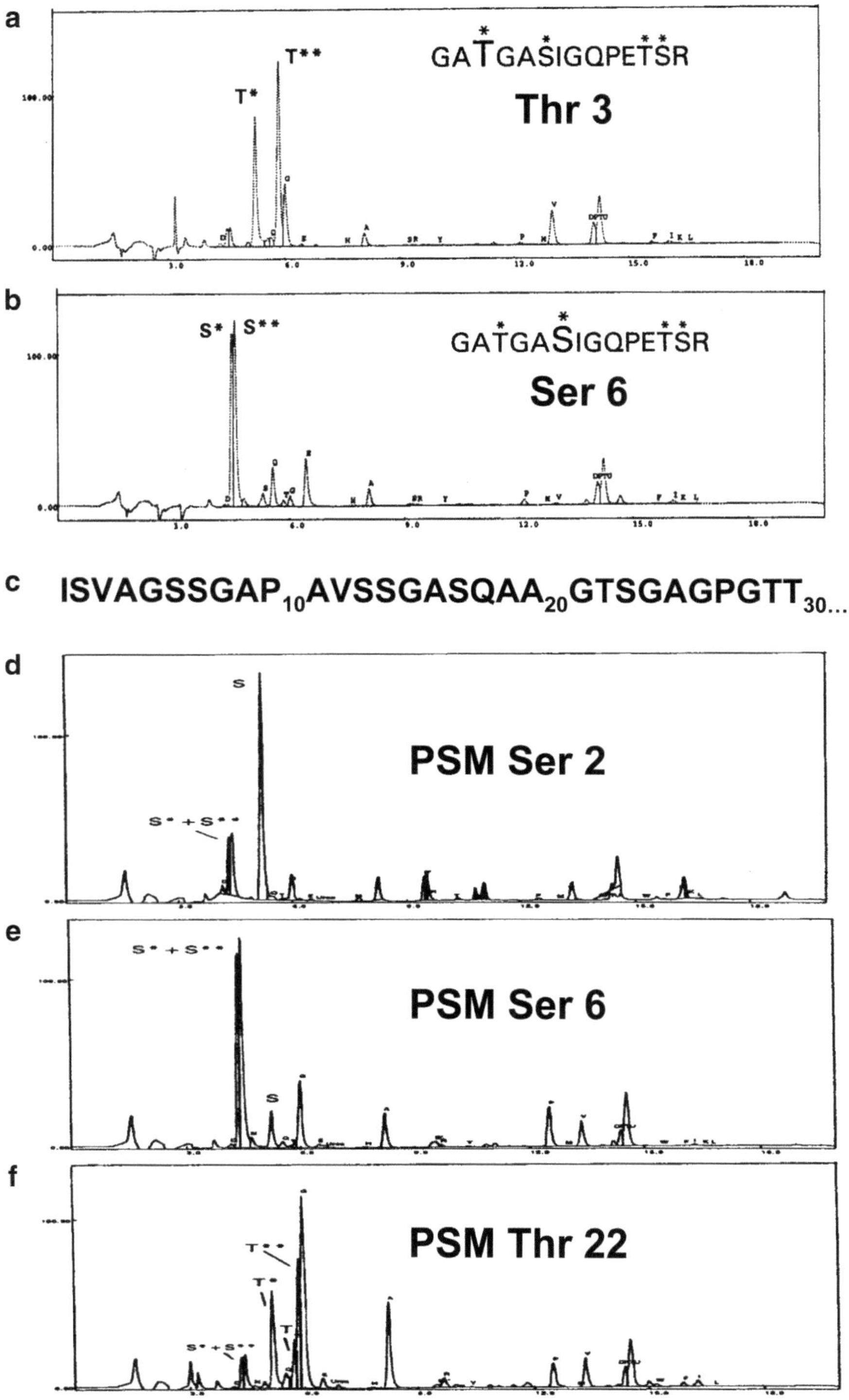

Fig. 2. Representative Edman degradation C18-PTH column chromatograms of α-GalNAc-O-glycosylated Ser and Thr residues in glycoproteins. (**a** and **b**) Chromatograms of fully glycosylated Thr 3 and Ser 6 from the PSM-derived glycopeptide GAT*GAS*IGQPETSR. (**c**) First 30 residues of the PSM tandem repeat. (**d–f**) Chromatograms of partially glycosylated Ser 2, Ser 6, and Thr 22 from the PSM tandem repeat. Conditions: Chromatograms were run on an Applied Biosystems Procise 494 protein sequencer using standard pulse liquid cycles and the Fast Normal I gradient program at a column temperature of 55°C. Approximately 3–10 nmol of glycopeptide were spotted.

α-GalNAc-O-Ser-PTH peak (labeled T*) and the other eluting between Thr-PTH and Gly-PTH (labeled T**). Similarly, the α-GalNAc-O-Ser-PTH peak appears, as a closely spaced doublet (labeled S* and S**), between the elution positions of Asp-PTH and Asn-PTH. The doublet nature of the glycosylated Ser and Thr PTH derivatives is due to the formation of two differently migrating diastereotopic derivatives upon conversion of the anilinothiazolinone to the PTH derivative. It has been observed that the ratios of the area of the two species remain relatively constant, typically 1:1 for Ser*:Ser** and ~45:55 for Thr*:Thr** (38). When sequencing long, highly glycosylated glycopeptides, cycle lag may result in the presence of overlapping PTH-Ser and T* in the same cycle. In this situation, one can use the area of the T** peak and known ratio of T* to T** to correct the area of the overlapping peak. We have found that the separation of PTH-Ser and T* peaks can be improved by decreasing the column temperature to 45°C (from 55°C), although at the expense of additional residue overlap in other regions of the chromatogram (see Note 14).

3. Once optimal C18-PTH column conditions are obtained, the TFMSA-treated glycopeptide of interest can be sequenced to determine its site-specific glycosylation pattern utilizing standard pulse liquid sequencing cycles and optimized C18-PTH analyzer conditions. PTH chromatograms for variably glycosylated Ser 2, Ser 6, and Thr 21 in the PSM tandem repeat are given in Fig. 2c–e. After assigning the PTH peaks and determining their relative areas via the sequencer software, the data is transferred to an Excel or Lotus 123 spreadsheet for further manipulation. (If necessary, peak areas are corrected for integration errors by cutting and weighing). This involves performing baseline corrections (particularly for long glycopeptide sequences) due to cycle lag and preview and the inclusion of response factors (obtained in the calibration step). The baseline corrections are typically performed by subtracting a five-cycle minimum value running average across the entire sequencing run for each PTH peak in the chromatogram (4). This effectively flattens the plot of residue area versus cycle number as shown in Fig. 3. Alternatively, for shorter peptides (~20 residues), a simple lag correction can also be used that provides nearly identical results. The percent of glycosylation is obtained from the corrected areas of the unique PTH peaks in the chromatogram: S and S*+S**, and T and T*+T**. The glycosylation patterns of the PSM and CSM tandem repeats obtained by this approach can be found in Fig. 3 and refs. 6, 8. Edman sequencing is a very straightforward approach for quantifying the glycosylation of ppGalNAc T peptide and glycopeptide substrates that contain multiple potential sites of

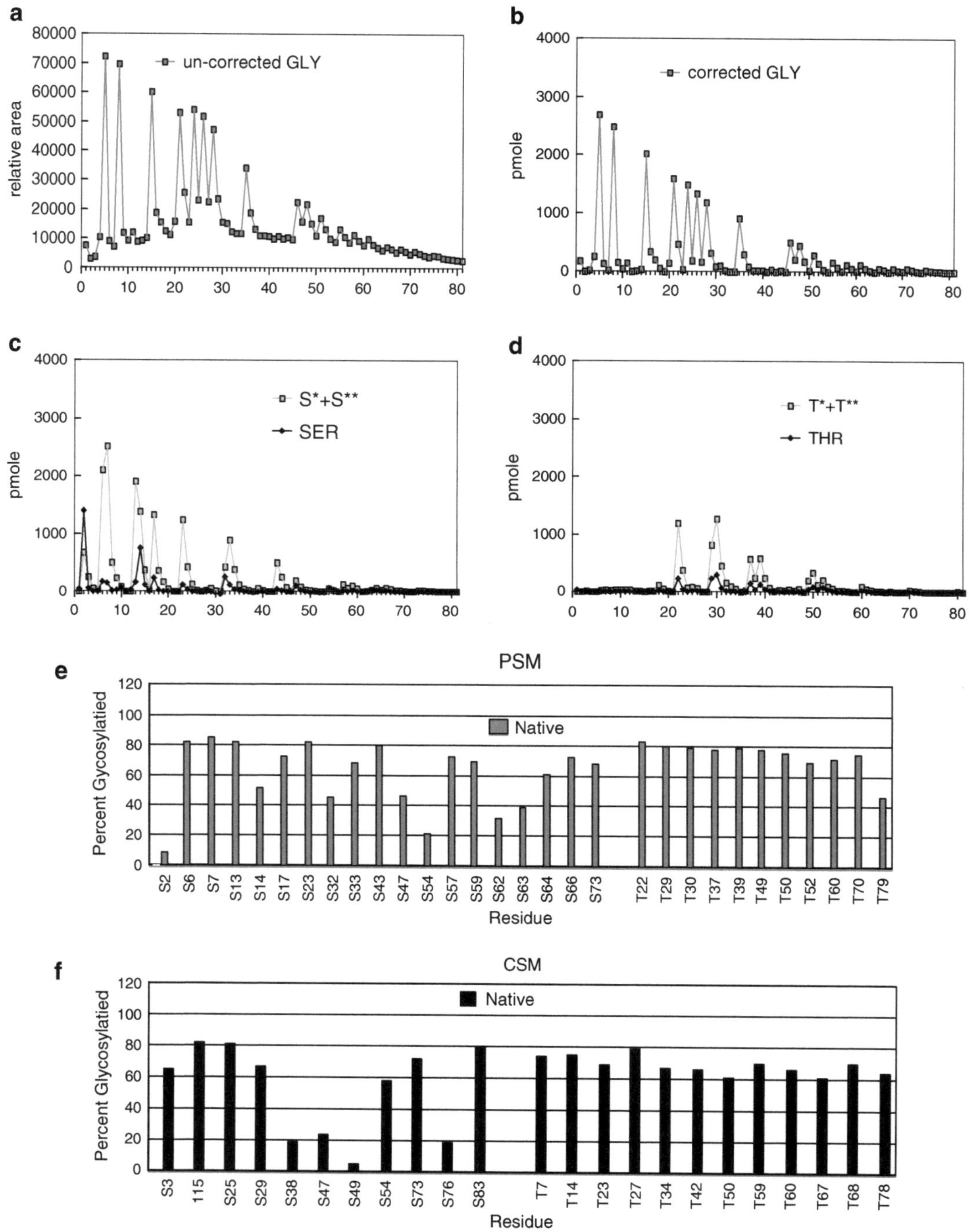

Fig. 3. Representative data analysis of the Edman sequencing of the PSM tandem repeat. (**a**) Plot of the uncorrected response (area) vs. cycle number for the PTH-Gly peak. (**b**) Plot of the corrected pico mole area of the PTH-Gly peak after baseline correction. (**c** and **d**) Plots of the corrected pico mole areas of the glycosylated and nonglycosylated (S*, S**, T*, T**) PTH derivatives of Ser and Thr, respectively, from which the percent of glycosylation can be determined. (**e** and **f**) Plots of the percent α-GalNAc glycosylation of the Ser and Thr residues of the PSM and CSM tandem repeats (6, 8). The sequences of the PSM and CSM tandem repeats are found in Fig. 6d and h.

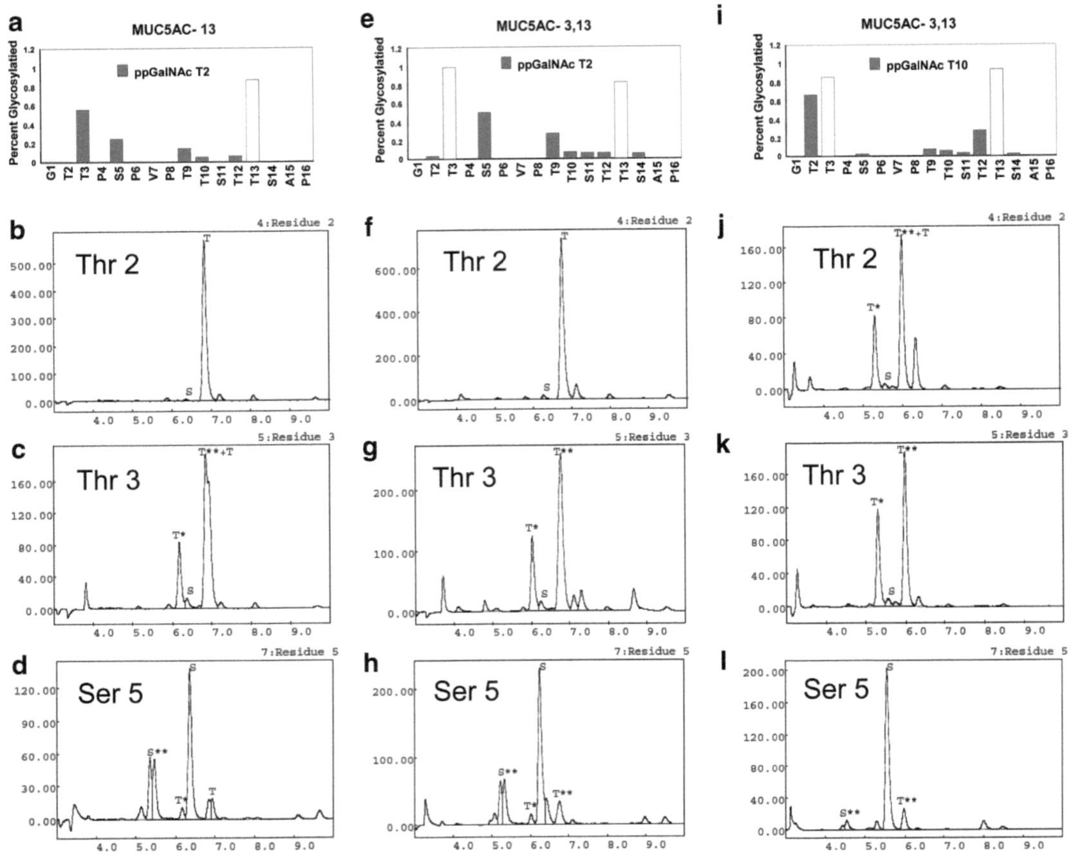

Fig. 4. Edman sequence analysis of the glycosylation of human MUC5AC tandem repeat glycopeptides by ppGalNAc T2 and T10. Left column (**a–d**): Analysis of MUC5AC-13 (GTTPSPVPTTSTT*SAP) after glycosylation by ppGalNAc T2. Center column (**e–h**): Analysis of MUC5AC-3, 13 (GTT*PSPVPTTSTT*SAP) after glycosylation by ppGalNAc T2. Right column (**i–l**): Analysis of MUC5AC-3, 13 after glycosylation by ppGalNAc T10. Top panels (**a**, **e**, and **i**): Plot extent of glycosylation determined by Edman sequencing. Bottom three panels display portions of the PTH chromatograms for Cycles 2, 3, and 5 representing Thr 2, Thr 3, and Ser 5, respectively; see ref. 12 for details.

glycosylation. An example is given in Fig. 4 for the glycosylation by ppGalNAc T2 and T10 of two glycopeptides derived from the human MUC5AC tandem repeat that contain up to nine potential sites of glycosylation (11).

3.2. Site Prediction of O-Glycosylation Sites Based on Random Peptide Substrates

Systematic studies of the peptide substrate specificity of the different isoforms in the family of ppGalNAc T transferases are lacking. Recently, we developed an approach using oriented random peptide substrates, containing a single central Thr (or Ser) residue, see Table 1, that can be used to obtain this information in a relatively straightforward manner (6, 32, 33). The most critical step in this procedure is the isolation of the product glycopeptide by lectin affinity chromatography which is described in detail below. Our initial studies suggest that the obtained random peptide substrate preference data correlate reasonably well with experimentally observed glycosylation patterns reported for other peptides and hence may be useful in predicting isoform-specific sites of glycosylation and the design of optimal and selective isoform-specific substrates (6).

Key to the approach is the use of lectin affinity chromatography to isolate glycosylated random peptide from nonglycosylated peptide and the subsequent use of Edman amino acid sequencing to quantify the amino acid composition at each randomized position flanking the site of glycosylation. By comparing the mole fraction composition of the isolated random glycopeptide to the composition of the input random peptide, positional-specific enhancement factors can be obtained. This approach represents the first attempt to systematically quantify the peptide substrate preferences of the ppGalNAc transferase family. The design of the random peptide substrates utilized in this approach (Table 1) addresses several technical issues, the most important being the ability to separate the random (glyco)peptide from buffers and free GalNAc on gel filtration. The GAGA flanking sequences, in addition to increasing peptide size, are used as sequence identification markers, thereby allowing an assessment of (glyco)peptide purity and the extent of (glyco)peptide degradation. Moreover, the GAGA sequences represent a relatively extended peptide conformation, thereby eliminating end effects likely to be present with shorter peptide substrates. Using the series of random peptides

Table 1
Oriented Random Peptide Substrates for Obtaining ppGalNAc Transferase Amino Acid Residue Specificities: Random Peptide Substrates P-I, P-III, P-IV, P-V, P-VI, and P-VII

No. of unique C18-PTH[a]

Peptide	Sequence	No. of unique sequences (reference)	C18-PTH Column Temp
	GAGAXXXTXXXAGAGK		
P-I	X=GAPRENYV	2.6×10^5 (31)	45
P-III	X=GAPRDQKFI	5.3×10^5 (31)	55
P-IV	X=GAPREQHL	2.6×10^5 (31)	55
P-V	X=GAPRDNKM(W)[b]	5.3×10^5 (31)	45
	GAGAXXXXXTXXXXXAGAGK		
P-VI	X=G,A,P,V,L,Y,E,Q,R,H	10.0×10^9 (33)	55
P-VII	X=G,A,P,I,M,F,D,N,R,K	10.0×10^9 (33)	45

[a]Edman Sequencer C-18 PTH column temperature used for optimal separation of glycosylated Ser and Thr residues

[b]The Trp (W) residues are not detected on Edman sequencing of this peptide due to oxidation; therefore, W is not included in the analysis (31) (see Note 16)

given in Table 1, enhancement values for all amino acid residues, except Trp, Cys, Thr, and Ser, can be obtained for residues +/− 3 or +/−5 positions from the site of glycosylation depending on which series of random peptide substrate is used (see Note 15).

3.2.1. ppGalNAc T Random Peptide Substrate Preferences: Isolation of α-GalNAc-Glycopeptide on Lectin Chromatography

1. Random peptide is glycosylated by ppGalNAc T utilizing appropriate reaction conditions for the transferase. Typically, 1 mg of random peptide is incubated in 0.1 mL of reaction buffers containing 10 mM $MnCl_2$, 1–2 mM UDP-GalNAc (^3H-labeled 0.1 μCi), and appropriate buffer, pH 6.5–7.5. Protease inhibitors (1%, Sigma P8340 and P8849) (see Note 6) and 2-mercaptoethanol (4–10 mM) may be optionally included. The source of transferase may be media supernatant, affinity-tagged transferase bound to affinity beads, or purified transferase. The reaction is quenched by adding 2× volume of 250 mM EDTA. It is important that the amount of transferase used gives less than 5% glycopeptide product in a typical overnight incubation based on ^3H incorporation into the random peptide (see Note 17). Conversions greater than 10% are undesirable. Unreacted UDP-GalNAc is removed by passing the diluted (to 4 mL) reaction mix through a washed 2–3 mL Dowex 1-X8 anion-exchange column (see Note 18). After lyophilizing, the random (glyco)peptide is separated from free GalNAc and buffers by gel-filtration chromatography on Sephadex G-10 (0.7 × 113-cm column), each fraction monitored by ^3H and absorbance at 220 and/or 280 nm. The radiolabeled (glyco)peptide peak is pooled and lyophylized.
2. Random glycopeptides are separated from nonglycosylated random peptides by passage across a Ser/Thr-O-α-GalNAc (Tn-antigen)-specific, mixed-bed, immobilized lectin column containing the lectins, SJA, SBA, HPA, and VVA, equilibrated with 1× lectin buffer as follows. The lyophylized random (glyco)peptide from G10 chromatography is dissolved in 1 mL water (from which 0.1 mL is removed as the prelectin control) and made 1× lectin buffer by adding 0.1 mL of 10× lectin buffer. After adding 10 μL each of protease inhibitors P8340 and P8849, the sample is loaded on the lectin column and allowed to slowly enter the gel. After stopping the flow for 10–15 min to allow binding of glycopeptides, column washing is begun at a flow rate of ~8 mL/h while collecting ~3.5-mL fractions on a fraction collector. It is imperative to allow a very extensive wash to prevent contamination of the bound glycopeptide by nonglycosylated peptide. After collecting 25–30 fractions (90–110 mL total volume), bound glycopeptide is released from the lectin column with 5 mL of 10 mM GalNAc in buffer and an additional 10–15 fractions collected. Subsequently, the column is reequilibrated with an overnight wash (~250 mL) with 1× lectin buffer—after which the column is fully functional

for a subsequent run. Fractions are monitored (see Note 19) at 220 nm for peptide and added GalNAc and by scintillation counting 50–100 μL for ^{3}H-labeled glycopeptide (see Note 20). GalNAc-released ^{3}H-glycopeptide fractions (see Note 21) are pooled, lyophilized, and separated from salt, buffer, and GalNAc by gel-filtration chromatography on Sephadex G10 as performed above. The radiolabeled glycopeptide peak [called the post-lectin sample (see Note 22)] is pooled and lyophylized. Prior to Edman sequencing, both the prelectin and post-lectin (glyco)peptides are lyophylized exhaustively from a small volume of water 3–4 times to remove trace acetic acid (see Note 23).

3. The pre- and post-lectin random peptide samples are pulse liquid Edman sequenced as described in Subheading 3.1.2, except that standard proline cleavage cycles are utilized for the randomized X residue positions to ensure complete cleavage. The C18-PTH column temperatures given in Table 1 can be used to reduce overlap of the T* and T** lag with other residues in subsequent cycles. Typically, the entire post-lectin sample is sequenced while only a fraction of the prelectin sample is sequenced, adjusted to match the peak intensities of the post-lectin sequencing run (see Notes 24 and 16). Peak areas provided by the ABI Procise software (and if necessary corrected for integration errors by cutting and weighing) are transferred to Excel or Lotus 123 spreadsheets and processed to eliminate preview and lag either by the average minimum value baseline correction approach described (4) or by mathematically correcting for lag and preview using spreadsheet software (unpublished). After correcting for PTH recovery based on the PTH standards or other suitable approach, the mole fraction of each randomized residue is calculated at each X position of the peptide for both the pre- and post-lectin (glyco) peptides. Enhancement factors are obtained by dividing the post-lectin glycopeptide mole fraction by the control prelectin random (glyco)peptide mole fraction at each random cycle position for each amino acid residue present. These amino acid residue-specific enhancement factors are taken as specific indicators of transferase preference, with values greater than 1 indicating an increased preference and values below 1 a decreased preference for the given residue. We have found that the enhancement values obtained for a given amino acid residue are typically very similar regardless of the peptide substrate in Table 2 that is used (6, 33). Therefore, the enhancement factors can be averaged across the different peptides to obtain residue-specific enhancement factors. The obtained enhancement values for ppGalNAc T1, T2, and T10 are listed in Table 2 and plotted in Fig. 5 for comparison (6, 32, 33) (see Note 25).

Table 2
ppGalNAc T1, T2, and T10 random peptide-derived enhancement values[a,b]

	−5	−4	−3	−2	−1	0	+1	+2	+3	+4	+5
ppGalNAc T1:											
G	1.15	0.98	0.90	1.15	0.98		0.63	1.67	0.86	0.93	0.89
A	1.27	1.07	1.09	1.30	1.42		0.97	1.45	0.86	0.94	0.85
P	1.06	1.18	1.00	1.21	2.04		2.24	1.16	3.01	1.66	1.33
V	1.04	1.16	1.20	1.12	1.85		1.44	0.87	0.72	1.06	1.35
I	0.65	0.75	1.03	0.68	1.08		1.01	0.59	0.53	0.91	1.19
L	0.75	0.83	0.90	0.84	0.23		0.50	0.47	0.46	0.81	0.99
M	0.95	0.99	1.08	1.03	0.63		1.14	0.88	0.85	1.05	1.14
F	0.62	0.73	1.09	1.00	1.14		0.39	0.54	0.42	0.75	0.68
Y	1.14	1.33	1.19	1.01	1.06		0.52	0.75	1.89	1.31	1.16
E	0.96	0.91	1.10	0.81	0.32		1.49	0.92	0.46	0.89	0.92
D	0.95	0.99	1.21	0.90	0.57		1.31	0.85	0.86	0.93	0.97
Q	1.07	1.09	1.17	0.87	0.64		1.20	0.80	0.50	0.89	0.92
N	1.01	1.05	0.97	0.90	0.56		0.46	0.51	0.78	1.00	0.95
R	0.74	0.80	0.83	0.75	0.60		0.56	0.94	0.57	0.74	0.89
K	0.82	0.81	0.60	0.50	0.32		0.41	0.82	0.46	1.01	1.14
H	0.84	0.97	1.08	0.93	0.75		0.60	0.71	0.59	0.73	0.85
S*	–	0.40	0.39	0.26	0.16		0.28	0.97	0.39	0.61	–
ppGalNAc T2:											
G	1.03	1.11	0.72	1.74	0.99		0.74	1.61	1.16	1.02	0.99
A	1.08	1.06	1.06	1.02	1.31		1.10	1.29	0.99	1.00	0.92
P	1.23	1.20	1.90	1.00	4.23		2.28	1.21	2.84	1.63	1.34
V	1.08	1.02	1.68	0.79	0.77		1.30	0.97	0.88	1.00	1.23
I	0.87	0.81	1.76	0.59	0.40		0.84	0.81	0.74	0.86	0.98
L	0.86	0.87	1.10	0.75	0.18		0.59	0.65	0.56	0.85	0.96
M	0.98	1.02	1.02	0.92	0.38		1.00	1.01	0.68	1.06	1.11
F	0.71	0.74	1.05	0.95	0.23		0.45	0.74	0.59	0.75	0.73
Y	1.00	1.08	0.81	1.10	0.43		0.53	0.65	1.15	1.13	1.09
E	0.82	0.85	0.64	0.81	0.29		1.00	1.02	0.62	0.85	0.99
D	1.04	1.01	0.92	0.88	0.70		1.13	0.94	1.01	0.99	0.98
Q	0.99	1.06	0.88	1.01	0.26		1.16	1.03	0.59	0.89	0.98
N	1.05	1.04	0.73	0.71	0.33		0.52	0.58	0.51	0.87	0.82
R	0.99	0.92	0.62	0.83	0.50		0.72	0.70	0.52	0.74	0.88
K	0.84	0.84	0.55	0.71	0.32		0.60	0.69	0.49	0.82	1.08
H	0.93	0.93	0.67	0.78	0.34		0.52	0.63	0.50	0.75	0.85
S*	–	1.58	0.60	0.76	0.50		0.43	0.75	0.61	0.57	–
ppGalNAc T10:											
G	1.86	1.68	1.44	1.58	1.52		1.04	1.03	1.09	1.08	1.34
A	1.32	1.17	1.12	1.16	1.01		1.08	1.56	1.20	1.12	0.98
P	0.87	0.85	0.82	0.89	1.35		1.53	1.67	0.94	1.03	1.00
V	0.65	0.73	0.91	0.70	0.77		0.83	0.83	1.08	1.01	1.09
I	0.51	0.66	0.89	0.66	0.61		0.72	0.63	0.75	0.80	0.87
L	0.56	0.69	0.98	0.76	0.85		0.61	0.50	0.78	0.72	0.77
M	0.68	0.76	0.88	1.05	1.37		1.13	0.60	1.04	0.91	0.94
F	0.60	0.68	1.05	1.01	1.08		0.67	0.56	0.87	0.64	0.78
Y	0.87	1.04	1.09	1.41	1.32		0.88	0.76	1.16	1.03	1.13
E	0.98	0.95	0.92	0.73	0.50		0.98	0.65	0.83	0.85	1.00

(continued)

Table 2 (continued)

	−5	−4	−3	−2	−1	0	+1	+2	+3	+4	+5
D	1.64	1.41	1.26	1.03	0.77		0.92	0.84	1.35	1.12	1.10
Q	1.01	1.01	0.99	0.88	0.73		1.31	0.73	1.07	0.96	0.99
N	0.85	0.75	0.70	0.84	0.52		0.57	0.54	0.95	0.75	0.96
R	1.23	1.10	1.04	1.06	1.19		1.39	1.26	1.09	1.54	1.28
K	0.71	0.95	0.90	0.76	0.93		0.91	0.79	0.67	0.76	0.64
H	0.83	0.65	0.69	0.74	0.68		0.50	0.47	0.88	0.82	0.75
S*	–	0.72	0.62	0.61	0.38		8.58	2.21	1.29	1.52	–

[a] Column heading numbers represent positions relative to the site of glycosylation (position 0)

[b] Enhancement values represent the ratio of the mole fraction of the isolated random glycopeptide over the initial random peptide for each residue as determined by Edman sequencing. S* represents enhancement factors forα-GalNAc-O-Ser (see Note 26). Values from refs. 31–33.

3.3. Use of TFMSA and Lectin Chromatography for Glycoproteomics Studies

Although beyond the scope of this section, we have also begun treating bulk tissue extracts with the mild TFMSA followed by exhaustive trypsin treatment to obtain α-GalNAc-glycopeptides for isolation on lectin chromatography. It is anticipated that this approach will yield a series of O-linked glycopeptides for subsequent glycoproteomics analysis of bulk tissues of ppGalNAc T transferase knockout animals. After further isolation, the glycopeptides may be analyzed by Edman sequencing or the MS approaches described in other chapters of this issue.

3.4. Use of Random Peptides Substrate Preferences to Predict Isoform-Specific Sites of Glycosylation

As can be seen from Fig. 5, the most selective substrate positions for ppGalNAc T1 and T2 (31, 33) are the three residues N-terminal of the site of glycosylation (i.e., positions −3, −2, and −1, in Fig. 5) with the −1 position displaying the largest differences between isoforms: ppGalNAc T1 preferring Pro and Val while ppGalNAc T2 almost exclusively preferring Pro. Interestingly, the three C-terminal residues (positions +1, +2, and +3) appear to have very similar patterns for ppGalNAc T1 and T2, with Pro, Gly, and Pro being the most preferred residues, respectively. Unpublished results from several other ppGalNAc Ts (see Note 25) show similar trends, the N-terminal positions showing the greatest differences while the C-terminal positions displaying the Pro Gly Pro pattern. Of further note is the observation that the outer flanking residues at positions −5, −4, +4, and +5 display enhancement values that are very close to 1 for almost all of the amino acid residues (Fig. 5). This suggests that the major determinants in the peptide recognized by the ppGalNAc Ts characterized to date are within the three N- and C-terminal residues flanking the site of glycosylation. These findings are consistent with previous O-glycosylation prediction approaches (13, 14). Therefore, only the enhancement values for the three residues in the N and C direction flanking the site of glycosylation need to be utilized in assessing glycosylation site propensity. Note that ppGalNAc T10 shows very few specific

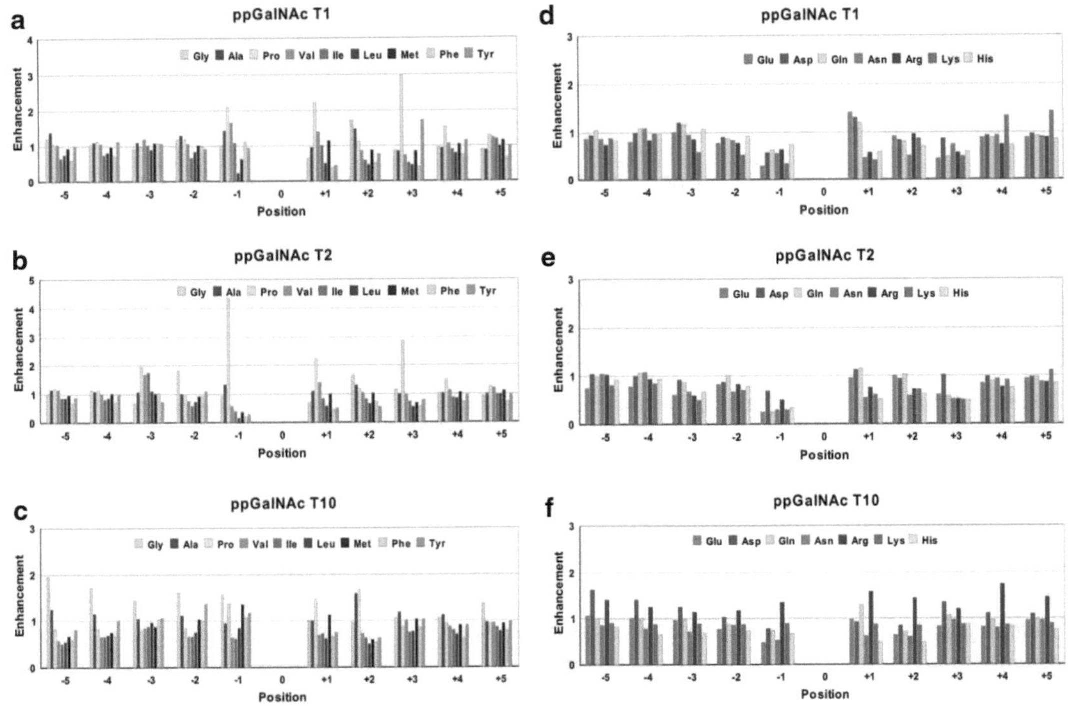

Fig. 5. Plots of hydrophobic (*left*) and hydrophilic (*right*) enhancement values for ppGalNAc T1, T2, and T10; see refs. 31–33 and Table 2.

peptide residue enhancements in keeping with the observation that this isoform is most active against GalNAc-O-glycopeptide substrates (40, 41) (see Note 26).

We have found that the simple product of the flanking enhancement values surrounding a given site of glycosylation roughly correlates with the ability (or rate) of the site to be glycosylated by a given transferase (31) (unpublished). Thus, we have proposed that enhancement product values may be used to rank sites of potential glycosylation, thereby predicting the order of glycosylation of substrates with multiple acceptor sites. This is shown in Table 3a, b, where the enhancement products are compared to the glycosylation of an MUC1 peptide and a wild-type and mutant gC-1 peptide from HSV-1 (42) (however, see Note 27). Also listed in Table 3c are the optimal and most selective ppGalNAc T1 and T2 peptide sequences derived from the random peptide data along with the experimental-determined Kcat/Km values for each (31). We have also plotted in Fig. 6 the ppGalNAc T1 and T2 enhancement products obtained for the PSM and CSM tandem repeats for comparison to their experimentally determined ppGalNAc T1 and T2 glycosylation pattern (7, 11). In addition, the Net-O-Glyc Web-based predictions (14) for mucin-type O-glycosylation are given for comparison. It is clear from the figure that the enhancement value products roughly approximate the isoform-specific glycosylation patterns for most of the residues. In contrast, the Net-O-Glyc

Table 3
Enhancement value products predict experiment glycosylation site order (a and b) and design of isoform-specific substrates (c)

(a) Glycosylation of MUC1 tandem repeat (relative order of glycosylation) [31, 53]:			
MUC1:	PAPGS T_6	APPAHGV T_{14}	SAPD T_{19} R
ppGalNAc T1 Exp [53]	2nd	1st	nd
ppGalNAc T1 product[a]	3.9	10.0	0.7
ppGalNAc T2 Exp [53]	1st	2nd	nd
ppGalNAc T2 product[a]	12.5	3.3	0.5

(b) Glycosylation of wild-type (wt) and mutant gC-1 peptide from HSV-1 by ppGalNAc T2 (relative order of glycosylation) [42]:				
wt peptide:	PKNN T_5	T_6 PAK	S_{10} GRP	T_{14} KPPGP
ppGalNAc T2 Exp sites	–	–	2nd	1st
ppGalNAc T2 product[a]	0.15	0.75	0.91	5.21
Mutant peptide (K9A,R12A):	PKNN T_5	T_6 PA*A*	S_{10} G*A*P	T_{14} KPPGP
ppGalNAc T2 Exp sites	–	3rd	2nd	1st
ppGalNAc T2 product[a]	0.15	1.51	6.88	6.41

(c) Design of Optimal and Selective ppGalNAc T1 and T2 substrates [31]:					
		Enhancement product[a,c]		Kcat/Km	
	Peptide[a]	T1	T2	T1	T2
Optimal for:					
ppGalNAc T1	R1-FFPTPGP-R2	45	87	4.3	17
ppGalNAc T2	R1-PGPTPGP-R2	28	209	0.6	39
Selective for:					
ppGalNAc T1	R1-HVFTERY-R2	43	0.5	0.41	nd
ppGalNAc T2	R1-PGPTAIG-R2	1.2	19	nd	0.14

[a]Products of enhancement values for positions −3 to +3 obtained from Table 2. Product values greater than 1 suggest favorable glycosylation, less than 1 unfavorable glycosylation
[b]R1= GAGA-, R2= -AGAGK
[c]Note that listed enhancement products are from ref. 31 and differ from those using the updated enhancement values given in Table 2

predictions predict all of the sites to be nearly equally O-glycosylated (G-score values >0.5 being predictive) (14). Presently, the reported use of the enhancement products to predict glycosylation sites is limited (12); therefore, to date, no specific cutoff values have been determined, although, by the nature of the approach, product values greater than 1 would be expected to indicate favorable glycosylation while the larger the value the greater the probability for glycosylation. Enhancement product values significantly less than 1 would, therefore, be expected to be very poor targets.

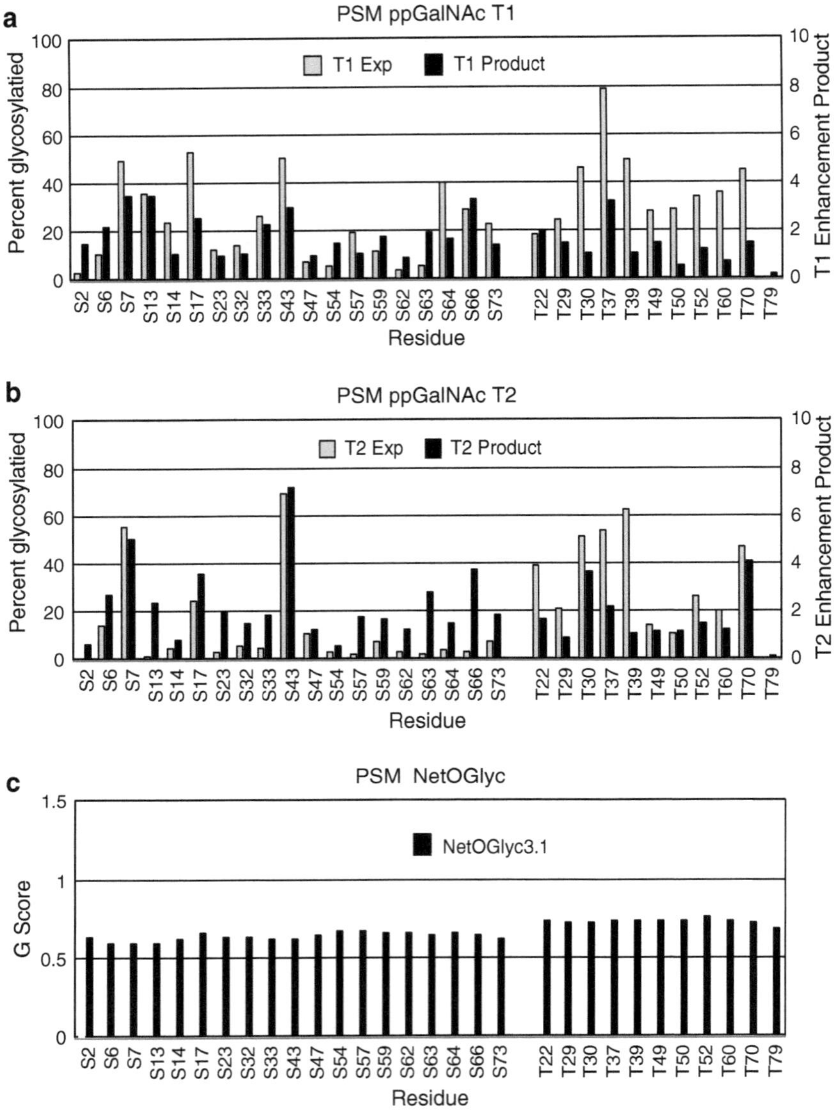

d

ISVAGSSGAP$_{10}$AVSSGASQAA$_{20}$GTSGAGPGTT$_{30}$
ASSVGVTETA$_{40}$RPSVAGSGTT$_{50}$GTVSGASGST$_{60}$
GSSSGSPGAT$_{70}$GASIGQPETS$_{80}$R

Fig. 6. Enhancement value products roughly correspond to the observed in vitro glycosylation of the PSM and CSM tandem repeats. (**a–d**) PSM tandem repeat and (**e–h**) CSM tandem repeat. (**a** and **e**) Enhancement products for ppGalNAc T1 and observed ppGalNAc T1 glycosylation. (**b** and **f**) Enhancement products for ppGalNAc T2 and observed ppGalNAc T2 glycosylation. (**c** and **g**) Net-O-glc predictions for the PSM and CSM tandem repeats, respectively. (**d** and **h**) Tandem repeat sequences of PSM and CSM, respectively. Experimental glycosylation data obtained from Edman sequencing; see ref. 7, 8.

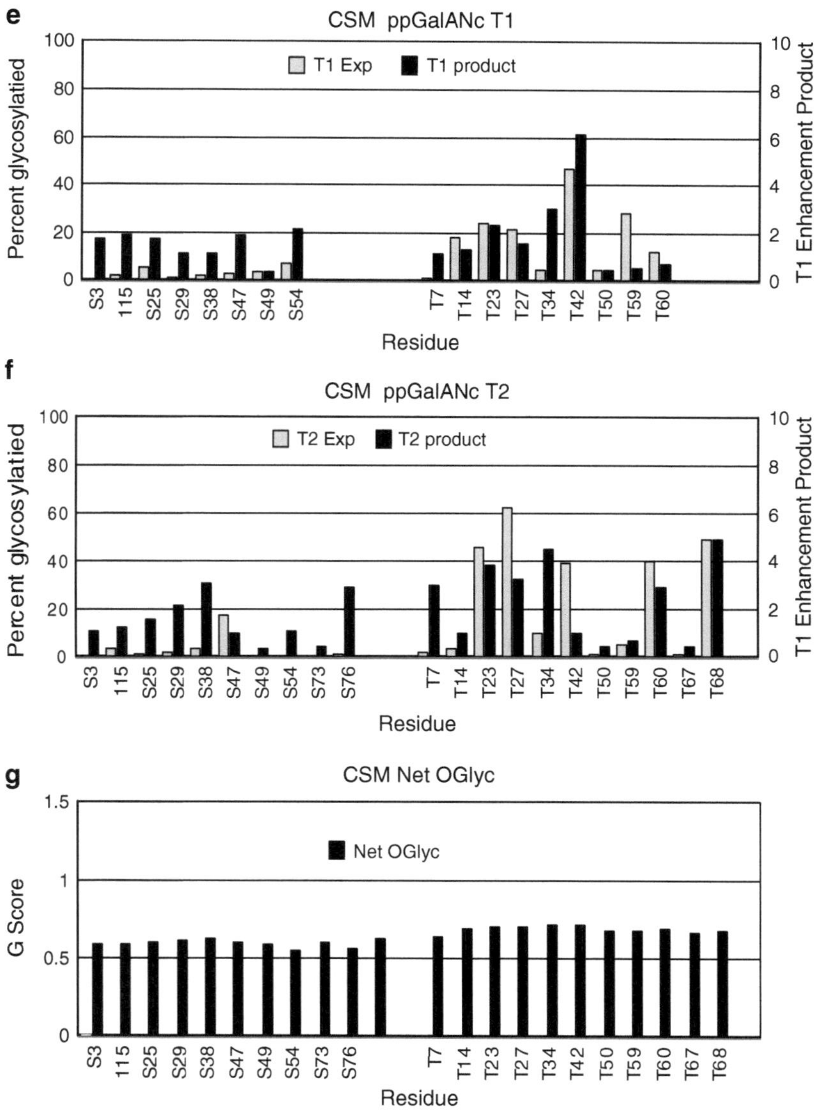

h AGSVGRTAGG$_{10}$PGFTSPGRVA$_{20}$GGTGSPTASA$_{30}$
RGGTPGGSEG$_{40}$FTAAPGSEST$_{50}$GGHSGAPGTT$_{60}$
LAGRAGTTLG$_{70}$PRSEPSGTGV$_{80}$GGSPVATTL

Fig. 6. (continued)

4. Notes

1. The so-called O-glycanase (endo-α-*N*-acetylgalactosaminidase) from *Diplococcus pneumoniae* (renamed *Streptococcus pneumoniae*) only cleaves the unsubstituted β-Gal(1–3) α-GalNAc-O-Ser/Thr disaccharide, leaving all other mucin-type glycan structures intact (43, 44).
2. There are two Web sites available predicting sites of O-glycosylation:

   OGPET v1.0 (Rafael Torres Jr., Igor C. Almeida, Yash Dayal and Ming-Ying Leung: unpublished): http://129.108.112.23/OGPET/; NetOGlyc 3.1 Server (14): http://www.cbs.dtu.dk/services/NetOGlyc/.
3. Structurally, the ppGalNAc Ts consist of an N-terminal catalytic domain tethered by a short linker to a C-terminal ricin-like lectin domain containing three recognizable carbohydrate-binding sites (see ref. 45). Some members of the ppGalNAc T family prefer substrates which have been previously modified with O-linked GalNAc on nearby Ser/Thr residues, hence having the so-called glycopeptide, or filling-in activities, i.e., ppGalNAc T7 and T10 (40, 41, 46). Others simply posses altered preferences against glycopeptide substrates, i.e., ppGalNAc T2 and T4 (11, 47–49), or may be inhibited by neighboring glycosylation, i.e., ppGalNAc T1 and T2 (6, 8, 41). These latter transferases have been called early or initiating transferases, preferring nonglycosylated over glycosylated substrates. The roles of the catalytic and lectin domains on modulating ppGalNAc T peptide and glycopeptide specificity are not fully understood (see Notes 26 and 27).
4. Commercial mucin preparations are commonly relatively crude and may require additional purification via gel filtration or isopycnic density-gradient centrifugation (54) prior to TFMSA treatment depending on final use of the partially deglycosylated mucin. However, since the TFMSA approach is successful on very crude mucin preparations, further purification may not be necessary depending on subsequent isolation steps.
5. It is very important that the synthesis of the random peptides is performed under conditions to allow complete coupling at each random cycle in order to minimize the appearance of the central Thr (or Ser) residue into adjacent positions due to synthesis lag. Therefore, the random peptides must be Edman sequenced prior to use to confirm their suitability for use.
6. Combined protease inhibitors P8340 and P8849 contain phenylmethylsulfonyl fluoride (PMFS), phosphoramidon, E-64 (trans epoxysuccinyl L-lucylamido(4-guanidino)-butane),

4-(2-aminoethyl) benzenesulfonyl fluoride (AEBSF), pepstatin A, bestatin, leupeptin, and aprotinin.

7. Sialic acid is typically removed after an overnight incubation of mucin (100–400 mg, 2–10 mg/mL) in buffer containing 1 M NaCl, 0.1 M sodium acetate, pH 5.0, and 0.04% sodium azide at 37°C utilizing 5–10 U of *Clostridium perfringens* neuraminidase, type VIII from Sigma–Aldrich (St. Louis, MO).
8. Use care when purging so as not to blow the lyophilized sample out of the tube!
9. Be sure to dry off the outside of the TFMSA/anisole tube with paper towel to prevent ethanol or condensate from entering the mucin-containing tube during reagent transfer.
10. The reaction can be temporally stopped by placing in dry ice overnight and resumed the next day by returning it to an ice-water bath.
11. On the initial dialysis, the container should be tightly capped to reduce pyridine fumes.
12. In contrast to α-GalNAc-O-Ser/Thr, the βGalNAc-O-Ser/Thr residues are completely lost on Edman sequencing; see author's correction for reference (45).
13. The transfer of longer O-linked glycans to the flask requires alterations in the transfer solvents to a more hydrophylic solvent containing methanol; see ref. 50–52.
14. Best results are typically obtained with fresh C18-PTH columns which give the sharpest peaks with minimal overlap.
15. We have also shown that for ppGalNAc T1 and T2 an analogous peptide to P-I synthesized with a Ser acceptor site gives similar results to the Thr-containing peptide P-I (31).
16. Sequence determinations of peptide P-V revealed the near-complete loss of Trp in the pre- and post-lectin sequence determinations. It was determined that the Trp residue is lost after the initial Sephadex G10 gel filtration step due to oxidation. Multiple efforts to inhibit the presumed Trp oxidation at this step have not been successful. Therefore, enhancement values for Trp have not been determined.
17. Incorporation is measured from the ^{3}H content of the peptide peak eluted from Sephadex G10 chromatography. ppGalNAc T hydrolyzes UDP-GalNAc; therefore, ^{3}H eluting from the Dowex resin is a combination of hydrolysis and transfer to peptide; see ref. 6, 7.
18. It is important to remove all UDP-GalNAc as it interferes with the lectin column isolation of glycopeptide. The sample must be diluted to 4 mL prior to loading on Dowex since high concentrations of EDTA interfere with UDP-GalNAc binding.

19. It is important that the fractions are monitored in reverse elution order to eliminate potential contamination of glycopeptide fractions with free peptide.
20. The unbound pass through peak containing unglycosylated peptide can be pooled and rechromatographed on Sephadex G10 for reuse as substrate.
21. The glycopeptide peak usually appears 3–4 fractions after the initial addition of GalNAc.
22. There is usually very little absorbance detected at 220 or 280 nm for the glycopeptide peak.
23. Nonbinding radiolabeled material has been observed in the pass through fraction of the lectin chromatography on occasion. In one instance, this was due to the radiolabeled UDP-GalNAc not being fully *N*-acetylated. Although the deacetylated ^{3}H-GalNAc was transferred by the ppGalNAc T to random peptide substrate, the resultant deacetylated GalNAc glycopeptide failed to bind the lectin; see supplement materials of 32 for details. In another instance, unreacted UDP-GalNAc was not fully removed on Dowex. Since UDP-GalNAc does not bind the lectin column, any UDP-GalNAc (which is not readily detected on Sephadex G10 chromatography as comigrates near the random peptide peak) also appears in the nonglycosylated peptide pass through fraction.
24. We have observed that the Lys residue content of peptides P-III and P-V decreases relative to other residues when low amounts of (glyco)peptide are sequenced (typically, <0.1 nmol). To compensate for this behavior, both the prelectin and post-lectin sequence determinations are performed using similar amounts of peptide, where identical losses of Lys occur.
25. A manuscript reporting preferences for ppGalNAc T3, T5, and T12 is under preparation from the Gerken laboratory.
26. ppGalNAc T10 is known as a glycopeptide preferring transferase, showing its greatest activity against glycopeptide substrates (40, 41, 46). Recently, we have utilized a random peptide substrate containing Ser-O-α-GalNAc in the randomized X positions and the *N*-azido derivative of UDP GalNAc, UDP-GalNAz (32), as the donor to address the glycopeptide specificity of ppGalNAc T10 as well as ppGalNAc T1 and T2. The detailed methods of this approach are beyond the scope of this chapter but are described in sufficient detail in ref. 32. Nevertheless, the obtained preferences for Ser-O-α-GalNAc show ppGalNAc T10 to possess a very strong (approximately eightfold) enhancement for a Ser/Thr-O-α-GalNAc at the +1 position of the site of glycosylation. In contrast, ppGalNAc T1 and T2 show predominantly inhibitory effects for neighboring Ser-O-α-GalNAc residues, consistent with our earlier findings

(7, 8). These data clearly demonstrate that ppGalNAc T10 possesses a Ser/Thr-O-α-GalNAc binding site on its catalytic domain, whereas ppGalNAc T1 and T2 do not.

27. Confounding the predictions is the observation that several of the ppGalNAc Ts (i.e., T2 and T4) can have their substrate specificity significantly altered by prior substrate GalNAc glycosylation. For example, it has been shown that the presence of the lectin domain of ppGalNAc T2 significantly shifts the preferred sites of glycosylation on glycopeptide substrates (see Fig. 4) (11, 40, 46, 49). As ppGalNAc T7 and T10 prefer to transfer α-GalNAc to glycopeptide acceptors, it has been widely assumed that their C-terminal lectin domains would play significant roles in this activity; however, recent studies (11, 32) have shown that indeed ppGalNAc T10's catalytic domain is responsible for its glycopeptide specificity, as shown by the enhancement values given in Table 2.

Notes Added in Proof

The manuscript in Note 25 has been published: Gerken, T.A., Jamison, O., Perrine, C.L., Collette, J.C., Moinova, H., Ravi, L., Markowitz, S.D., Shen, W., Patel, H., and Tabak, L.A. (2011) Emerging paradigms for the initiation of mucin type protein O-glycosylation by the polypeptide GalNAc transferase (ppGalNAcT) family of glycosyltransferases, *J. Biol. Chem 286*, 14493–14507.

The enhancement value product calculations described in the text, can now be performed at the Isoform Specific O-Glycosylation Prediction (ISOGlyP) web site: http://isoglyp.utep.edu/.

Acknowledgments

This work was supported by the National Institutes of Health, National Cancer Institute, R01-CA 78834, to TAG. The assistance of Oliver Jamison in performing these studies is also acknowledged.

References

1. Ten Hagen, K. G., Fritz, T. A., and Tabak, L. A. (2003) "All in the Family" - The UDP-GalNAc:polypeptide N-acetylgalactosaminyl-transferases, *Glycobiology 13*, 1R–16R.
2. Hollingsworth, M. A. and Swanson, B. J. (2004) Mucins in cancer: protection and control of the cell surface, *Nat Rev Cancer 4*, 45–60.

3. Brockhausen, I. (2007) Biosynthesis of Mucin-Type O-Glycans, *Comprehensive Glycoscience From Chemistry to Systems Biology 3*, 33–59.
4. Gerken, T. A., Owens, C. L., and Pasumarthy, M. (1997) Determination of the site-specific O-glycosylation pattern of the porcine submaxillary mucin tandem repeat glycopeptide: Model proposed for the polypeptide:GalNAc transferase peptide binding site, *J. Biol. Chem. 272*, 9709–9719.
5. Gerken, T. A., Owens, C. L., and Pasumarthy, M. (1998) Site-specific core 1 O-glycosylation pattern of the porcine submaxillary gland mucin tandem repeat: Evidence for the modulation of glycan length by peptide sequence, *J. Biol. Chem. 273*, 26580–26588.
6. Gerken, T. A., Gilmore, M., and Zhang, J. (2002) Determination of the site-specific oligosaccharide distribution of the O-glycans attached to the porcine submaxillary mucin tandem repeat: Further evidence for the modulation of O-glycan side chain structures by peptide sequence, *J. Biol. Chem 277*, 7736–7751.
7. Gerken, T. A. (2004) Kinetic modeling confirms the biosynthesis of mucin Core 1 (β-Gal(1–3) α-GalNAc-*O*-Ser/Thr) *O*-glycan structures are modulated by neighboring glycosylation effects, *Biochemistry 43*, 4137–4142.
8. Gerken, T. A., Tep, C., and Rarick, J. (2004) Role of Peptide Sequence and Neighboring Residue Glycosylation on the Substrate Specificity of the Uridine 5'-Diphosphate-α-*N*-acetylgalactosamine: Polypeptide *N*-acetylgalactosaminyl Transferases T1 and T2: Kinetic Modeling of the Porcine and Canine Submaxillary Gland Mucin Tandem Repeats, *Biochemistry 43*, 9888–9900.
9. Muller, S., Goletz, S., Packer, N., Gooley, A., Lawson, A. M., and Hanisch, F. G. (1997) Localization of O-glycosylation sites on glycopeptide fragments from lactation-associated MUC1. All putative sites within the tandem repeat are glycosylation targets in vivo, *J. Biol. Chem. 272*, 24780–24793.
10. Takeuchi, H., Kato, K., Hassan, H., Clausen, H., and Irimura, T. (2002) O-GalNAc incorporation into a cluster acceptor site of three consecutive threonines: Distinct specificity of GalNAc-transferase isoforms, *Eur J Biochem 269*, 6173–6183.
11. Raman, J., Fritz, T. A., Gerken, T. A., Jamison, O., Live, D., Lu, M., and Tabak, L. A. (2008) The catalytic and lectin domains of UDP-GalNAc :Polypeptide alpha-N-Acetylgalactosaminyltransferase function in concert to direct glycosylation site selection, *J. Biol. Chem 283*, 22942–22951.
12. Miwa, H. E., Gerken, T. A., Jamison, O., and Tabak, L. A. (2010) Isoform-specific O-glycosylation of osteopontin and bone sialoprotein by polypeptide N-acetylgalactosaminyltransferase-1, *J. Biol. Chem 285*, 1208–1219.
13. Elhammer, A. P., Poorman, R. A., Brown, E., Maggiora, L. L., Hoogerheide, J. G., and Kezdy, F. J. (1993) The specificity of UDP-GalNAc:polypeptide N-acetylgalactosaminyltransferase as inferred from a database of in vivo substrates and from the in vitro glycosylation of proteins and peptides, *J. Biol. Chem 268*, 10029–10038.
14. Gupta, R., Birch, H., Rapacki, K., Brunak, S., and Hansen, J. E. (1999) O-GLYCBASE version 4.0: a revised database of O-glycosylated proteins, *Nucl. Acids Res. 27*, 370–372.
15. Ten Hagen, K. G. and Tran, D. T. (2002) A UDP-GalNAc: polypeptide N-acetylgalactosaminyltransferase is essential for viability in Drosophila melanogaster, *J. Biol. Chem 277*, 22616–22622.
16. Schwientek, T. J., Bennett, E. P., Flores, C., Thacker, J., Hollman, M., Reis, C. A., Behrens, J., Mandel, U., Keck, B., Schafer, M. A., Hazelmann, K., Zubarev, R., Roepstorff, P., Hollingsworth, M. A., and Clausen, H. (2002) Functional conservation of subfamilies of putative UDP-N-acetylgalactosamine: Polypeptide N-acetylgalactosaminyltransferases in drosophila, C. elegans and mammals: One subfamily comprised of l(2)35Aa is essential in drosophila, *J. Biol. Chem 277*, 22623–22638.
17. Zhang, L., Zhang, Y., and Ten Hagen, K. G. (2008) A mucin-type O-glycosyltransferase modulates cell adhesion during Drosophila development, *J. Biol. Chem 283*, 34076–43086.
18. Tian, E. and Ten Hagen, K. G. (2007) A UDP-GalNAc:Polypeptide N-Acetylgalactosaminyltransferase Is Required for Epithelial Tube Formation, *J. Biol. Chem 282*, 606–614.
19. Zhang, L., Tran, D. T., and Ten Hagen, K. G. (2010) An O-Glycosyltransferase Promotes Cell Adhesion during Development by Influencing Secretion of an Extracellular Matrix Integrin Ligand, *J. Biol. Chem 285*, 19491–19501.
20. Kato, K., Jeanneau, C., Tarp, M. A., Benet-Pages, A., Lorenz-Depiereux, B., Bennett, E. P., Mandel, U., Strom, T. M., and Clausen, H. (2006) Polypeptide GaINAc-transferase T3 and familial tumoral calcinosis: Secretion of FGF23 requires O-glycosylation, *J. Biol. Chem 281*, 18370–18377.
21. Wagner, K. W., Punnoose, E. A., Januario, T., Lawrence, D. A., Pitti, R. M., Lancaster, K., Lee, D., von Goetz, M., Yee, S. F., Totpal, K.,

Huw, L., Katta, V., Cavet, G., Hymowitz, S. G., Amler, L., and Ashkenazi, A. (2007) Death-receptor O-glycosylation controls tumor-cell sensitivity to the proapoptotic ligand Apo2L/TRAIL, *Nat Med 13*, 1070–1077.

22. Herr, P., Korniychuk, G., Yamamoto, Y., Grubisic, K., and Oelgeschlager, M. (2008) Regulation of TGF-beta signalling by N-acetyl galactosaminyltransferase-like 1, *Development 135*, 1813–1822.
23. Nakamura, N., Toba, S., Hirai, M., Morishita, S., Mikami, T., Konishi, M., Itoh, N., and Kurosaka, A. (2005) Cloning and Expression of a Brain-Specific Putative UDP-GalNAc: Polypeptide N-Acetylgalactosaminyltransferase Gene, *Biological & Pharmaceutical Bulletin 28*, 429–433.
24. Merla, G., Ucla, C., Guipponi, M., and Reymond, A. (2002) Identification of additional transcripts in the Williams-Beuren syndrome critical region, *Human Genetics 110*, 429–438.
25. Simmons, A. D., Musy, M. M., Lopes, C. S., Hwang, L. Y., Yang, Y. P., and Lovett, M. (1999) A direct interaction between EXT proteins and glycosyltransferases is defective in hereditary multiple exostoses, *Hum. Mol. Genet. 8*, 2155–2164.
26. Wood, L. D., Parsons, D. W., Jones, S., Lin, J., Sjoblom, T., Leary, R. J., Shen, D., Boca, S. M., Barber, T., Ptak, J., Silliman, N., Szabo, S., Dezso, Z., Ustyanksky, V., Nikolskaya, T., Nikolsky, Y., Karchin, R., Wilson, P. A., Kaminker, J. S., Zhang, Z., Croshaw, R., Willis, J., Dawson, D., Shipitsin, M., Willson, J. K. V., Sukumar, S., Polyak, K., Park, B. H., Pethiyagoda, C. L., Pant, P. V. K., Ballinger, D. G., Sparks, A. B., Hartigan, J., Smith, D. R., Suh, E., Papadopoulos, N., Buckhaults, P., Markowitz, S. D., Parmigiani, G., Kinzler, K. W., Velculescu, V. E., and Vogelstein, B. (2007) The Genomic Landscapes of Human Breast and Colorectal Cancers, *Science 318*, 1108–1113.
27. Guda, K., Moinova, H., He, J., Jamison, O., Ravi, L., Natale, L., Lutterbaugh, J., Lawrence, E., Lewis, S., Willson, J. K. V., Lowe, J. B., Wiesner, G. L., Parmigiani, G., Barnholtz-Sloan, J., Dawson, D. W., Velculescu, V. E., Kinzler, K. W., Papadopoulos, N., Vogelstein, B., Willis, J., Gerken, T. A., and Markowitz, S. D. (2009) Inactivating germ-line and somatic mutations in polypeptide N-acetylgalactosaminyltransferase 12 in human colon cancers, *PNAS 109*, 12921–12925.
28. Willer, C. J., Sanna, S., Jackson, A. U., Scuteri, A., Bonnycastle, L. L., Clarke, R., Heath, S. C., Timpson, N. J., Najjar, S. S., Stringham, H. M., Strait, J., Duren, W. L., Maschio, A., Busonero, F., Mulas, A., Albai, G., Swift, A. J., Morken, M. A., Narisu, N., Bennett, D., Parish, S., Shen, H., Galan, P., Meneton, P., Hercberg, S., Zelenika, D., Chen, W. M., Li, Y., Scott, L. J., Scheet, P. A., Sundvall, J., Watanabe, R. M., Nagaraja, R., Ebrahim, S., Lawlor, D. A., Ben Shlomo, Y., Davey-Smith, G., Shuldiner, A. R., Collins, R., Bergman, R. N., Uda, M., Tuomilehto, J., Cao, A., Collins, F. S., Lakatta, E., Lathrop, G. M., Boehnke, M., Schlessinger, D., Mohlke, K. L., and Abecasis, G. R. (2008) Newly identified loci that influence lipid concentrations and risk of coronary artery disease, *Nat Genet 40*, 161–169.
29. Kathiresan, S., Melander, O., Guiducci, C., Surti, A., Burtt, N. P., Rieder, M. J., Cooper, G. M., Roos, C., Voight, B. F., Havulinna, A. S., Wahlstrand, B., Hedner, T., Corella, D., Tai, E. S., Ordovas, J. M., Berglund, G., Vartiainen, E., Jousilahti, P., Hedblad, B., Taskinen, M. R., Newton-Cheh, C., Salomaa, V., Peltonen, L., Groop, L., Altshuler, D. M., and Orho-Melander, M. (2008) Six new loci associated with blood low-density lipoprotein cholesterol, high-density lipoprotein cholesterol or triglycerides in humans, *Nat Genet 40*, 189–197.
30. Tabak, L. A. (2010) The role of mucin-type O-glycans in eukaryotic development, *Seminars in Cell & Developmental Biology 21*, 616–621.
31. Gerken, T. A., Raman, J., Fritz, T. A., and Jamison, O. (2006) Identification of common and unique peptide substrate preferences for the UDP-GalNAc: polypeptide alpha -N-acetylgalactosaminyltransferases T1 & T2 (ppGalNAc T1 & T2) derived from oriented random peptide substrates, *J. Biol. Chem 281*, 32403–32416.
32. Perrine, C. L., Ganguli, A., Wu, P., Bertozzi, C. R., Fritz, T. A., Raman, J., Tabak, L. A., and Gerken, T. A. (2009) The glycopeptide preferring polypeptide- GalNAc transferase-10 (ppGalNAcT10),involvedinmucintype-O-glycosylation, has a unique GalNAc-O-Ser/Thr binding site in its catalytic domain not found in ppGalNAc T1 or T2, *J. Biol. Chem 284*, 20387–20397.
33. Gerken, T. A., Ten Hagen, K. G., and Jamison, O. (2008) Conservation of peptide acceptor preferences between Drosophila and mammalian polypeptide-GalNAc transferase orthologue pairs, *Glycobiology 18*, 861–870.
34. Perrine, C., Ju, T., Cummings, R. D., and Gerken, T. A. (2009) Systematic determination of the peptide acceptor preferences for the human UDP-Gal: glycoprotein-{alpha}-GalNAc {beta}3 galactosyltranferase (T-synthase), *Glycobiology 19*, 321–328.
35. Edge, A. S. B. Deglycosylation of glycoproteins with trifluoromethanesulphonic acid:

elucidation of molecular structure and function. Biochem J 376, 339–350. 2003.

36. Gerken, T. A., Gupta, R., and Jentoft, N. (1992) A novel approach for chemically deglycosylating O-linked glycoproteins: The deglycosylation of submaxillary and respiratory mucins, *Biochemistry 31*, 639–648.

37. Gerken, T. A. and Dearborn, D. G. (1984) Carbon-13 NMR studies of native and modified ovine submaxillary mucin, *Biochemistry 23*, 1485–1497.

38. Gerken, T. A., Owens, C. L., and Pasumarthy, M. (1997) in *Techniques in Glycobiology* (Townsend, R. R. and Hotchkiss, A. T., Eds.) pp 247–269, Marcel Dekker, Inc, New York.

39. Abernethy, J. L., Wang, Y., Eckhardt, A. E., and Hill, R. L. (1992) in *Techniques in Protein Chemistry III* (Angeletti, R. H., Ed.) Academic Press, New York.

40. Cheng, L., Tachibana, K., Zhang, Y., Guo, J., Kahori, T. K., Kameyama, A., Wang, H., Hiruma, T., Iwasaki, H., Togayachi, A., Kudo, T., and Narimatsu, H. (2002) Characterization of a novel human UDP-GalNAc transferase, pp-GalNAc-T10, *FEBS Lett 531*, 115–121.

41. Pratt, M. R., Hang, H. C., Ten Hagen, K. G., Rarick, J., Gerken, T. A., Tabak, L. A., and Bertozzi, C. R. (2004) Deconvoluting the Functions of Polypeptide *N*-α-Acetylgalactosaminyltransferase Family Members by Glycopeptide Substrate Profiling, *Chem. Biol. 11*, 1009–1016.

42. Mardberg, K., Nystrom, K., Tarp, M. A., Trybala, E., Clausen, H., Bergstrom, T., and Olofsson, S. (2004) Basic amino acids as modulators of an O-linked glycosylation signal of the herpes simplex virus type 1 glycoprotein gC: functional roles in viral infectivity, *Glycobiology 14*, 571–581.

43. Brooks, M. M. and Savage, A. V. the substrate specificity of the enzyme endo-alpha-N-acetyl-D-galactosaminidase from Diplococcus pneumonia. Glycoconj. J. 14, 183–190. 1997.

44. Iwase, H. and Hotta, K. Release of O-linked glycoprotein glycans by endo-alpha-N-acetyl-D-Galactosaminidase. Methods in Molecular Biology 14, 151–159. 1993.

45. Fritz, T. A., Raman, J., and Tabak, L. A. (2006) Dynamic association between the catalytic and lectin domains of human UDP-GalNAc:polypeptide alpha -N-acetylgalactosaminyltransferase-2, *J. Biol. Chem 281*, 8613–8619.

46. Bennett, E. P., Hassan, H., Hollingsworth, M. A., and Clausen, H. (1999) A novel human UDP-N-acetyl-D-galactosamine:polypeptide N-acetylgalactosaminyltransferase, GalNAc-T7, with specificity for partial GalNAc-glycosylated acceptor substrates, *FEBS Lett. 460*, 226–230.

47. Hassan, H., Reis, C. A., Bennett, E. P., Mirgorodskaya, E., Roepstorff, P., Hollingsworth, M. A., Burchell, J., Taylor-Papadimitriou, J., and Clausen, H. (2000) The lectin domain of UDP-N-acetyl-D-galactosamine: polypeptide N-acetylgalactosaminyltransferase-T4 directs its glycopeptide specificities, *J. Biol. Chem 275*, 38197–38205.

48. Hanisch, F. G., Reis, C. A., Clausen, H., and Paulsen, H. (2001) Evidence for glycosylation-dependent activities of polypeptide N-acetylgalactosaminyltransferases rGalNAc-T2 and -T4 on mucin glycopeptides, *Glycobiology 11*, 731–740.

49. Wandall, H. H., Irazoqui, F., Tarp, M. A., Bennett, E. P., Mandel, U., Takeuchi, H., Kato, K., Irimura, T., Suryanarayanan, G., Hollingsworth, M. A., and Clausen, H. (2007) The lectin domains of polypeptide GalNAc-transferases exhibit carbohydrate-binding specificity for GalNAc: lectin binding to GalNAc-glycopeptide substrates is required for high density GalNAc-O-glycosylation, *Glycobiology 17*, 374–387.

50. Gooley, A. A., Ou, K., Russell, J., Wilkins, M. R., Sanchez, J. C., Hochstrasser, D. F., and Williams, K. L. (1997) A role for Edman degradation in proteome studies, *Electrophoresis 18*, 1068–1072.

51. Gooley, A. A. and Williams, K. L. (1997) How to find, identify and quantitate the sugars on proteins, *Nature 385*, 557–559.

52. Zachara, N. E. and Gooley, A. A. (2000) Identification of glycosylation sites in mucin peptides by edman degradation, *Methods in Molecular Biology 125*, 121–128.

53. Wandall, H. H., Hassan, H., Mirgorodskaya, E., Kristensen, A. K., Roepstorff, P., Bennett, E. P., Nielsen, P. A., Hollingsworth, M. A., Burchell, J., Taylor-Papadimitriou, J., and Clausen, H. (1997) Substrate specificities of three members of the human UDP-N-acetyl-alpha-D-galactosamine:Polypeptide N-acetylgalactosaminyltransferase family, GalNAc-T1, -T2, and -T3, *J. Biol. Chem. 272*, 23503–23514.

54. Davies, J. R. and Carlstedt, I. (2000) Isolation of large gel-forming mucins. *Methods in Molecular Biology 125*, 3–13.

Chapter 6

Analysis of Assembly of Secreted Mucins

Malin E.V. Johansson and Gunnar C. Hansson

Abstract

Studies of assembly and secretion of gel-forming mucins are complex. The pulse-chase methods for mucins described here include metabolic radiolabeling and labeling in animals with azido-GalNAc. The labeled mucins are analyzed by composite agarose-polyacrylamide gel electrophoresis and autoradiography or by mucus-preserving tissue fixation and Click-iT® chemistry.

Key words: Mucus, Mucin, Glycan, Intestine, Pulse chase, Agarose-polyacrylamide gel electrophoresis

1. Introduction

The assembly of gel-forming mucins is difficult to study due to their large size and complex multimeric nature. For example, the MUC2 mucin has a 5,000 amino acid apomucin that initially forms disulfide-linked dimers that after the addition of O-glycans reach a size of 2.5 MDa. Available methods limit the studies of mucin assembly to the early biosynthetic steps. Radioactive labeling has successfully been used to follow these early processes by pulse-chase studies. Since the secreted oligomerizing mucins are rich in cysteines, this amino acid is well-suited as a labeled building block. The early biosynthesis of MUC2 has been studied by this approach in LS174T colon cancer cells (1–4). Biosynthesis of MUC5AC has also been studied using pulse chase in HT29 cells (5). The latter steps as well as the complete biosynthesis of these large molecules have been assessed by expressing plasmids encoding recombinant proteins containing, for example, the terminal cysteine-rich domains. Using this approach, we have studied the MUC2 mucin assembly and shown the formation of C-terminal dimers and N-terminal trimers (6, 7). Porcine submaxillary mucin (PSM) was also shown to be assembled into C-terminal dimers and N-terminal trimers as for MUC2 (8, 9).

Michael A. McGuckin and David J. Thornton (eds.), *Mucins: Methods and Protocols*, Methods in Molecular Biology, vol. 842, DOI 10.1007/978-1-61779-513-8_6, © Springer Science+Business Media, LLC 2012

The formation of the complete mucus can only be studied in whole biological systems, where all functions of the tissue are present, including assembly and secretion into the appropriate environment. We have characterized the mucus in colon of mouse and shown that it consists of two different layers (10). A way to study the later stages of biosynthesis and turnover of mucins is to take advantage of the highly glycosylated nature of these molecules. Azido-glycan incorporation and labeling of these with Click-it chemistry make in vivo labeling possible without using large amounts of radioactivity as discussed previously (11). As every O-glycan on a mucin contains at least one GalNAc, labeling with GalNAz is an efficient way to label mucins produced by cells in culture or live animals.

2. Materials

2.1. Radioactive Metabolic Labeling of Mucins in Tissue Culture

1. Methionine and cysteine free cell culture medium (Dulbecco's Modified Eagles Medium (DMEM), Catalogue No. 21013024, Invitrogen) supplemented with 2 mM L-glutamine and 10% (v/v) fetal bovine serum.
2. Radioactive labeling reagent [^{35}S] Met, [^{35}S] Cys mix (EXPRE^{35}S^{35}S Protein labeling mix, EasyTag™, 11 mCi/mL, Catalogue No. NEG772, Perkin Elmer).
3. Sterile PBS.
4. Chase medium (IMDM or DMEM) supplemented with 15 μg/mL L-methionine and 25 μg/mL L-cysteine.

2.2. Preparation of Cell Lysate

1. Safety equipment, including protection shield, glasses, waste container, and gloves.
2. Tray with ice.
3. Ice-cold phosphate-buffered saline (PBS).
4. Lysis buffer (50 mM Tris–HCl, pH 7.9, 150 mM NaCl, 1% (v/v) Triton X-100) with 1× Complete protease inhibitor mix, EDTA free (Roche).
5. Cell scraper.
6. Sonicator (Soniprep 150, MSE or similar).

2.3. Immunoprecipitation of Labeled Mucins

1. Specific antibodies to mucins (against tandem repeat or glycosylation insensitive) or to tags of fusion proteins (see Note 1).
2. Magnetic beads (Dynabeads®, M280 anti-mouse or anti-rabbit, Invitrogen) or Protein G PLUS-agarose (Santa Cruz Biotechnology).
3. Magnetic holder for 1.5-mL tubes.

4. Microcentrifuge for 1.5-mL tubes.
5. Rotating or rocking device for microcentrifuge tubes.
6. Blocking buffer (0.1% (w/v) bovine serum albumin (BSA) in PBS).
7. Preimmune serum from the same species as the antiserum used.
8. Lysis buffer (50 mM Tris–HCl, pH 7.9, 150 mM NaCl, 1% (v/v) Triton X-100).
9. Wash buffer (10 mM Tris–HCl, pH 7.5, 2 mM EDTA, 0.1% (w/v) SDS, 0.1% (v/v) Triton X-100).
10. Reducing sample buffer (0.75 M Tris–HCl, pH 8.1, 2% (w/v) SDS, 0.01% (w/v) bromophenol blue, 60% (v/v) glycerol, 100 mM DTT).
11. Heating block (95°C).

2.4. Polyacrylamide Gel Electrophoresis

Protocols for polyacrylamide gel electrophoresis can be found in many laboratory manuals.

2.5. Composite Agarose-Polyacrylamide Gel Electrophoresis

1. Agarose (Type 1-B: LowEEO, Sigma).
2. Acrylamide (40% (w/v) solution, acrylamide/bisacrylamide 19:1, BioRad).
3. Ammonium persulfate (APS, BioRad, 40% (w/v) solution in ddH_2O, stored at −20°C).
4. TEMED (BioRad).
5. 5× Tris–HCl buffer (1.875 M Tris–HCl, pH 8.1).
6. Glycerol (50% (v/v) solution).
7. Casting equipment (BioRad Miniprotean or equivalent).
8. Large and small glass plate, two 1.5-mm spacers, two 0.75-mm combs.
9. Lower gel solution (0.08 g agarose, 1.6 mL 5× Tris–HCl buffer, pH 8.1, 1.6 mL 50% (v/v) glycerol, 1.6 mL ddH_2O).
10. Upper gel solution (0.04 g agarose, 1.6 mL 5× Tris–HCl buffer, pH 8.1, 6.4 mL ddH_2O).
11. Microwave oven.
12. Oven (60°C).
13. Magnetic stirrer, gradient mixer, and peristaltic pump (SG Gradient Makers and P-1 pump GE Healthcare or equivalent).
14. Gel electrophoresis running equipment (BioRad or equivalent) and power supply.
15. Running buffer (192 mM boric acid, 1 mM EDTA, 0.1% (w/v) SDS, pH 7.6 (set with Tris base)).

2.6. Fixation of Gel and Autoradiography

1. Fixation solution (30% (v/v) ethanol, 10% (v/v) acetic acid).
2. Signal amplification solution (Amplify fluorographic reagent, GE Healthcare).
3. Filter paper (3MM) and plastic wrap.
4. Gel dryer (DrygelSr, Hoefer).
5. Autoradiography exposure cassette.
6. Autoradiography film (Kodak® Biomax, Sigma–Aldrich).
7. Autoradiography film developer (XOMAT 1000, Kodak) with developer and fix solutions (Kodak® RP X-Omat, Sigma–Aldrich).

2.7. In Vivo GalNAz Labeling of O-Glycans in Mouse

1. Tetraacetylated azido galactosamine (Click-iT® GalNAz, Invitrogen): 5.2 mg dissolved in 100 μL DMSO and diluted to 1 mL in PBS.
2. Intraperitoneal injection material (needle, 25 G, and syringe).

2.8. Methanol-Carnoy's Fixation

1. Dissection instruments for mouse tissue.
2. Tubes with 5 mL methanol-Carnoy's fixative (60% (v/v) dry methanol, 30% (v/v) chloroform, 10% (v/v) glacial acetic acid).
3. Methanol, absolute ethanol, absolute ethanol/xylene, xylene.
4. Paraffin embedding and tissue sectioning service.

2.9. TAMRA Detection of GalNAz Containing Glycoproteins in Tissue Sections

1. Oven: 60 and 50°C.
2. Xylene substitute (Sigma–Aldrich).
3. Ethanol solutions (100, 95, 70, 50, 30% (v/v)).
4. PAP pen used to make circles around tissue sections. It produces a thin hydrophobic film creating a water-repellent barrier.
5. Click-iT® Tetramethylrhodamine (TAMRA) Protein Analysis Detection Kit (Invitrogen).
6. Humid incubation chamber for glass slides.

3. Methods

3.1. Radioactive Labeling of Mucin-Producing Cell Lines

Keep the lysate and/or medium on ice to minimize degradation of the proteins.

1. Grow the appropriate cells expressing gel-forming mucins or fusion proteins of their domains to reach confluence in a Petri dish (6 cm in diameter).
2. The medium is removed, the attached cells are rinsed with PBS, and 2 mL of cysteine- and methionine-free cell culture

medium is added. Incubate the cells under normal conditions (37°C, 5% CO_2, 95% humidity) for 2 h.

3. The labeling is started by adding 10 μL of EXPRE^{35}S^{35}S Protein labeling mix to the culture and the solution is gently mixed with a pipette. Use appropriate safety shields to protect yourself from the radioactive reagent.
4. Incubate the cells for the desired labeling time under normal cell culture conditions.
5. Remove and discard the medium (highly radioactive! see Note 2). Rinse the cells with PBS and add 2 mL new medium supplemented with extra cysteine and methionine. This commences the chase period, where the labeling interval is ended by competition of the unlabeled amino acids during protein synthesis.
6. Chase the labeled proteins for desired time by incubating the cell at normal conditions (see Note 3).
7. Remove the medium and lyse the cells (see Subheading 3.2). The medium can also be used for analyzing secreted material and should be supplemented with protease inhibitors and used directly for immunoprecipitation.

3.2. Preparation of Cell Lysate

1. Place the tissue culture dish on a tray with ice to slow down metabolism.
2. Remove the medium which can be used to study secreted proteins.
3. Wash the labeled and chased cells with ice-cold PBS twice to reduce background from unincorporated radioactive amino acids.
4. Add 1 mL lysis buffer per 6-cm-diameter Petri dish and let it spread out on the whole cell surface.
5. Use a cell scraper to scrape the cells from the plastic and collect the material in one area, where it can be removed using a pipette. Transfer the lysate to a 1.5-mL microcentrifuge tube on ice.
6. Complete the lysis by sonicating the solution three times for 20 s at the amplitude of 16 μm (see Note 4).
7. The sample can be used for immunoprecipitation or directly for electrophoresis (see Note 5).

3.3. Immunoprecipitation of Labeled Mucins

1. Immunoprecipitation is difficult with molecules as large as the mucins. Two different methods of capturing the mucin with antibody are presented, where the first method mainly has been used with fusion proteins and antibodies against tags and the second method with full-length mucins and specific antisera. Different antibodies detect different forms of the molecules, and antibodies specific to the tandem repeat sequences do not

detect the mature O-glycosylated mucin while antisera specific for the terminal domains are usually able to do so (see Note 1).

3.3.1. Immunoprecipitation Using Magnetic Beads

The nonspecific binding to the beads is reduced by an initial blocking step and the specific antibody/antiserum is bound to the beads before incubation with the lysate. This is an efficient way to bind antibody to all beads. If the antigen-antibody reaction is the initial reaction the amount of beads should be increased.

1. Resuspend the magnetic beads slurry (Dynal Cat. M280 anti-mouse or anti-rabbit Dynabeads®) and transfer 50 μL (3.5×10^7 beads) to a microcentrifuge tube.
2. Nonspecific binding is reduced by washing the beads three times in blocking buffer. This is performed by resuspending the beads in the blocking buffer and after 5 min the tube is placed in the magnetic holder to trap the beads to the side of the tube. Discard the supernatant.
3. The beads are resuspended in wash buffer with 2–15 μg antibody in a total volume of 50 μL and incubated with mixing at room temperature for 2 h (see Note 5).
4. The beads are washed in lysis buffer three times using the magnetic holder.
5. Add the cell lysate from Subheading 3.2 and incubate the samples with rotation or rocking at 4°C overnight. The labeled medium can also be used in this way for immunoprecipitation (see Notes 2 and 6).
6. Pellet the beads using the magnet and wash the beads in lysis buffer three times and in wash buffer three times (see Note 7). The supernatants contain radioactive material.
7. Remove the supernatant of the last wash and add 20 μL reducing sample buffer to the beads. Boil the sample at 95°C for 5 min and place it on ice. The tube should be sealed with a cap lock during boiling to circumvent opening (popping) of the tubes and spreading of the radioactive material. Put the tube in the magnetic holder and transfer the supernatant to a new tube. The sample can now be analyzed by electrophoresis.

3.3.2. Immunoprecipitation Using Protein G Agarose

1. Preclear the lysate by adding 1 μg of nonspecific immunoglobulins from the same species as the host of the primary antibody and incubate at 4°C for 1 h (see Note 8).
2. Mix the protein G agarose solution and add 20 μL to the lysate with an additional incubation at 4°C for 30 min.
3. Spin the sample for 5 min ($1{,}000 \times g$) at 4°C. Transfer the supernatant to a new tube and discard the pellet.

4. Add 0.5–5 μg of primary antibody to the supernatant and incubate in a rotating or rocking device at 4°C overnight (see Note 5).
5. Add 30 μL Protein G PLUS Agarose (mixed to a slurry) and incubate at 4°C for 2 h.
6. Centrifuge the sample at 1,000 × *g* for 5 min at 4°C and discard the supernatant (see Note 2).
7. Wash the pellet three times in lysis buffer and three times in wash buffer and remove the supernatant. A final centrifugation of the pellet can be performed to remove any remaining wash buffer (see Note 7).
8. Add 20 μL reducing sample buffer to the pellet and mix well. Incubate at 95°C for 5 min and place on ice. The tube should be sealed with a cap-lock seal during heating.
9. Centrifuge the sample for 5 min and transfer the supernatant to a new tube. The sample can now be analyzed by electrophoresis.

3.4. Polyacrylamide Gel Electrophoresis

The long gel-forming mucins cannot be analyzed by conventional SDS-PAGE and other methods have, thus, been developed; see Subheading 3.5. If the analysis is to be done of smaller proteins, such as the non O-glycosylated apomucin precursors, or for smaller recombinant mucin parts in the size of 10–500 kDa, conventional SDS-PAGE using standard techniques is recommended to achieve the best separation.

3.5. Composite Agarose-Polyacryl Amide Gel Electrophoresis

The following protocol for analysis of mature mucins with composite agarose-polyacrylamide gel electrophoresis was originally adapted for mucins by Dr. Niclas G. Karlsson and coworkers (12).

1. The gel casting equipment (glass plates and spacers) is mounted and placed in a 60°C oven, together with a gradient mixer and a pump. The pump is connected to the gradient mixer and a tube from the pump is placed in between the glass plates of the casting equipment.
2. A gel containing agarose (0.5–1% gradient), acrylamide (0–6%), and glycerol (0–10%) is prepared from the lower and upper gel solutions. Both solutions are boiled in a microwave oven until the agarose is melted and then immediately placed in the 60°C oven.
3. Allow the lower gel solution to cool down to ~60°C and add 1.2 mL 40% (w/v) acrylamide. Swirl gently to mix.
4. The upper and lower gel solutions (5 mL of each) are added to each chamber of the gradient mixer, respectively. APS (2 μL) and TEMED (2 μL) are added to the two chambers, respectively, and the solutions are mixed. The lower solution is stirred with a magnet on a magnetic stirrer.

5. The pump is started on maximum rate (10 on a GE Healthcare P-1 pump) and the two valves of the gradient mixer are opened. The gel must be casted within 2 min. The gel is filled all the way up to the top of the glass plates.
6. Two combs (0.75 mm) are placed together into the gel which is allowed to polymerize at RT for at least 6 h or for 1 h in RT followed by +4°C overnight (see Note 9). The gels can be stored (with combs) at +4°C covered with wet blotting paper and wrapped in plastic wrap for at least a week.
7. If the polymerized gel around the combs has shrunk, the gel can be reconstituted with the upper gel solution reheated in a microwave oven.
8. Prior to electrophoresis, remove the combs carefully, place the gel in the electrophoresis equipment, and add running buffer. The wells are carefully washed with running buffer prior to use. Fill the outer and inner container with running buffer to transfer the heat generated more efficiently.
9. The electrophoresis container is placed on ice in a cold room (+4°C). After samples are loaded, the gel is run for 200 mAh/gel (10–20 mA for 20–10 h) (see Note 10).

3.6. Detection by Autoradiography

1. The gel equipment is disassembled and the gel is placed in fixation solution for 1 h at room temperature.
2. The fixative is replaced by Amplify fluorographic reagent and incubated for 30 min at room temperature.
3. The gel is placed on a filter paper and covered with plastic wrap.
4. Dry the gel in a gel dryer with the filter paper facing down and the vacuum is started. The agarose-containing gels only permit low temperatures to avoid melting during the drying process; thus, maximum 45°C is recommended. Usually, the drying takes 4–6 h.
5. The dry gel is placed in an autoradiography cassette and the gel is overlaid by a film in a dark room (see Note 11).
6. The film is exposed in −80°C for a time that varies depending on the efficiency of immunoprecipitation, level of incorporation, and relative abundance of the mucins studied. The film can be developed after 1–2 days and a new film can be added if necessary.
7. The films are developed using an automated film developer or manually in developer, fixative, and water.

The result of radioactive labeling and immunoprecipitation of lysate from LS174T cells expressing endogenous MUC2 and a transfected MUC2 N-terminal construct is shown in Fig. 1.

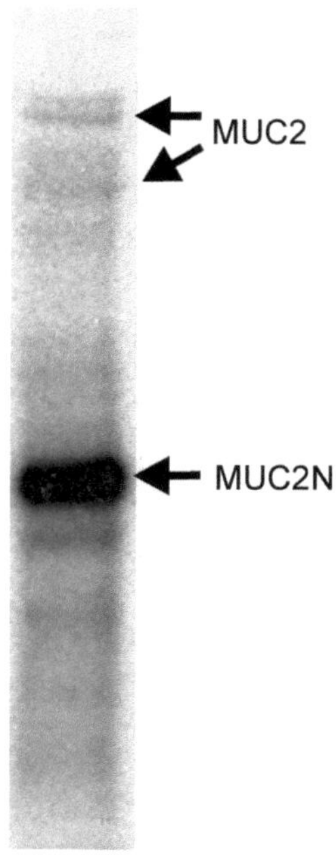

Fig. 1. Radioactive labeling of LS174T cells expressing MUC2 (MUC2) and an N-terminal construct of MUC2 (MUC2N) with a chase of 30 min. Immunoprecipitation was performed using an MUC2 N-terminal-specific antisera (αMUC2N3) (1) and the samples were analyzed by composite agarose-polyacrylamide gel electrophoresis. The construct is produced at higher rates during this short period of labeling compared to the endogenously expressed MUC2.

3.7. In Vivo GalNAz Labeling of O-Glycans in Mouse

1. Wild-type mice are injected intraperitoneally with 2.6 mg GalNAz in 0.5 mL PBS with 10% (v/v) DMSO.
2. The animals are kept in their normal environment for the duration of the labeling (see Note 12).
3. The labeling is terminated by sacrificing the animals according to local regulations. Avoid stressing the animals as intestinal emptying limits the sampling of fecal-filled distal colon (see Note 13).

3.8. Fixation of Tissue in Methanol-Carnoy's Fixative

A water-free fixation of the intestine containing fecal material preserves the mucus and the method is described in Chapter 13.

3.9. TAMRA Detection of GalNAz Containing Glycoproteins in Tissue Sections

The method to visualize mucus by GalNAz labeling on tissue sections has been used successfully on mouse colon as shown in Fig. 2.

1. Dewax the samples by an initial incubation at 60°C for 10 min and two additional incubations in xylene substitute solution. Hydrate the sample in solutions with decreasing concentration of ethanol (100, 95, 70, 50, 30% (v/v)). Incubate the slide for 5 min in each bath.

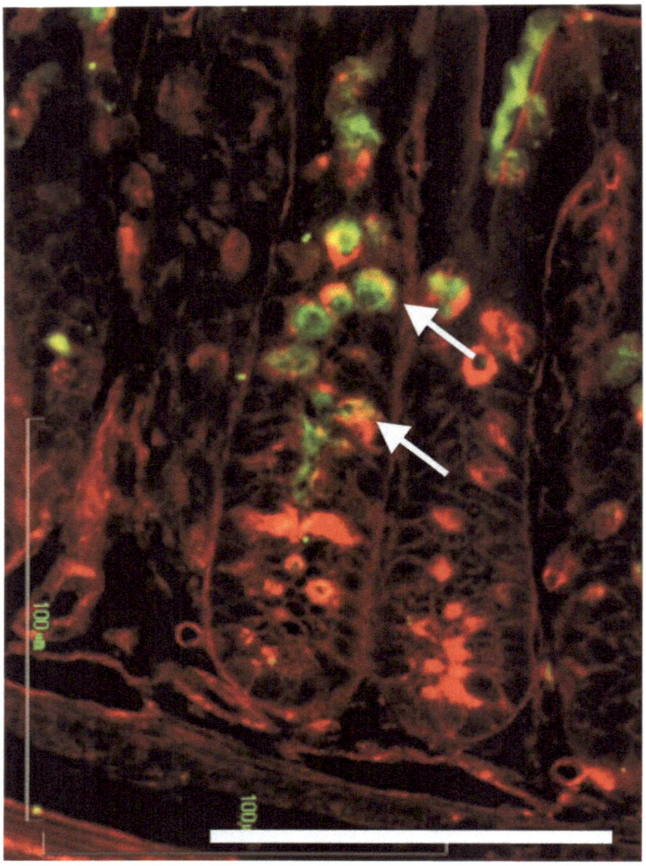

Fig. 2. Colon sections from a mouse labeled with GalNAz for 7 h were detected using the click-iT® reaction with TAMRA-alkyne which is seen as a red staining. This is especially observed in the mucus-filled goblet cells (indicated by *arrow*). The scale bar is 100 μm.

2. Wash in PBS and mark around the sections with a PAP pen.
3. The Click-iT® TAMRA reaction is performed by following the supplied protocol in a small scale. Prepare the TAMRA alkyne solution (component A and B) and also components D and E. Make a master mix containing 5 μL of the TAMRA alkyne solution, 3 μL ddH_2O, and 0.5 μL of component C and mix and 0.5 μL of component D per section to stain. Mix and incubate for 3 min and add 9 μL to each section. Add 1 μL of component E and mix again. The reaction turns bright orange at this step.
4. The reaction is allowed to continue for 1–2 h in the dark at RT in a humid chamber.
5. The reaction mix is removed and the samples are washed in PBS. If a combination with immunostaining is desired, continue with incubation in blocking and antibody solutions (see Note 14). Mount the sections using Prolong antifade mounting

media and allow to set at room temperature in the dark. Analyze the sections using a fluorescent or confocal microscope (see Note 15).

4. Notes

1. As monoclonal antibodies only recognize a single epitope, they can be very specific, but its reactivity is more sensitive to modifications and conformation of the antigen. Polyclonal sera on the other hand contain many different antibody specificities, also on the same antigen, and generally these antisera result in strong multivalent binding, but at the same time also give more unspecific background. The choice of antibody depends on what is available and the questions asked. Gel-forming mucins are secreted and the newly synthesized molecules are folded with formation of the many disulfide bonds in the endoplasmic reticulum. N-glycans of the high mannose type are attached in the ER. Antibodies directed to the protein core of the mucin domains only detect this non-O-glycosylated form found in the endoplasmic reticulum. The mucin biosynthesis continues with translocation into the Golgi apparatus, where N-glycans are matured and numerous O-glycans attached. This material is not further detected with antibodies reacting with the protein core of the mucin domains, but can on the other hand be detected with antibodies directed against the nonglycosylated domains. Lectins or antibodies against glycan epitopes can also be used to immunoprecipitate mucins, but these do not detect a specific protein.
2. Radioactive samples must be handled with care and personal protective equipment, as protective clothing, face shield, and gloves, should be used. The waste should be discarded according to local regulations. The work area and equipment as well as hands should be monitored for radioactive contaminations by a portable meter (GM tube). Contaminations should be cleaned immediately with water and detergent.
3. As an example, the precursor apo-MUC2 mucin is produced by the cell line, LS174T, at detectable levels as fast as after 10 min after labeling and the O-glycosylated mucin after 1 h of labeling. The labeled mucins are stored within the cells for several days (3, 13). Fusion proteins are also possible to detect as precursors in the ER within minutes after labeling. The duration of the mucin in the cells depends on if the cells are able to store the mucins in the regulated secretory pathway or if the expressed mucin is constitutively secreted resulting in a disappearance from the cells within a few hours. The N-terminus of

MUC2 in CHO cells has almost disappeared after a chase of 3 h followed by a 15-min pulse (7).

4. Sonication might damage the large gel-forming mucin networks and milder solubilization techniques as homogenization can be used.
5. The specificity of the IP should be tested by comparison to an identical IP performed with an unspecific control using an irrelevant antibody of the same isotype and with the same concentration.
6. Nonreduced mucins are large, fragile, and sticky, so mechanical stress, filtration, and centrifugation must be avoided. The MUC2 mucin complex is partly insoluble (14) and due to this centrifugation is omitted at this step, but if only soluble forms are to be studied then centrifugation at low speed is a way to reduce nonspecific background.
7. Different stringency wash solutions are required for different antibody–sample combinations. Washing in series with analysis of the removed supernatants by electrophoresis can reveal appropriate conditions for sufficient specific binding with most of the background removed.
8. The same volume of serum from nonimmunized animals can be used as control when a polyclonal serum is used as primary antibody.
9. The use of two thinner combs makes it easier to remove them without breaking the fragile agarose wells.
10. Load the samples from the bottom of the well to allow loading of large volumes. The dye running at the front of the electrophoresis should run out of the gel and the molecular mass markers of 250 kDa shall be found in the lower third of the gel.
11. Standard cassettes can be used as the signal already is enhanced by the Amplify reagent. Double-emulsion films with clear background as Kodak® Biomax MS have the highest sensitivity.
12. Initial GalNAz incorporation can be detected after 1 h. Some mucins are still labeled after 24 h.
13. Cervical dislocation is a preferred method, causes less stress, is fast, and retains the intestinal luminal material.
14. Omit antigen retrieval procedures, where the temperature is increased, as this results in a weak TAMRA stain. Compatibility with other procedures to regain antigens has not been tested.
15. Care must be taken to remove bubbles from the samples at mounting. Apply a pressure on the cover glass, but take care not to disrupt the tissue or move the loosely fixed luminal material or mucus layer.

Acknowledgments

This work was supported by the Swedish Research Council (no. 7461, 21027, and 342-2004-4434), The Swedish Cancer Foundation, The Knut and Alice Wallenberg Foundation (KAW2007.0118), IngaBritt and Arne Lundberg Foundation, Sahlgren's University Hospital (LUA-ALF), EU-FP7 IBDase (no. 200931), Wilhelm and Martina Lundgren's Foundation, The Swedish CF Foundation, Torsten och Ragnar Söderbergs Stiftelser, and The Swedish Foundation for Strategic Research—The Mucosal Immunobiology and Vaccine Center (MIVAC) and the Mucus-Bacteria-Colitis Center (MBC) of the Innate Immunity Program (2010–2014).

References

1. Asker, N., Axelsson, M. A. B., Olofsson, S. O., and Hansson, G. C. (1998) *J. Biol. Chem. 273*, 18857–18863.
2. Asker, N., Baeckstrom, D., Axelsson, M. A., Carlstedt, I., and Hansson, G. C. (1995) *Biochem. J.* ***308*** *(Pt 3)*, 873–880.
3. McGuckin, M. A., Devine, P. L., and Ward, B. G. (1996) *Biochem. Cell Biol.* 74, 87–93.
4. Van Klinken, B. J., Einerhand, A. W., Buller, H. A., and Dekker, J. (1998) *Glycobiology 8*, 67–75.
5. Sheehan, J. K., Kirkham, S., Howard, M., Woodman, P., Kutay, S., Brazeau, C., Buckley, J., and Thornton, D. J. (2004) *J. Biol. Chem. 279*, 15698–15705.
6. Lidell, M. E., Johansson, M. E. V., Mörgelin, M., Asker, N., Gum, J. R., Kim, Y. S., and Hansson, G. C. (2003) *Biochem. J. 372*, 335–345.
7. Godl, K., Johansson, M. E. V., Karlsson, H., Morgelin, M., Lidell, M. E., Olson, F. J., Gum, J. R., Kim, Y. S., and Hansson, G. C. (2002) *J. Biol. Chem. 277*, 47248–47256.
8. Perez-Vilar, J. and Hill, R. L. (1998) *J. Biol. Chem.* 273, 6982–6988.
9. Perez-Vilar, J., Eckhardt, A. E., de Luca, A., and Hill, R. L. (1998) *J. Biol. Chem. 273*, 14442–14449.
10. Johansson, M. E., Phillipson, M., Petersson, J., Velcich, A., Holm, L., and Hansson, G. C. (2008) *Proc. Natl. Acad. Sci. USA* ***105***, 15064–15069.
11. Laughlin, S. T. and Bertozzi, C. R. (2009) *Proc. Natl. Acad. Sci. USA* 106, 12–17.
12. Schulz, B. L., Packer, N. H., and Karlsson, N. G. (2002) *Anal. Chem. 74*, 6088–6097.
13. Axelsson, M. A. B., Asker, N., and Hansson, G. C. (1998) *J. Biol. Chem. 273*, 18864–18870.
14. Carlstedt, I., Herrmann, A., Karlsson, H., Sheehan, J. K., Fransson, L., and Hansson, G. C. (1993) *J. Biol. Chem.* ***268***, 18771–18781.

Chapter 7

MUC1 Membrane Trafficking: Protocols for Assessing Biosynthetic Delivery, Endocytosis, Recycling, and Release Through Exosomes

Franz-Georg Hanisch, Carol L. Kinlough, Simon Staubach, and Rebecca P. Hughey

Abstract

MUC1 is normally apical in polarized epithelial cells but is aberrantly localized in tumor cells. To better understand the mechanism of this altered localization as well as the normal functions of MUC1, we are focused on characterizing the features of MUC1 that regulate the membrane trafficking of this mucin-like transmembrane protein. Previous studies using heterologous expression of MUC1 in CHO and MDCK cells revealed that trafficking to the cell surface as well as endocytosis and recycling is modulated by glycosylation, palmitoylation, and docking of adaptor protein complexes. Protocols for assessing MUC1 trafficking have utilized membrane-impermeant cell surface biotinylation and subsequent stripping with reducing reagents, such as MESNA. The cumulative data have been used for computer modeling and calculation of rate constants. As MUC1 is released through trafficking to exosomes, we have devised protocols for the affinity isolation of MUC1-containing lipid rafts from nanovesicular subpopulations to perform proteomic mapping of protein constituents in these sorting platforms. Our studies to date have shown that plasma membranous MUC1 traffics via lipid raft-associated pathways to exosomes, which are independent of caveolin-1 or dynamin, but dependent on flotillin.

Key words: MUC1, Transmembrane mucin, Glycoprotein membrane trafficking, Endocytosis, Recycling, Exosomes, Metabolic labeling, Immunoprecipitation, Computer modeling, Lipid rafts, Proteomics

1. Introduction

MUC1 is a multifunctional glycoprotein expressed at the apical surface of polarized epithelia. The function of MUC1 as a cell surface sensor that promotes cell growth and survival is overshadowed by its name and classification as a mucin. Like all mucins, MUC1 does provide cell surface protection from bacteria, viruses, and other insults from the environment, but MUC1 also functions in

Michael A. McGuckin and David J. Thornton (eds.), *Mucins: Methods and Protocols*, Methods in Molecular Biology, vol. 842, DOI 10.1007/978-1-61779-513-8_7, © Springer Science+Business Media, LLC 2012

modulating EGF receptor activity, signal transduction through docking cytoplasmic proteins, and modifying transcription factor activity after nuclear targeting (for review, see refs. 1–3).

1.1. Apical Targeting of MUC1

As MUC1 is a transmembrane protein, we have focused on characterizing the features of the protein that modulate its membrane trafficking with emphasis on surface delivery from the biosynthetic pathway, endocytosis, recycling, and delivery to exosomes. Glycosylation of MUC1 does play a key role in some aspects of MUC1 membrane trafficking. For example, MUC1 traverses apical recycling endosomes along the biosynthetic pathway in polarized MDCK cells which is a route used by apically expressed proteins with glycan-dependent targeting signals, and our recent studies indicate that MUCI apical targeting is dependent on glycosylation of the mucin-like repeats with core O-glycans (4, 5). We also observed that MUC1 synthesized with truncated O-glycans in glycosylation-defective CHO cells was delivered to the cell surface with slower kinetics when compared to MUC1 with normal glycosylation (6). However, the presence of truncated glycans on MUC1 also enhanced its clathrin-mediated endocytosis in these same cells by twofold, and replacement of the entire ectodomain with that of the model protein Tac enhanced MUC1 endocytosis by almost threefold (6, 7).

To avoid the complications of glycan-dependent trafficking of MUC1, we used the Tac-MUC1 chimera to characterize the role of the cytoplasmic domain in endocytosis and recycling. We found that efficient endocytosis requires cytoplasmic binding of the adaptor complex 2 (AP2) at a YHPM motif and Grb2 at a tyrosine-phosphorylated YTNP motif while efficient recycling back to the cell surface requires dual palmitoylation of cysteines at a CQC motif and binding of adaptor complex 1 (AP1) at the YHPM motif (7, 8). All of the experiments that were carried out to assess MUC1 delivery to the cell surface, MUC1 endocytosis, and MUC1 recycling relied on the ability to modify lysine residues on cell surface MUC1 or chimera with biotin using a membrane-impermeant reagent sulfosuccinimidyl 2-(biotinamido)-ethyl-1,3-dithiopropionate (sulfo-NHS-SS-biotin). Biotinylated MUC1 can then be recovered from immunoprecipitates (IPs) with avidin conjugated to agarose. The fraction of the total MUC1 IP that is subsequently recovered with avidin beads represents the fraction of cellular MUC1 that was initially on the cell surface. The disulfide bond between the lysine-reactive NHS group and the biotin also allows efficient release of the biotin from the modified protein by treatment with reducing reagents, such as β-mercaptoethanol (β ME), which is commonly used in SDS-gel sample buffer. However, treatment of the biotinylated cell surface with the membrane-impermeant reducing agent, sodium-2-mercaptoethanesulfonic acid (MESNA), also releases biotin from lysines modified with sulfo-NHS-SS-biotin without toxic effects on the cell. For this reason, biotinylated cells can be returned to culture and biotinylated proteins can continue their normal membrane trafficking with intermittent treatment of the cells with MESNA to strip cell surface biotin while protecting intracellular biotin. A portion of this chapter describes the protocols for cell surface biotinylation and MESNA stripping of biotin in order to measure membrane trafficking of MUC1.

1.2. Exosomal Targeting of MUC1

MUC1 is synthesized as a single peptide but exhibits autocatalytic cleavage while still in the endoplasmic reticulum to produce two subunits that remain stably associated (9, 10). Shedding of MUC1 from cells has been attributed both to dissociation of the subunits and proteolytic cleavage (11, 12). We and others have obtained evidence for an additional "shedding" route of MUC1 via exosomal export (13, 14) (see Fig. 1). Transfected MCF-7 breast cancer cells were demonstrated to release exosomes that contained not

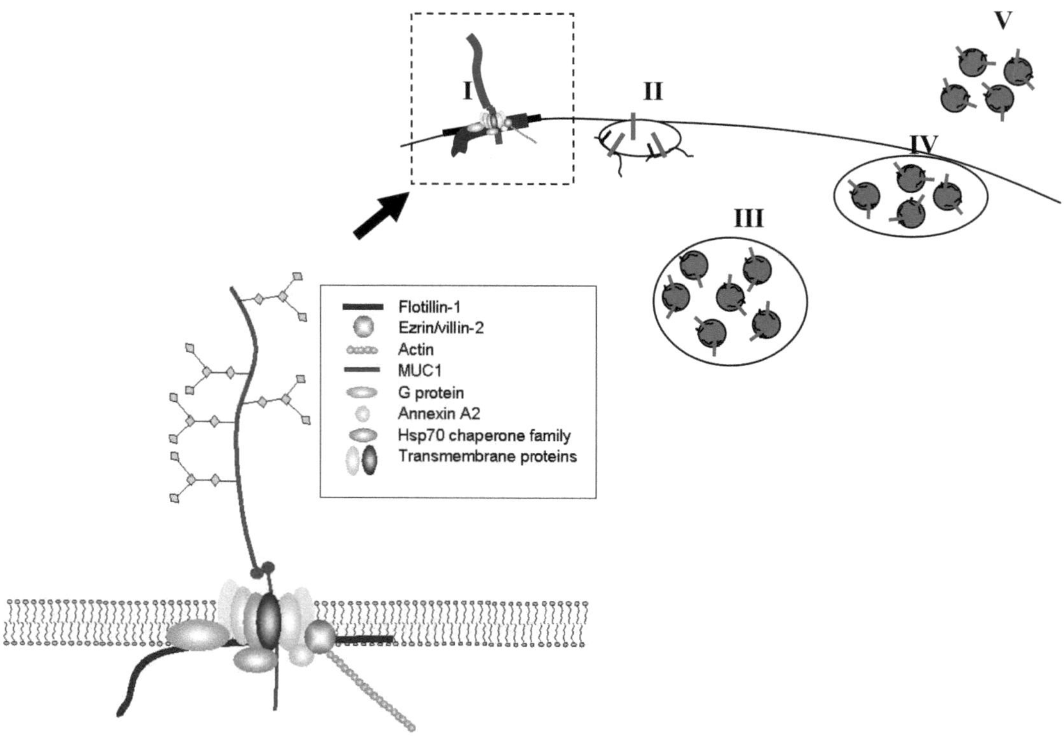

Fig. 1. MUC1 delivery to exosomes. MUC1-positive lipid rafts with a characteristic protein composition and flotillin-1 as a scaffolding protein (step I) are endocytosed to form early endosomes (step II). Exosomes originate by a second invagination of the endosomal membrane, which results in the formation of multivesicular bodies (MVBs) (step III). After membrane fusion of MVBs with the plasma membrane (step IV), the exosomes are released into the extracellular space (step V). The extracellular topology of the heavily O-glycosylated MUC1 ectodomain as well as the protein composition of the respective lipid rafts are retained in exosomal membranes.

only endogenous MUC1, but also the recombinant myc-tagged MUC1 fusion protein (14). Exosomes are nanovesicles released by leukocytes and epithelial cells. Their function is still a matter of debate, but they may play a key role in homeostasis and adaptation of plasma membrane glycoprotein patterns. Hence, the removal of de novo-synthesized plasma membrane proteins and immaturely glycosylated (undersialylated or unprocessed) species via lysosomal degradation or via exosomal export might be a mechanism by which cells regulate the exposure of a uniquely glycosylated protein in apical membranes.

Trafficking of MUC1 is dependent on the formation of specific sorting platforms, which represent subdomains of the plasma and Golgi complex membranes. The simple fluid mosaic model of plasma membranes has evolved to include more advanced concepts taking into account the high dynamicity of protein traffic and the observation that membranes exhibit subdomains or lipid rafts with characteristic protein and lipid compositions. These rafts can be regarded as sorting platforms for targeted transport of transmembrane and GPI-anchored proteins and hence their protein compositions are highly fluctuating. We observed that MUC1 formed an integral component of such lipid rafts in both plasma membranes and released exosomes and followed trafficking pathways which were independent of clathrin, caveolin-1, or dynamin, but dependent on flotillin-1 or -2 (14). In the second portion of this chapter, we describe a strategy that allows us to isolate specific subpopulations of exosomal lipid rafts that contain a myc-tagged version of recombinant MUC1 (MUC1-M2) as a bait protein. The affinity-purified protein fractions can be used in proteomic studies aimed at molecular characterization of MUC1-containing rafts, which serve as sorting platforms in cellular trafficking.

2. Materials

2.1. CHO and MDCK Cell Culture, Transfection/Infection, and Lysis

1. Chinese Hamster ovary (CHO) cells from M. Kreiger at Massachusetts Institute of Technology or Madin Darby canine kidney (MDCK) cells from G. Apodaca at University of Pittsburgh are maintained on 60-mm plastic dishes in DMEM/F12/FBS (Dulbecco's minimum Eagle's medium (DMEM) and Ham's F12 (1:1) with 5% fetal bovine serum (FBS)) at 37°C in a humidified incubator with 5% CO_2.
2. Nonpolarized CHO and MDCK cells are plated on 12-well plastic plates for endocytosis or recycling assays. Polarized cultures of MDCK cells are grown on 12-mm permeable supports with 0.4 μm membranes in 12-well plates for 4–5 days before use (e.g., Costar, Corning, NY).

3. MUC1 expression is achieved by stable transfection of cells with MUC1 cDNA subcloned into pcDNA3.1 (Invitrogen, San Diego, CA) by growth in 0.5 mg/mL G418 and isolation of clones or by infection of cells with an adenovirus (AV-MUC1 used at m.o.i. of 100) prepared by subcloning MUC1 cDNA into the pAdlox vector.
4. Detergent cell lysis buffer: 60 mM *n*-octyl β-D-glucopyranoside, 0.1% SDS, 150 mM NaCl, 10 mM HEPES-NaCl, pH 7.4.

2.2. SDS-PAGE and Immunoblotting

1. PAGE sample buffer: BioRad SDS sample buffer with fresh 5% β ME.
2. Criterion™ Precast Gels (4–15% Tris–HCL, 1.0 mm) and a BioRad Mini Protean Gel system (BioRad, Hercules, CA).
3. Nitrocellulose (e.g., 0.45 μm from BioRad) with a BioRad transfer apparatus.
4. Blocking buffer: 10% nonfat dry milk predissolved with vigorous stirring for 20 min in phosphate-buffered saline (PBS: 100 mM NaCl, 80 mM Na_2HPO_4, 20 mM NaH_2PO_4).
5. Kodak TR-Phosphorimager screen with an Imager and Quantity One software (Biorad Laboratories).
6. MUC1 Antibodies: VU-3C6 and B27.29 mouse monoclonal antibodies (15).

2.3. Biotinylation-Based Assays

1. HEPES-buffered saline (HBS): 150 mM NaCl, 10 mM HEPES, pH 7.4. The buffer is used alone or with addition of Triton X-100 (1% (v/v)) or SDS (1 or 0.01% (w/v)).
2. Dulbecco's PBS with calcium and magnesium ($DPBS^{++}$) (Mediatech, Inc, Manassas, VA).
3. DMEM lacking Met and Cys (Sigma, St. Louis, MO).
4. (^{35}S)Met/Cys (Easy Tag Express-(^{35}S)Protein Labeling Mixture, Perkin Elmer, Waltham, MA).
5. TEA-buffered saline (TBS): 10 mM triethanolamine–HCl, pH 7.6, 137 mM NaCl, and 1 mM $CaCl_2$. Prepare by dissolving 66 μL TEA in 50 mL water and adjust pH to 7.6 with HCl.
6. Stripping buffer: 50 mM Trizma–HCl, pH, 8.6, 100 mM NaCl, 1 mM EDTA, 0.2% (w/v) BSA can be stored as a stock at 4°C and is used with or without addition of 100 mM MESNA. The MESNA is added fresh just before each experiment and chilled on ice for at least 1 h before use.
7. Sulfo-NHS-SS-biotin is prepared fresh by dissolving 1 mg in 5 μL DMSO and storing the tube on ice until needed (~1 h). The sample is then thawed when needed and the 5 μL directly added to 1 mL TBS.

8. Iodoacetic acid (120 mM) is prepared fresh in $DPBS^{++}$ and chilled on ice for at least 1 h before use.
9. HEPES-buffered MEM: Minimum Essential Medium Eagle (Sigma) containing 20 mM HEPES, pH 7.4, 0.6% BSA (w/v), 4 mM $NaHCO_3$. Prepare using a stock solution of 1 M HEPES, pH 7.6.
10. Software for determining rate constants for endocytosis and recycling: IGOR Pro 4.19 (WaveMetrics, Lake Oswego, OR).

2.4. MCF-7 Cell Culture and Transfection

1. Culture requirements: DMEM, 10% FBS, penicillin/streptomycin (100×), puromycin (5 μg/mL), PBS, trypsin, 300 cm^2 culture flasks.
2. Transfection reagent: Superfect (Qiagen, Hilden, Germany).

2.5. Chemicals and Buffers for Exosome and Lipid Raft Isolation

1. Sucrose, Protease inhibitor cocktail (e.g., Sigma–Aldrich P8340 plus 1 mM phenylmethanesulfonyl fluoride (PMSF), highest purity Triton X-100 (Roth, Karlsruhe, Germany).
2. Buffer μMAC isolation kit (Miltenyi Biotec, Bergisch Gladbach, Germany):

   Washing Buffer 2: 20 mM Tris–HCl, pH 7.5.

   SDS-elution buffer: 50 mL Tris–HCl, pH 6.8, 50 mM DTT, 1% SDS (w/v), 1 mM EDTA, 0.005% bromophenol blue (w/v), 10% glycerol (v/v).

3. Methods

3.1. MUC1 Delivery to the Surface of Polarized Epithelial Cells

3.1.1. Metabolic Labeling (Pulse-Chase Protocol)

1. Cells expressing MUC1 and growing on permeable supports are washed twice, top and bottom, with 0.5 mL sterile prewarmed (37°C) DMEM lacking Met and Cys. The medium in the lower chamber is removed first and replaced last to avoid pressure on the monolayer from below and thereby disruption of the tight junctions.
2. Cells are *starved* for Met and Cys by incubation in the same media for 30 min at 37°C in 5% CO_2.
3. Cells are metabolically labeled by incubation in 0.5 mL DMEM lacking Met and Cys with 50–75 μCi of (^{35}S)Met/Cys included only in the lower chamber. Pulse labeling is usually for 30 min at 37°C in 5% CO_2.
4. Medium is removed from the bottom chamber and then the top chamber and disposed of appropriately as radioactive waste. Cells are washed twice, top and bottom, with 0.5 mL sterile prewarmed (37°C) $DPBS^{++}$.
5. Cells are chased for 15–120 min in DMEM/F12/FBS at 37°C in 5% CO_2.

3.1.2. Cell Surface Biotinylation

1. Cells (in 12-well dishes on permeable supports) are moved to a metal plate (9 × 13 × 0.25 in.) sitting on ice in a rectangular ice pan and gently washed, top and bottom, four times with 0.5 mL ice-cold DPBS^{++}. Each wash remains on the cells for 5 min and medium in the lower chamber is always removed first and replaced last. The cells remain on the metal plate on ice for the entire protocol.
2. The top or bottom surface of the polarized cells is incubated for 10–15 min on ice with 1 mg/mL sulfo-NHS-SS-biotin in 0.5 mL TBS to label the apical or basolateral surface, respectively. The opposite surface of the cells being treated with sulfo-NHS-SS-biotin is incubated with TBS lacking the biotin reagent.
3. The biotinylation reaction is quenched by washing the top and bottom of the cells three times with 0.5 mL DMEM/F12/FBS. Each *wash* remains on the cells for 3 min.

3.1.3. Immunoprecipitation and Recovery of Biotinylated MUC1

1. Permeable supports are removed, cells are solubilized, and supernatants are moved to a clean tube as described in Subheading 3.5.
2. Antibodies (0.5 μg VU-3C6 or B27.29) and 30 μL of a 50% slurry of protein-G conjugated to Sepharose 4B (Invitrogen, Carlsbad, CA) are added to the supernatant. Tubes are vortexed and centrifuged in a microcentrifuge for 10 s at top speed to pellet the Sepharose.
3. Cell extracts with antibodies and protein G beads are incubated at 4°C overnight with end-over-end mixing.
4. Tubes are centrifuged in a microcentrifuge twice for 30 s (twice in a row without removing tubes from the centrifuge) at 9,300 × *g* to pellet the protein G beads. The supernatant is removed and the pellet is washed with (1) 1 mL 1% (v/v) Triton X-100 in HBS, (2) 1 mL 0.01% (w/v) SDS in HBS, and (3) 1 mL HBS. After the last wash, all the liquid is removed from the pellet with a Hamilton syringe (Gastight #1710, 100 μL with a 22S-gauge needle) (see Note 1). The immunoprecipitated protein is recovered by heating the Sepharose pellet with 80 μL 1% SDS in HBS for 2 min at 90–100°C. The tubes are vortexed and centrifuged in a microcentrifuge twice for 30 s at top speed to pellet the Sepharose, and the supernatant is moved to a clean tube using a Hamilton syringe.
5. A 20-μL aliquot of supernatant is removed and saved at 4°C (this represents 25% of the *total* immunoprecipitate). One milliliter of 0.5% (v/v) Triton X-100 in HBS is added to the remaining 60 μL to dilute out the SDS.
6. Avidin conjugated to agarose (30 μL) (Pierce, Rockford, IL) is added to the SDS-Triton sample and incubated at 4°C overnight with end-over-end mixing.

7. The biotinylated proteins bound to the avidin-conjugated beads are washed with (1) 1 mL 1% Triton X-100 in HBS and then (2) 1 mL HBS.
8. The biotinylated proteins are eluted from the avidin-conjugated beads by adding 30 μL PAGE sample buffer with 5% βME and heating for 3.5 min at 90–100°C.
9. The avidin-conjugated beads are pelleted after vortexing by centrifugation for 30 s in a microcentrifuge at top speed and the supernatant is loaded onto the SDS-gel using a Hamilton syringe. Aliquots representing the total IP are also loaded onto the SDS-gel for comparison and to calculate the fraction biotinylated.
10. Densitometric data are used to calculate the percent of the total MUC1 found on the surface (percent biotinylated) at each time point.

3.2. MUC1 Endocytosis in Polarized Epithelial Cells

1. Endocytosis of MUC1 is carried out in *triplicate* at each time point by using only three wells on each 12-well plate for each time point. A separate 12-well plate is used for each time point, including the zero time point and the control plate (to determine total biotinylated).
2. Cells growing on plastic or permeable supports in 12-well dishes are biotinylated as described in Subheading 3.1. All subsequent washes are 0.5 mL unless otherwise noted. For permeable supports, use 0.5 mL on the top and the bottom.
3. To allow endocytosis of the biotinylated cell-surface MUC1, plates of cells are rapidly warmed to 37°C on a metal plate (9 × 13 × 0.25 in.) in a circulating water bath by the addition of prewarmed HEPES-buffered MEM.
4. The plate of cells remains in the 37°C bath for 1–30 min.
5. To stop endocytosis, plates of cells are rapidly cooled on the metal plate on ice by two quick washes with ice-cold $DPBS^{++}$.
6. The plate of cells for the zero time point remains on ice and is quickly rinsed once with ice-cold HEPES-buffered MEM (10 s) and then rapidly rinsed twice with $DPBS^{++}$ (see Note 2). Biotin remaining on the cell surface is stripped by washing the cells twice for 20 min each with ice-cold membrane-impermeant 100 mM MESNA in stripping buffer.
7. The residual MESNA on the cells is quenched by washing the cell surface once with 120 mM iodoacetic acid in $DPBS^{++}$ on ice for 10 min.
8. The plate of cells representing the *total* biotinylated MUC1 remains on ice in $DPBS^{++}$ and is not treated with MESNA or iodoacetic acid.
9. Cells are extracted as described in Subheading 3.5.

10. MUC1 is immunoprecipitated and total and biotinylated MUC1 recovered as described in Subheadings 3.1.2 and 3.1.3.
11. MUC1 is analyzed after SDS-PAGE by quantitative immunoblotting with a BioRad Versadoc. (^{35}S)MUC1 from metabolically labeled cells is analyzed after SDS-PAGE with a BioRad Imager.
12. The percent endocytosis of MUC1 is calculated by dividing the amount of biotinylated MUC1 remaining after MESNA treatment by the amount of biotinylated MUC1 recovered without MESNA treatment (control). Background values obtained at $t=0$ are subtracted from each time point and are typically 1–5% of *total* MUC1 (control). An example of the raw and processed data for one experiment is shown in Fig. 2.
13. Data from each experiment are plotted as percent internalized (mean ± SD, $n=3$), whereas for publications data are calculated from three distinct experiments (mean ± SEM). Rate constants for endocytosis are calculated by computer modeling using both endocytosis and recycling data (see below) (see Note 3).

3.3. MUC1 Recycling in Polarized Epithelial Cells

1. Plates of cells are biotinylated on ice, moved to a 37°C water bath for 5–10 min to allow endocytosis, and returned to ice to strip surface biotin with MESNA as described in Subheading 3.1. Only one 20-min MESNA incubation/wash is done at this time followed by one quench with 120 mM iodoacetic acid in DPBS^{++} for 10 min. Cells are then washed rapidly twice with DPBS^{++}.
2. To measure recycling, plates of cells are rapidly warmed to 37°C a second time by the addition of prewarmed HEPES-buffered MEM.
3. The plate of cells remains in the 37°C bath for 1.5–10 min. At the end of each time point, the plates are returned to ice and rapidly washed twice with ice-cold DPBS^{++}. All time points are held on ice until the entire time course is completed.
4. Biotin on proteins returning to the cell surface is stripped by washing the cells once for 20 min with 0.5 mL ice-cold MESNA in stripping buffer as described in Subheading 3.2.
5. MESNA is quenched by washing the cell surface once with 120 mM iodoacetic acid in DPBS^{++} on ice for 10 min.
6. The plate of cells representing the total biotinylated MUC1 remains on ice and is not treated with MESNA or iodoacetic acid.
7. The plate of cells representing the $t=0$ time point of recycling and the one representing the total 5–10-min endocytosis point remain on ice after the initial endocytosis protocol, including the MESNA and iodoacetic acid wash (i.e., these cells are treated a second time with MESNA and iodoacetic acid while on ice).

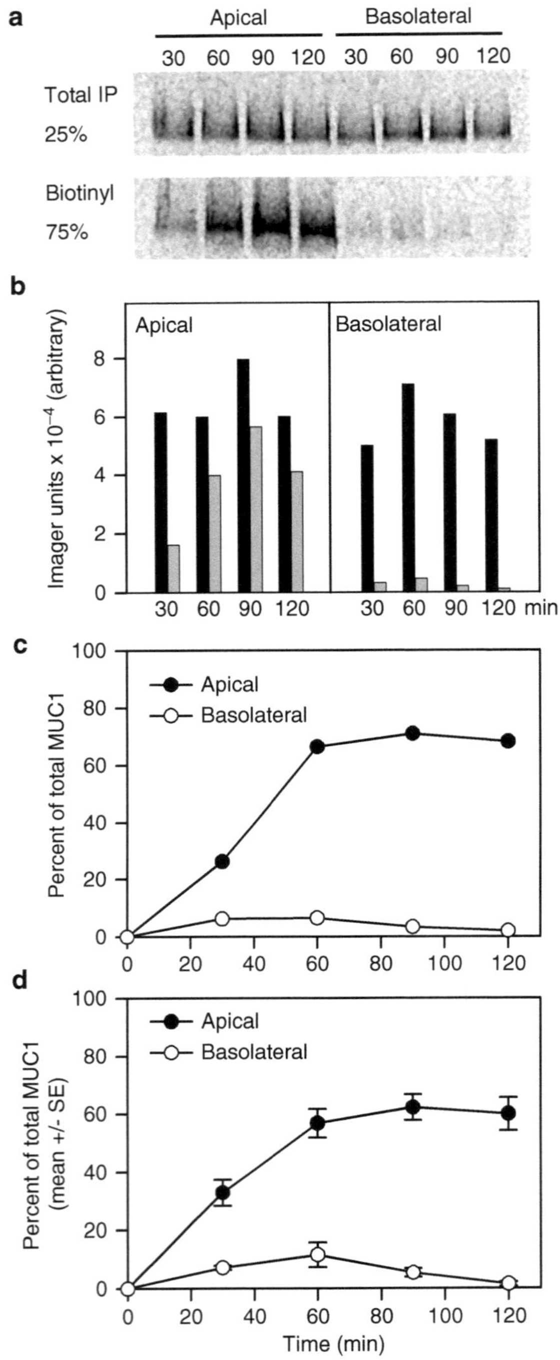

Fig. 2. Example of a biotinylation experiment that follows MUC1 biosynthetic delivery to the cell surface. Polarized MDCK cells expressing MUC1 and growing on permeable supports were metabolically labeled with [^{35}S]Met/Cys for 30 min and chased for 30, 60, 90, or 120 min. Plates of cells were moved to ice for treatment with sulfo-NHS-SS-biotin to biotinylate the cell surface proteins on the apical (top) or basolateral (bottom) surface. Cells were extracted with detergent and biotinylated MUC1 was recovered with avidin beads from anti-MUC1 immunoprecipitates for SDS-PAGE. A representative BioRad Imager profile of an SDS-gel is shown in (**a**) and the corresponding data are presented in (**b**) (biotinylated MUC1; *gray bars*). Twenty-five percent of the immunoprecipitate was saved to calculate the percent of total MUC1 (*black bars*) reaching the cell surface at each time point as shown in (**c**). Data from three experiments are combined and presented in (**d**) for publication quality (mean and SE).

8. Cells are extracted as described in Subheading 3.5.
9. MUC1 is immunoprecipitated and *total* and biotinylated MUC1 recovered as described in Subheading 3.1.2.
10. MUC1 is analyzed after SDS-PAGE by quantitative immunoblotting with a BioRad Versadoc. (^{35}S)MUC1 from metabolically labeled cells is analyzed after SDS-PAGE with a BioRad Imager.
11. The data is plotted as the *percent* of biotinylated MUC1 remaining after subtracting the $t=0$ endocytosis time point.
12. Data from each experiment are plotted as percent internalized (mean, $n=1$ or 2) while published data are calculated from three experiments (mean ± SEM). Rate constants for endocytosis are calculated by computer modeling using recycling data (see below). An example of the raw and processed data for a TacMUC1 mutant is shown in Fig. 3.

3.4. Calculating Rate Constants Based on Endocytosis and Recycling Data (see Fig. 4) (see Note 4)

1. Experimental data in the format of percent biotinylated for each time point of endocytosis and recycling are collected for computer modeling using IGOR Pro 4.19.
2. The assay can be simulated using formulas in Fig. 4, where A is MUC1 at the cell surface and B and C are intracellular pools of MUC1 available and unavailable, respectively, for recycling back to the cell surface.
3. Data can be fit simultaneously with kinetic rate constants as global parameters and initial values (percent biotinylated) as local parameters.

3.5. Extraction of Cells for Immunoprecipitation and PAGE

1. After experimental protocols are complete as described in Subheadings 3.1–3.3, cells growing on 12-well plastic dishes are extracted into 0.2 mL detergent cell lysis buffer for 20 min at room temperature on a rotating shaker. Cell extracts are moved to a 1.5-mL conical plastic tube (with a snap cap) and centrifuged at ~20,000 × g for 7 min in a microcentrifuge at 4°C. Supernatants are moved to a clean conical tube.
2. Cells growing on 12-mm Costar supports are solubilized by first removing the filter from the plastic insert with a new razor blade and placing it inside of a 1.5-mL conical plastic tube with 0.25 mL detergent cell lysis buffer for end-over-end rotation on a wheel at room temperature for 20 min. Filters are removed from the tubes with tweezers before centrifugation (leaving all liquid behind) and moving supernatants to a clean tube.

3.6. SDS-PAGE and Immunoblotting

1. All samples from protocols described in Subheadings 3.1–3.3 are analyzed by SDS-gel electrophoresis using 4–15% Criterion™ Precast Gels.
2. Proteins on gels with samples for immunoblotting are transferred electrophoretically to nitrocellulose with a BioRad transfer

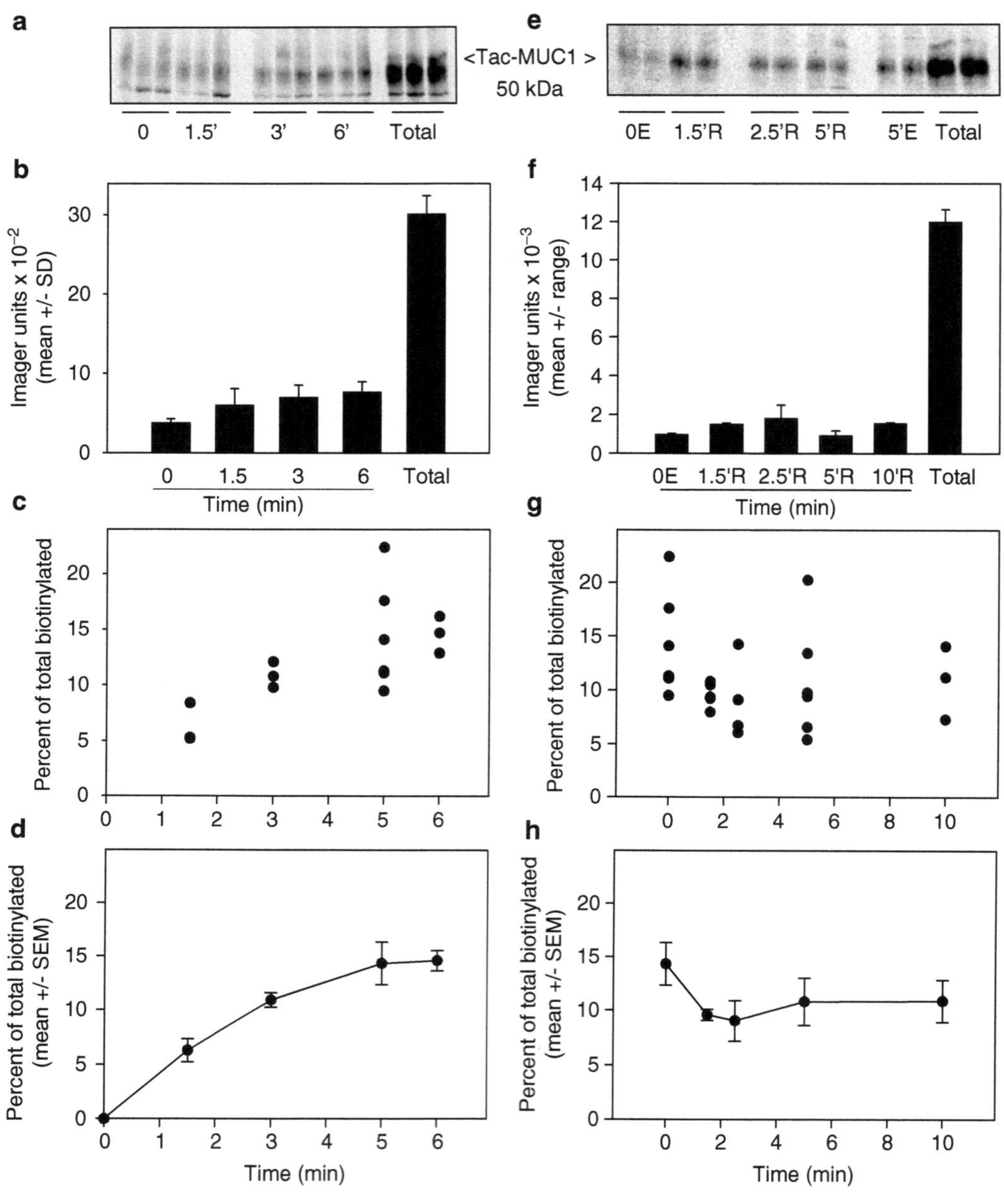

Fig. 3. Examples of endocytosis and recycling experiments. CHO cells expressing a chimera Tac-MUC with ectodomain of Tac and transmembrane and cytoplasmic tail of MUC1 (mutant CQC/AQA) were metabolically labeled with (^{35}S)Met/Cys for 30 min and chased for 90 min. The cell surface was biotinylated on ice and endocytosis was carried out for 0, 1.5, 3, 5, or 6 min as indicated (**a–d**) before stripping surface biotin with MESNA. Cells were extracted with detergent and biotinylated Tac-MUC1 was recovered with avidin beads from anti-Tac immunoprecipitates for SDS-PAGE. A representative BioRad Imager profile of an SDS-gel is shown in **a**, and the corresponding data from triplicate samples are shown in **b**. Data from multiple experiments are presented in **c** after subtraction of the zero time point (background) and calculating the percent of the total biotinylated Tac-MUC1 recovered at each time point. Recycling of biotinylated (^{35}S)Tac-MUC1 was followed after 5-min endocytosis (**e–h**) and a representative profile of an SDS-gel is shown in **e** with the corresponding data from duplicate samples shown in **f**. Data from multiple experiments is presented in **g** after subtraction of the zero time point (background is 0E, zero endocytosis) and calculating the percent of the total biotinylated Tac-MUC1 recovered at each time point. Note that the data at the zero time point of recycling are the same as the 5-min endocytosis data (5′E in **e** and **f**). The endocytosis and recycling data from **c** and **g** (mean and SEM) are used for computer modeling to calculate the rate constant for endocytosis ($k_1 = 0.2 \pm 0.08$), recycling ($k_2 = 0.3 \pm 0.12$), and degradation ($k_3 = 0.13 \pm 0.09$). The data from **d** and **h** can also be plotted with the best-fitting curve generated by the computer (see our published data in 8).

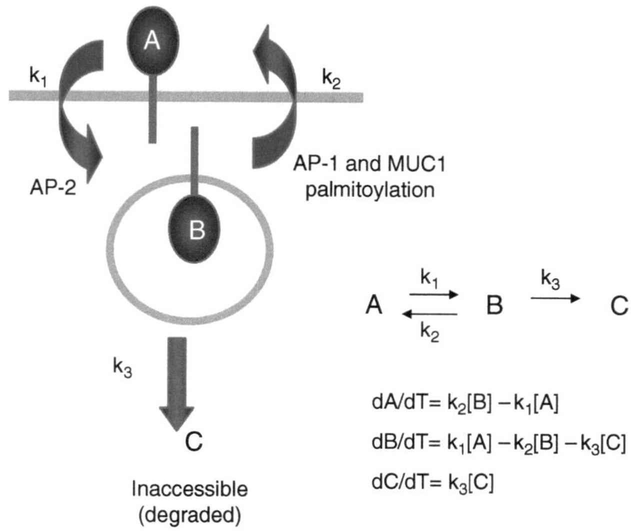

Fig. 4. Calculating rate constants based on endocytosis and recycling data. Experimental data in the format of percent biotinylated for each time point of endocytosis and recycling are collected for computer modeling using IGOR Pro 4.19 (WaveMetrics, Lake Oswego, OR). The assay was simulated with the model on the *left* using differential equations on the *right*, where A is MUC1 at the cell surface, and B and C are intracellular pools of MUC1 available and unavailable (degraded), respectively, for recycling back to the cell surface. Data were fit simultaneously with kinetic rate constants as global parameters and initial values (percent biotinylated) as local parameters. Rate constant for endocytosis is k_1, for recycling is k_2, and degradation is k_3.

apparatus and the nitrocellulose is blocked for at least 30 min in blocking buffer.

3. Gels with radioactive samples are dried and placed with a Kodak TR-Phosphorimager screen for 1–14 days before analyzing with a BioRad Imager and Quantity One software. After SDS-PAGE, gels with radioactive samples can also be transferred to nitrocellulose and air dried for 1 h before placing with the phosphorimager screen. After analyzing the radioactive bands, the nitrocellulose can be blocked with blocking buffer for 30 min and subjected to immunoblotting.
4. MUC1 expressed in CHO cells is immunoblotted (and immunoprecipitated) with VU-3C6 while MUC1 expressed in MDCK cells is immunoblotted (and immunoprecipitated) with B27.29 (see Note 5).

3.7. Exosome Isolation from Cell Culture Supernatants

1. It is essential to deplete exosomes from FBS used for cell culture to avoid contamination of MCF-7-derived human nanovesicles with bovine exosomes.
   (a) Centrifuge 50 mL of FBS in 70-mL centrifuge bottles in a fixed angle rotor at 114,000 × *g* for 90 min at 4°C (e.g., Beckmann Ti45 rotor at 40,000 rpm).

(b) Discard the FBS-derived exosomal pellet and use the exosome-free FBS supernatant for preparation of exosome cell culture medium: DMEM, 10% (v/v) FCS, penicillin, streptomycin, puromycin.

2. Collect the cell culture supernatant every 2–3 days.
3. Centrifuge six aliquots of 55 mL supernatant in 70-mL centrifuge bottles with caps in a fixed angle rotor (e.g., Beckman Ti45) in four progressive steps at 4°C:
   (a) 1,200 × *g*—10 min
   (b) 10,000 × *g*—30 min
   (c) 10,000 × *g*—30 min
   (d) 125,000 × *g*—90 min
4. Discard the final supernatant and all but the last pellet. Resuspend the final exosomal pellet in a centrifuge bottle in 2–3 mL cold 1% Triton X-100 in PBS.

3.8. Extraction of Lipid Rafts from Exosomes

1. Solubilize pellets of exosomes in ice-cold 1% Triton X-100 in PBS by pipetting up and down with an Eppendorf pipette (100 μL tip, 30 s) with intermittant vortexing for 2 min. Use visual inspection to be sure that the pellet is fully resuspended (use an additional 10-min incubation on ice if needed). The exosomes are lysed in cold Triton X-100 to separate membranous raft protein from contaminating cytosolic proteins captured during invagination of the multivesicular body membrane from the limiting membrane.
2. Wash exosomal lipid rafts with PBS by an additional centrifugation at 125,000 × *g* for 90 min at 4°C.
3. Continue with the protocol described in Subheading 3.10. Resuspend the lipid pellet finally in 1 mL cold Triton X-100 in PBS and follow the protocol for μMAC isolation of myc-tagged protein in Subheading 3.11, steps 2a–2i.

3.9. Lipid Raft Extraction from Cellular Membranes of Cultured Cells

1. Wash approximately 3×10^8 cells in four culture flasks (each 300 cm^2) twice with warm PBS (37°C).
2. Detach cells using 4 mL of 10% trypsin for 5 min at 37°C and combine cells from 300-cm^2 flasks into 50-mL conical tubes.
3. Wash cells five times with cold PBS by centrifugation at 180 × *g* at 4°C (3×10^8 cells should form approximately a 1.5-mL pellet).
4. Divide cells during the last washing step into two equal portions in 15-mL tubes.
5. Add cold PBS with the protease inhibitor cocktail up to a volume of 1.8 mL and disperse the cell pellet by vortexing.
6. Add 200 μL of 10% (v/v) Triton X-100 in PBS to 1.8 mL to a final concentration of 1% Triton X-100 on ice and vortex.

7. Incubate cells by rotation at 4°C for 1 h.
8. Add 1.8 g sucrose to each 15-mL tube to obtain a concentration of 45% sucrose, and fill to a volume of 4 mL. To dissolve the sugar, rotate the tubes for another 30 min at 4°C, and check that the final volume is 4 mL.

3.10. Lipid Raft Isolation by Density Gradient Centrifugation

1. Place the cell lysate/Triton X-100/sucrose solution at the bottom of a centrifuge tube for a swinging rotor (e.g., Beckmann SW40).
2. Place carefully 4 mL of 35% sucrose as bottom layer into the centrifuge tube and fill with 2–3 mL of 5% sucrose close to the top of the tube to obtain a discontinuous sucrose gradient (from top to bottom: 5/35/45% sucrose).
3. Centrifuge for 21 h at 29,000 × *g* at 4°C with slow acceleration and deceleration without braking.
4. After centrifugation, recover the lipid rafts that have floated to a light-scattering band between the 5 and 35% sucrose layer in a 3-mL volume using a 5-mL pipette and transfer it into a centrifugation tube for a fixed angle rotor (e.g., Beckmann 60Ti) and vortex. Use PBS to fill to the top of the tube (total volume 26 mL for 60Ti rotor).
5. Centrifuge for 90 min, at 114,000 × *g* at 4°C, to wash lipid rafts.
6. Discard the bulk of the supernatant and drain the remaining supernatant away from the lipid raft pellet for 5 min in the tube, bottom up, on Whatman filter paper.

3.11. Lipid Raft Isolation by Affinity Purification

1. To isolate lipid rafts containing myc-tagged MUC1, solubilize the pellet described in Subheading 3.10, step 6, in 1 mL cold Triton X-100 in PBS by pipetting up and down with an Eppendorf pipette (100 μL tip, 30 s) with intermittent vortexing.
2. Transfer the rafts suspended in 1 mL to a 1.5-mL Eppendorf tube and follow a modification of the protocol supplied in the instructions for μMAC isolation kit (Miltenyi Biotec):
   (a) Add 50 μL of anti-myc microbeads for binding to the myc-tagged bait protein to entrap specific subpopulations of lipid rafts and rotate for 30 min.
   (b) Equilibrate the paramagnetic bead column with Triton X-100 in PBS.
   (c) Pour the beads with bound lipid rafts (1 mL) into the column fixed within a magnetic holder and let the unbound raft fraction run through.
   (d) Wash the column five times: four times with cold 1% Triton X-100 in PBS and once with washing buffer 2.
   (e) Elute the lipid rafts with 100 μL SDS elution buffer at 95°C.

(f) Heat the eluted rafts at 70°C for 15 min to dissolve raft proteins.

(g) Separate raft proteins using 1D or 2D gel electrophoresis (E) (see Note 6).

(h) After 1D-electrophoretic separation: Reduce and alkylate proteins after excision from gels, digest with trypsin in-gel, and analyze by LC-ESI MS-MS for protein identification in databases (see Notes 7 and 8); see Albert et al. 16.

(i) Alternatively, after 2D separation: Pick spots from gels, digest in-gel with trypsin, and analyze by Peptide Mass Fingerprint analysis on a MALDI-mass spectrometer; see ref. 14.

4. Notes

1. This size of Hamilton allows removal of all the liquid while excluding the Sepharose or agarose beads when the tip is placed at the bottom of the tube in the middle of the pellet.
2. Prechill the pipettes for the zero time points before use to ensure that washes and samples remain ice cold at all times. This significantly reduces background.
3. Experiments designed to follow MUC1 delivery to the surface within the biosynthetic pathway are carried out as a single well because there is only one biotinylation step (and no MESNA steps). The single points are clearly sufficient, as a comparison of the representative experiment in Fig. 2c and the cumulative data ($n=3$) in Fig. 2d agree quite well.
4. The reader likely noted from Fig. 3 that the data generated from recycling experiments appears to have more scatter than the data produced from endocytosis experiments (compare Fig. 3c, g). This is likely due to two factors: First, the recycling experiment requires two rounds of warming (endocytosis and recycling) and two rounds of stripping biotin with MESNA treatment on ice that adds to variation in the results. Secondly, the recycling experiment is set up in duplicate (or sometimes as a single well) rather than triplicate to minimize the amount of time needed to wash each sample in order to achieve the necessary rapid warming or cooling. The number of time points for endocytosis and recycling is also limited by the availability of space for 12-well dishes (3×5 in. each) on the metal plates (9×13 in. each) for the biotinylation and MESNA stripping steps (on ice) and in the 37°C water bath. Despite the scatter in the results, computer modeling of the data produces useful information.

5. The choice of antibody to use for isolating and detecting MUC1 should be determined experimentally for each cell type as the majority of anti-MUC1 antibodies are directed against the tandem repeat domain and are generally sensitive to cell-specific glycosylation of the tandem repeats (15).
6. For 2D separation, several repetitions of the affinity isolation step are necessary to obtain about 30–40 μg protein. Before 2D separation, use chloroform/methanol precipitation to get rid of SDS. Proteins are reduced and alkylated before application.
7. Peptide Mass Fingerprint analysis can only be performed on protein samples of low complexity, since the identification is based merely on the pattern of molecular masses of tryptic peptides. Hence, a prior 2D separation is necessary to obtain protein spots that might contain one or two protein species.
8. After 1D separation, about 10–12 fractions of the gel are cut from top to bottom. Each fraction may contain more than ten protein species. Accordingly, reliable protein identification is achievable only by a technical setup, where identification of proteins is based on molecular masses and MS/MS data for single peptides.

Acknowledgments

Part of the work was funded by the Deutsche Forschungsgemeinschaft Grant HA 2092/15-1 (to FGH), National Institutes of Health Grant DK054787 (to RPH), and Genzyme Renal Innovations Program (to RPH).

References

1. Singh PK and Hollingsworth MA (2006) Cell surface-associated mucins in signal transduction. *Trends Cell Biol* **16**, 467–476.
2. Hattrup CL and Gendler SJ (2008) Structure and function of the cell surface (tethered) mucins. *Annu Rev Physiol* **70**, 431–457.
3. Kufe DW (2009) Mucins in cancer: function, prognosis and therapy. *Nat Rev Cancer* **9**, 874–885.
4. Mattila P, Kinlough CL, Bruns JR, Weisz OA and Hughey RP (2009) MUC1 traverses apical recycling endosomes along the biosynthetic pathway. *Biol Chem* **390**, 551–556.
5. Kinlough CL, Poland PA, Gendler SJ, Mattila PE, Mo D, Weisz OA and Hughey RP (2011) Core-glycosylated mucin-like repeats from MUC1 are an apical targeting signal. *J Biol Chem* **286**, 3972–3981.
6. Altschuler Y, Kinlough CL, Poland PA, Bruns JB, Apodaca G, Weisz OA and Hughey RP (2000) Clathrin mediated endocytosis of MUC1 is modulated by its glycosylation state. *Mol Biol Cell* **11**, 819–831.
7. Kinlough CL, Poland PA, Bruns JB, Harkleroad KL and Hughey RP (2004) MUC1 membrane trafficking is modulated by multiple interactions. *J Biol Chem* **279**, 53071–53077.
8. Kinlough CL, McMahan RJ, Poland PA, Bruns JB, Harkleroad KL, Stremple RJ, Kashlan OB, Weixel KM, Weisz OA and Hughey RP (2006) MUC1 recycling is dependent on its palmitoylation. *J Biol Chem* **281**, 12112–12122.
9. Macao B, Johansson DG, Hansson GC and Hard T (2006) Autoproteolysis coupled to protein folding in the SEA domain of the membrane-bound MUC1 mucin. *Nat Struct Mol Biol* **13**, 71–76.

10. Levitin F, Stern O, Weiss M, Gil-Henn C, Ziv R, Prokocimer Z, Smorodinsky NI, Rubinstein DB and Wreschner DH (2005) The MUC1 SEA module is a self-cleaving domain. *J Biol Chem* **280**, 33374–33386.
11. Lillehoj EP, Han F and Kim KC (2003) Mutagenesis of a Gly-Ser cleavage site in MUC1 inhibits ectodomain shedding. *Biochem Biophys Res Commun* **307**, 743–749.
12. Thathiah A, Blobel CP and Carson DD (2003) Tumor necrosis factor-alpha converting enzyme/ADAM 17 mediates MUC1 shedding. *J Biol Chem* **278**, 3386–3394.
13. Cho JA, Yeo DJ, Son HY, Kim HW, Jung DS, Ko JK, Koh JS, Kim YN and Kim CW (2005) Exosomes: a new delivery system for tumor antigens in cancer immunotherapy. *Int J Cancer* **114**, 613–622.
14. Staubach S, Razawi H and Hanisch FG (2009) Proteomics of MUC1-containing lipid rafts from plasma membranes and exosomes of human breast carcinoma cells MCF-7. *Proteomics* **9**, 2820–2835.
15. Rye PD, Price MR (1998) ISOBM TD-4 International workshop on monoclonal antibodies against MUC1. *Tumor Biology* **19**, 1–152.
16. Albert TK, Laubinger W, Muller S, Hanisch FG, Kalinski T, Meyer F and Hoffmann W (2010) Human intestinal TFF3 forms disulfide-linked heteromers with the mucus-associated FCGBP protein and is released by hydrogen sulfide. *J Proteome Res* **9**, 3108–3117.

Chapter 8

Glycomic Work-Flow for Analysis of Mucin *O*-Linked Oligosaccharides

Catherine A. Hayes, Szilard Nemes, Samah Issa, Chunsheng Jin, and Niclas G. Karlsson

Abstract

The high-throughput analysis of the glycosylation of high molecular weight proteins, such as mucins, has been the aim of glycomics initiatives for the last decade. Here, we present a work-flow for the efficient and reproducible analysis of reduced oligosaccharides from a typical mucin sample. This work-flow can be applied to any similar samples of oligosaccharides. We include recently developed bioinformatic procedures for the statistical analysis of sample sets. These procedures can be applied in any laboratory environment, using free programs that are platform independent. The scripts are explained and can be adjusted to suit the individual experiment. Finally, a number of example results are given to highlight the use of the statistical analysis in a biological context.

Key words: Glycomics, O-linked oligosaccharides, AG-PAGE composite gel electrophoresis, Electroblotting, β-Elimination, Graphitized carbon LC–MS, Statistical clustering, Heat plots, Probability models, ROC curves.

1. Introduction

Mucin glycosylation has been shown to be of increasing importance in a number of different biological roles, such as structure, molecular recognition, protection of epithelial cells, and adhesion sites for microorganisms.

Mucins are large glycoproteins that are made up of PTS domains, consisting of tandem repeated peptide sequences containing proline (P), threonine (T), and serine (S) residues. These domains can be heavily *O*-glycosylated and increase the mass of the glycoprotein significantly. There have been 20 identified human mucins to date (UniProtKB database, http://www.uniprot.org/), and these are located in various mucosal epithelial tissues, such as the intestine, the respiratory tract, the eye, and the kidney.

Michael A. McGuckin and David J. Thornton (eds.), *Mucins: Methods and Protocols*, Methods in Molecular Biology, vol. 842,
DOI 10.1007/978-1-61779-513-8_8,

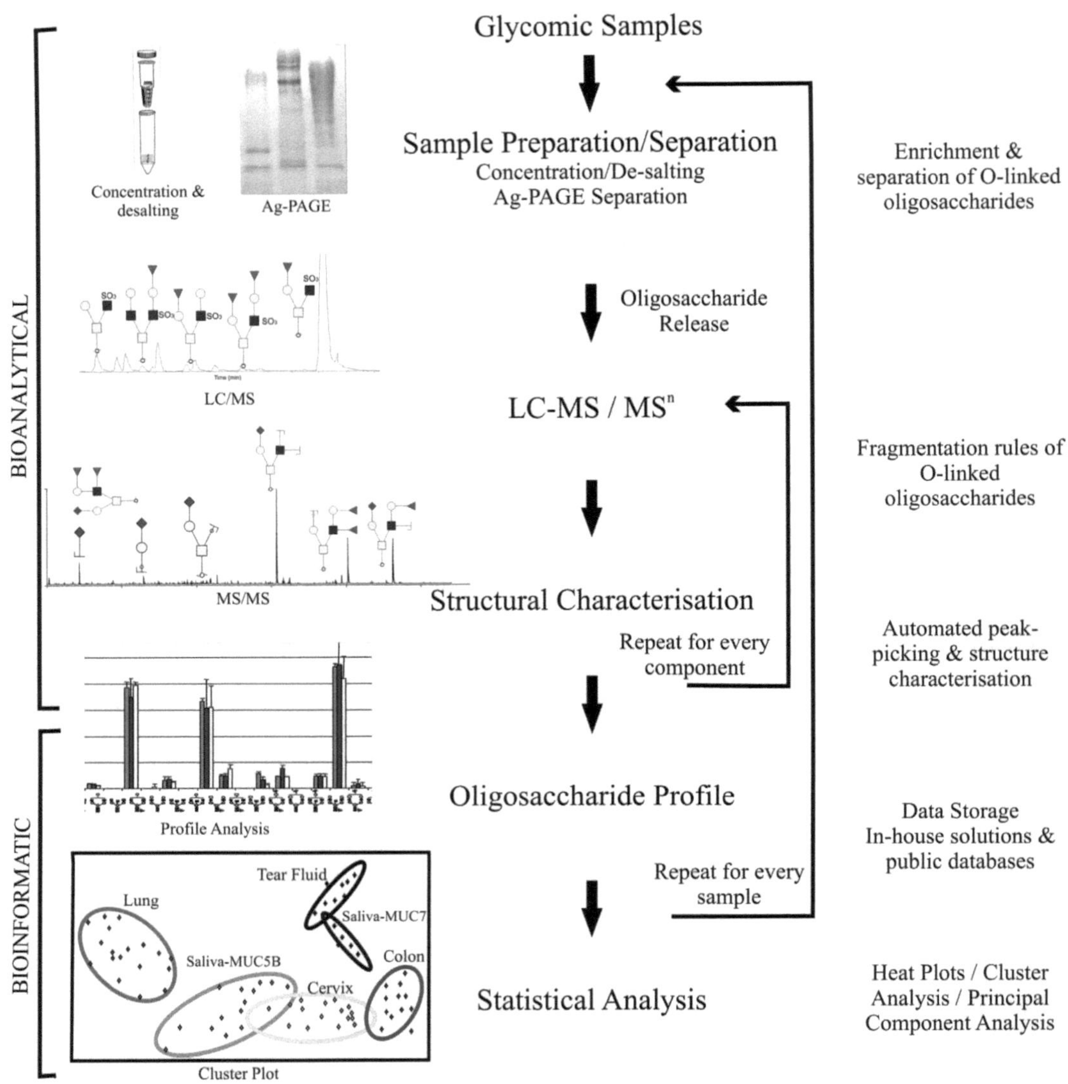

Fig. 1. Overview of the high-throughput work-flow for glycomic analysis of mucin samples.

Altered mucin glycosylation has been implicated in cancer (1), the inflammatory response (2, 3) as well as mucosal infection (4). This has important ramifications in many disease states, such as cystic fibrosis (5), ulcerative colitis (6), and rheumatoid arthritis (7), since altered glycosylation will influence the immune system with, for instance, altered interactions with the selectin-, siglec-, and the galectin families (8–10).

Mucins can roughly be divided into three main types, secreted gel-forming, secreted non-gel-forming and cell surface mucins (11). Because of their large size, it is not possible to separate these proteins on normal polyacrylamide gels. A composite gradient gel composed of a mixture of agarose and polyacrylamide has found to be the most efficient at this type of separation.

To gain a greater understanding of the role of mucins in the inflammatory response, it is necessary to query data sets rather than individual samples. Through our experience with mucins from diverse tissue types, we have optimized an analytical platform that can deal with high-throughput of sample sets of mucosal origin (the work-flow is depicted in Fig. 1). The work-flow presented here can be applied to any set of mucosal samples and will lead the researcher through experimental lab work, followed by interpretation, and finally statistical analysis of the data set.

2. Materials

All the reagents were of highest analytical grade and 18 MΩ/cm water is from a Milli-Q system (Millipore, Billerica, MA).

2.1. Sample Preparation

1. 2 M Tris–HCl, pH 7.8.
2. 20% (w/v) SDS.
3. 2× 1D AgPAGE sample buffer: 60% (v/v) glycerol, 0.75 M Tris–HCl, 0.01% (w/v), bromophenol blue, 2% (w/v) SDS.
4. 1 M dithiothreitol (DTT).
5. 1 M iodoacetamide (IAA).
6. Amicon ultra 0.5 mL 100 K spin filters (Millipore).

2.2. Ag-PAGE Composite Gels

2.2.1. Single Gel Casting

1. 5× Tris–HCl (1.875 mM, pH 8.1).
2. Low electro-osmotic grade agarose (<0.13 –mr).
3. 40% (w/v) acrylamide/bisacrylamide (2.6% C) (see Note 1).
4. 40% (w/v) ammonium persulphate (APS).
5. *N,N,N′,N′*-tetramethylethylenediamine (TEMED).
6. 50% (v/v) glycerol.
7. Gradient mixing chamber (15 mL) (Sigma-Aldrich Part no. Z34,039-1 or equivalent).
8. Peristaltic pump P-1 (Pharmacia Biotech, Piscataway, NJ or equivalent) with adjustable flow rate 0.6–500 mL/h.
9. Magnetic stirrer and magnetic bar for gradient mixer.
10. Casting equipment (Bio-Rad Miniprotean or equivalent) to include 1.0-mm spacer plates, short plates, casting frame, gel cassette assembly, casting stand, and 1-mm combs.
11. Oven.
12. Lower gel solution (for Miniprotean 1 mm spacer, for other formats increase or decrease volumes accordingly): agarose (0.08 g) is mixed with 5× Tris–HCl buffer, pH 8.1 (1.6 mL) and water (3.6 mL) (the lower solution is added to the lower chamber of the gradient mixer, or the chamber which feeds the pump first if the gel solution is applied from the top of the gel).

13. Upper gel solution (for Miniprotean 1 mm spacer, for other formats increase or decrease volumes accordingly): agarose (0.04 g) is mixed with 5× Tris–HCl buffer, pH 8.1 (1.6 mL) and water (6.4 mL) (the upper solution is added to the upper solution of the gradient mixer, or the chamber which feeds the pump last if the gel solution is added from the top of the gel).

2.2.2. Multigel Casting

1. Low electro-osmotic grade agarose (<0.13 –mr).
2. Urea (Electrophoresis grade).
3. 5× Tris–HCl (1.875 mM, pH 8.1).
4. 40% (w/v) acrylamide/bisacrylamide (2.6% C) (see Note 1).
5. 50% (v/v) glycerol.
6. *N*,*N*,*N*′,*N*′-tetramethylethylenediamine.
7. Multicasting equipment (Bio-Rad Mini-Protean 3 or equivalent) to include casting tank with lid, sealing gasket and tubing, front and back glass plates with 1.0-mm spacer plates, 1-mm combs and separation sheets.
8. Gradient mixing chamber (100 mL) (485 Gradient Former 165–4120, Bio-Rad or equivalent).
9. Peristaltic pump P-1 with adjustable flow rate 0.6–500 mL/h.
10. Plastic wrap.

2.2.3. Gel Running

1. Precast SDS-AgPAGE gel.
2. Gel electrophoresis equipment (Bio-Rad Miniprotean or equivalent) to include electrode assembly, clamping frame, mini-tank, and lid.
3. Power pack (Bio-Rad PowerPac Basic, 300 V, 400 mA, 75 W or equivalent).
4. 3× gel running buffer: 576 mM Boric acid, 3 mM EDTA, 0.3% (w/v) SDS, adjusted to pH 7.6 with Tris base.

2.3. Semidry Electroblotting of Mucins

1. Cathode buffer stock: 250 mM Tris–HCl and 400 mM α-amino-*n*-hexanoic acid in water.
2. Cathode buffer solution: 50 mL cathode buffer stock, 0.5 mL 10% (w/v) SDS made up to 500 mL in water.
3. Anode buffer stock: 250 mM Tris–HCl.
4. Anode buffer solution: 20 mL anode buffer stock, 40 mL methanol made up to 200 mL with water.
5. Ion trap stock: 3 M Tris made up to 500 mL in water.
6. Ion trap solution: 10 mL of ion trap stock made up to 100 mL in water.
7. Thick blotting paper cut to slightly larger than the gel (Bio-Rad).

8. Polyvinylidene fluoride (PVDF) membrane (Immobilion–P transfer membrane (Millipore) or equivalent) cut larger than the filter paper.
9. Methanol.
10. Semidry electroblotting apparatus (Bio-Rad Trans-blot SD semidry transfer cell).
11. Power pack (Bio-Rad PowerPac Basic, 300 V, 400 mA, 75 W).

2.4. Staining of Glycosylated Proteins on PVDF Membrane

1. Methanol.
2. Alcian blue solution: 25% (v/v) ethanol, 10% (v/v) acetic acid, 0.125% (w/v) alcian blue.
3. Destaining solution: 100% methanol.

2.5. Oligosaccharide Release

1. β-Elimination solution: 50 mM NaOH, 0.5 M $NaBH_4$.
2. Desalting solution: 1% glacial acetic acid in methanol.
3. AG 50W-X8 Cation Exchange Resin (Bio-Rad).
4. 1 M HCl.
5. P10 C18 Zip Tips (Millipore).

2.6. Column Packing

1. Fused silica capillary 250 μm internal diameter (i.d.), 370 μm outer diameter (o.d.) (Polymicro technologies, Phoenix, AZ).
2. 1/16 in. union (ZRU1.5, 0.25 mm bore, Vici AG, Schenkon, Switzerland).
3. Steel screen (0.5 μm or 1 μm pores, Vici AG).
4. PEEK tubing 20–25 mm, 400 μm i.d. (Upchurch Scientific, Oak Harbor, WA).
5. Hypercarb (graphitized carbon particles) 5 μm Particle size (Thermo Scientific, Waltham, MA).
6. Vespel ferrule (370 μm i.d., Vici AG).
7. Tetrahydrofuran (THF).
8. Methanol.
9. Ultrasonic bath.
10. Small magnetic bar (7 mm × 2 mm).
11. 2 mL borosilicate vial, 12 mm × 32 mm (Part no. VL-002, Western Fluid Engineering, Wildomer, CA or equivalent).
12. Compressed air.
13. Binary HPLC pump (Hewlett Packar 100 series or equivalent).
14. T-split (Valco-T, Vici AG).
15. Acetonitrile.
16. Water.
17. Nanobaume (Western Fluid Engineering) column packer.

2.7. Tuning of Mass Spectrometer

1. Maltopentaose solution (10 μg/mL) in 10 mM ammonium bicarbonate (40% (v/v) acetonitrile: 60% (v/v) water).
2. Ion trap mass spectrometer with standard electrospray interface (see Subheading 2.9).
3. 250-μL syringe.
4. Syringe pump.
5. PEEK tubing 20–25 mm, 400 μm i.d. (Upchurch Scientific).

2.8. Preparation of O-Linked Oligosaccharide Standards

1. Mucin from porcine stomach, Type II (PGM) (Sigma Aldrich, M1778).
2. β-Elimination solution: 50 mM NaOH, 1.0 M $NaBH_4$.
3. Desalting solution: 1% (v/v) glacial acetic acid in methanol.
4. AG 50 W-X8 Cation Exchange Resin (Bio-Rad).
5. 1 M HCl.

2.9. Graphitized Carbon LC–MS of Oligosaccharides

1. ESI–LC–MS system: Thermo LCQ Deca-XP Ion Trap mass spectrometer, Agilent XCT or equivalent, equipped with a standard electrospray module.
2. Auto-sampler (HTC-PAL, CTC Analytics, Zwingen, Switzerland or equivalent) with a 2-μL sample loop.
3. Graphitized carbon column: packed in-house (see Subheading 2.6) or alternatively a commercially available graphitized carbon column (e.g. Thermo Scientific 100 mm × 3.0 mm 5-μm Hypercarb column).
4. Mobile Phase A: 10 mM ammonium bicarbonate.
5. Mobile Phase B: 10 mM ammonium bicarbonate in 80% (v/v) acetonitrile.
6. Suitable control O-linked oligosaccharide standards: Porcine gastric mucin oligosaccharides.

2.10. LC–MS Data Interpretation

1. Peak picking: Automatic peak picking software: DeCyder MS Differential Analysis Software (GE Healthcare), ACD/MS Manager (Advanced Chemistry Development), or any of the free peak-picking softwares that are available (http://www.ms-utils.org/wiki/pmwiki.php/Main/SoftwareList peak picking/deconvolution).
2. Component identification: GlycoMod online tool (http://www.expasy.ch/tools/glycomod/).
3. Structure verification 1: GlycoWorkBench, downloadable java tool (http://www.glycoworkbench.org).
4. Structure verification 2: Database MS^2 matching and uploading of data, UniCarb-DB (http://www.unicarb-db.org/).

2.11. Statistical Analysis of Data

1. R statistical language (http://www.r-project.org/).
2. R script editor, for example WinEDT (http://www.winedt.com) or TinnR (http://www.sciviews.org/Tinn-R/).

3. Methods

3.1. Sample Preparation

1. Make up samples in 2 M Tris–HCl to give a final concentration of 0.32 M.
2. Add 20% SDS to a final concentration of 1% SDS.
3. Add 1 M DTT to a final concentration of 10 mM DTT.
4. Incubate at 90°C for approximately 20 min.
5. Add 1 M IAA to a final concentration of 25 mM IAA and incubate in the dark at room temperature for 20 min.
6. Add sample to filter of spin filter (maximum of 500 μL) and place in Eppendorf tube.
7. Spin samples at 14,000 × *g* for 20–60 min. This step concentrates the sample and removes salts.
8. Remove filter and place upside down in a clean, labelled Eppendorf tube. Spin at 1,000 × *g* for 2 min to collect sample.
9. Measure the volume of the remaining sample and make up samples in Ag-PAGE sample buffer, using water to bring volume to 20 μL (for loading on gel).
10. Heat for 5 min at 90°C prior to loading on gel.

3.2. Ag-PAGE Composite Gels

Gradient Ag-PAGE gels can be singly cast (one to two gels) or multicast (up to ten gels depending on equipment). The protocols below may need to be optimized in individual laboratories. Examples of mucins separated using these gels are shown in Fig. 2.

3.2.1. Single Gel Casting

1. Mount the gel casting equipment (glass plates, casting frame, and gel cassette assembly) and place in 60°C oven together with the gradient mixer and the pump. Connect the pump to the gradient mixer with the rubber tubing supplied. Place the gradient mixer on the magnetic stirrer plate placing the magnetic stirrer bar in the lower chamber (see Note 2).
2. Prepare the lower and upper gel solutions and heat in a microwave until the agarose has melted (see Note 3).
3. Transfer 5 mL of the hot upper solution to the upper chamber of the gradient mixer (see Note 4).
4. Immediately add 2 μL of 40% APS and 2 μL of TEMED (see Note 5).
5. Add 1.2 mL of 40% acrylamide to the lower solution and mix thoroughly.

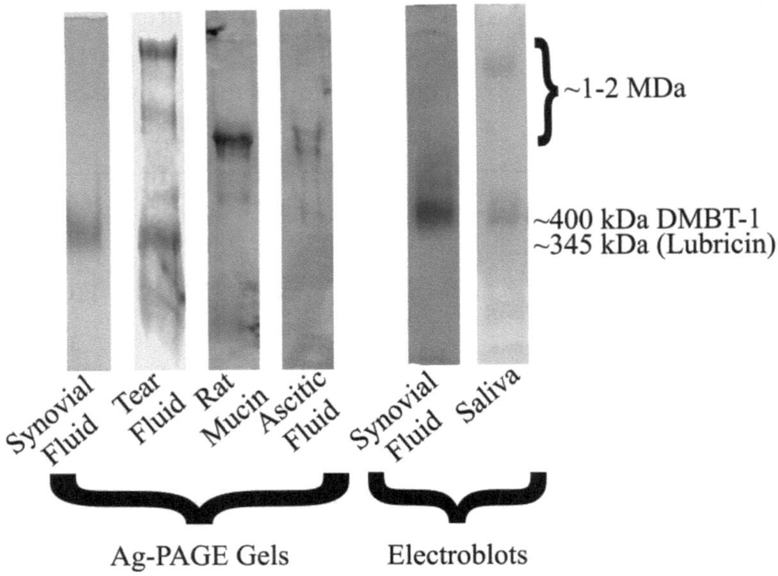

Fig. 2. Ag-PAGE composite gel and electro-blots of typical mucin samples. The gels were stained with Coomassie blue (protein staining) and the electro-blots with alcian blue (sugar staining).

6. Add 5 mL of the lower solution to the lower chamber of the gradient mixer.
7. Immediately add 2 μL of 40% APS and 2 μL of TEMED.
8. Switch on pump. Open the two valves of the gradient mixer and load the gradient from the top of the gel.
9. Add 1 mm comb (or a layer of isobutanol for preparative gels) and leave to polymerize at 50°C for 1–2 h (see Note 6). Gels can be stored at +4°C wrapped in a plastic film with wet paper for at least a week or wrapped in a plastic bag with gel buffer and kept in fridge for months (see Note 7).

3.2.2. Multigel Casting

1. Assemble the gel casting equipment (chamber, glass plates, separation plates).
2. Attach the casting lid tightly and attach the plastic tubing.
3. Check that the casting box is sealed by attaching the pump, pumping through water and checking for leaks.
4. Prepare gel solutions. The following is suitable for casting 13 gels; amounts can be varied as per necessary:

	Lower gel solution	Upper gel solution
Agarose (g)	0.55	0.65
5× Tris–HCl, pH 8.1 (mL)	11	13
50% Glycerol (mL)	11	0
Water	0	13

5. Repeatedly boil and stir solution until the agarose is completely dissolved.
6. Add the following components in order, to the same tubes as above:

	Lower gel solution	Upper gel solution
Urea (g)	13.2	15.6
40% Acrylamide (mL)	9.9	0
TEMED (μL)	37	43
Water	~10 mL (to final volume of 55 mL)	~30 mL (to final volume of 55 mL)

7. Make up a fresh solution of 40% APS.
8. Ensure that the gradient chambers are separated (internal tube is closed), the pump is set to maximum, the gradient pouring unit is raised and that the casting box is level.
9. Add 43 μL 40% APS to upper gel solution and invert several times to thoroughly dispense.
10. Transfer 60 mL of upper gel solution to gradient pourer unit (into the outlet chamber) and adjust the magnetic stirrer so that the solution is mixed well.
11. Add 37 μL 40% APS to lower gel solution and invert several times to dispense thoroughly.
12. Transfer 50 mL lower gel solution to gradient pourer unit (the non-outlet chamber).
13. Turn pump on to maximum speed.
14. Open internal door between gradient pourer unit chambers.
15. Pump until solution emerges from top of all gels.
16. Turn pump off.
17. Add combs beginning from front to back.
18. Clamp plastic tubing with two overlapping alligator clips.
19. Disconnect plastic tubing from pump.
20. Cover the casting box with plastic cling wrap.
21. Leave the casting box at room temperature for at least 1 h to polymerize acrylamide.
22. Carefully transfer the casting box to the cold room, taking care not to shake or disturb the solutions excessively. Leave at 4°C overnight.
23. To dissemble the casting box, place on bench on its back, unscrew and gently remove lid. Insert a scalpel blade and gently twist to separate the uppermost gel from the underlying glass.

24. Scrape off excess gel material and remove the plastic separator sheets.
25. Gels can be stored in 1× gel buffer in sealed bags at 4°C for up to 1 month.

3.2.3. Gel Running

1. Assemble gel running apparatus.
2. Add 1× composite gel running buffer to the inside chamber and halfway up in outside chamber.
3. Apply 20–40 μL of sample to each well and top up with running buffer.
4. Run at room temperature, and constant mA (10–15 mA) for 1 h. After 1 h, mA can be increased to 25–30 and run for another 1.5 h or until dye front runs off the bottom of the gel.

3.3. Semidry Electroblotting

1. Soak two filter papers in cathode buffer solution, two in anode buffer solution, and one in ion trap solution for 10 min.
2. Soak PVDF in methanol for 2 min, followed by 10 min in anode buffer solution.
3. Soak the gel in water for 2 min, followed by cathode buffer solution for a further 5 min.
4. Place the blot paper soaked in ion trap solution on a flat surface. Gently roll with a rolling tool to remove air bubbles.
5. Place the two blot papers soaked in anode buffer solution on top of the rolled blot paper, and roll to remove air.
6. Place the blot membrane on the stack (see Note 8).
7. Place the gel on the stack and ensure no air bubbles are trapped between the gel and the blot membrane.
8. Place the two blot papers soaked in cathode buffer solution on top of the stack and gently roll to remove air.
9. Transfer the stack to the electroblotting cassette and cover (see Note 9).
10. Run the gels at approximately 3–5 mA/cm^3 (of gel) for 45–60 min.
11. Remove blot, let air-dry, and store between two sheets of blotting paper. Alternatively continue staining with Alcian Blue, Fig. 3.

3.4. Staining of Glycosylated Proteins on PVDF Membrane

1. Wet the electroblotted membrane with methanol.
2. Add the blot to 20 mL of alcian blue solution, and shake for 7 min (see Note 10).
3. Pour off solution, and add 20 mL of destaining solution. Shake for 2 min. Repeat five times.
4. Air-dry the blot.
5. The blot can be imaged in visible reflected light (see Note 11).

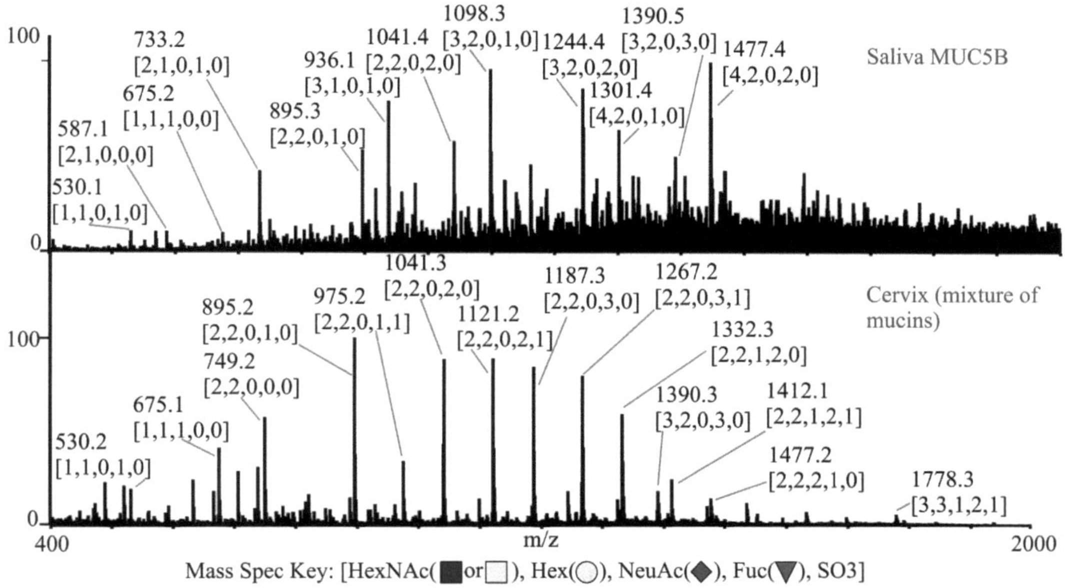

Fig. 3. Typical mass spectrum of released oligosaccharides of human mucin samples from cervix and saliva.

3.5. Oligosaccharide Release

1. Cut out relevant protein bands from alcian blue-stained PVDF membrane, cut into small pieces and place each into well of a microtitre plate.
2. Wet each band with ~20 μL methanol and remove excess methanol.
3. Add 20 μL β-elimination solution to each well (see Notes 12 and 13).
4. Cover tightly and incubate overnight (16 h) at 50°C.
5. Remove from oven, and add 1 μL glacial acetic acid to neutralize the reaction. Solutions should immediately fizz due to the production of H_2 gas.
6. Prepare cation exchange columns:
   (a) Wash AG 50W-X8 resin five times in equal volume of methanol, then five times in equal volume of water. Store at room temperature in an equal volume of methanol.
   (b) Pack columns by adding 25 μL of cation exchange suspension to Zip Tips (one per sample).
   (c) Place columns in 1.5-mL Eppendorf tubes and centrifuge for 10 s.
   (d) Wash columns with 2× 60 μL methanol followed by 2× 60 μL 1.0 M HCl, centrifuging for 10 s between each wash.
   (e) Remove the columns, discard the flow-through, and return to Eppendorf tubes.

(f) Wash columns with 2× 60 μL of water, centrifuging for 10 s after each wash.

(g) Remove the columns, discard the flow-through and place columns in new labelled 1.5-mL Eppendorf tubes.

7. Pipette solutions from the neutralized samples to the corresponding columns. Centrifuge for 10 s.
8. Wash wells with 2× 25 μL of water, load onto columns and centrifuge for 10 s after each wash.
9. Elute residual glycans from the columns by washing each with 30 μL of water. Centrifuge for 10 s.
10. Remove cation exchange columns from Eppendorf tubes, and dry the eluted glycan solutions to completion in a Speedivac at 45°C for approximately 60 min.
11. Remove white borate residues by addition of 50 μL desalting solution and drying to completion in a Speedivac at 45°C for approximately 8 min. Repeat five times (or until no white residue is visible).
12. Samples can be stored dry for at least 1 week at 4°C, and resuspended in water immediately prior to further analysis.

3.6. Column Packing

1. Connect a fused silica capillary of 15–18 cm length (250 μm i.d., 370 μm o.d.) to the 1/16 in. union containing the steel screen using 20–25 mm of PEEK tubing as a sleeve. Connect with a vespel ferrule a piece of fused silica to the opposite end of the union.
2. Prepare a slurry mixture of 15–20 mg of graphitized carbon in 1.2 mL methanol and sonicate in an ultrasonic bath for 5 min. Place the small magnet (7× 2 mm) in the slurry container for stirring during packing.
3. Place the slurry container in the Nanobaume pressure unit. Prepare the unit according to the manufactures' protocol. Once the pressure unit is set start the packing at approximately 50 bar and after 3–5 min increase to 120 bar. The packing takes about 40–45 min and the packing can be seen as the graphite particles are clearly visible through the fused silica column.
4. Turn off the pressure and allow the column to depressurize.
5. Connect to an HPLC pump via a T-split with a 50 cm × 50 μm fused silica capillary as splitter and wash with 80% (v/v) acetonitrile before connecting to the LC–MS. The LC is set to deliver a flow rate of 250 μL/min and the flow split in the T using a 50 cm × 50 μm of fused silica capillary as restrictor gives a flow through the 100 mm × 250 μm i.d. column packed with 5 μm graphite particle of about 5–10 μL/min.

3.7. Tuning of Mass Spectrometer

1. Fill the syringe with the maltopentaose standard.
2. Connect the syringe to the PEEK tubing and insert the syringe in the pump.
3. Pump at 4–5 μL/min and perform the tuning on the $(M–H)^-$-ion of *m/z* 827 in negative ion mode. Usually, the auto-tuning function is sufficient but the temperature and gas flow may have to be optimized manually.

3.8. Preparation of O-Linked Oligosaccharide Standards

1. Prepare a stock of PGM oligosaccharides by performing bulk β-elimination by dissolving 1 mg of mucin in 0.5 mL of β-elimination solution. Incubate at 50°C overnight.
2. Desalt sample in two times mEq of AG 50W-X8 (e.g. use 200 mg of AG 50W-X8 for 0.5 mL of β-elimination solution) (see Note 14). Stir or shake for 1 h.
3. Filter or decant the sample from the resin, washing two times with water to collect all the released oligosaccharides.
4. Lyophilize and desalt with five times washing with desalting solution.
5. Reconstitute samples in water to a concentration of 1 mg/mL and store at −20°C. Dilute 100 times before injection into mass spectrometer.

3.9. Graphitized Carbon LC–MS of Oligosaccharides

The following outlines the details and conditions for oligosaccharide LC–MS analysis using a Thermo LCQ Deca-XP Ion Trap or other suitable mass spectrometer. Note that ion signal optimization may be required.

1. Column details and preparation:
   (a) Column: Hypercarb—in-house preparation.
   (b) Mobile phases: A: 10 mM ammonium bicarbonate, B: 10 mM ammonium bicarbonate in 80% acetonitrile (v/v) (ACN).
   (c) Column equilibration: With flow rate at 0.250 mL/min with splitting giving an on-column rate of 5–10 μL/min or with split-less system giving 7 μL/min.
   (d) Start with 100% A for 30 min, then 100% B for 10 min and 100% A for 24 min.
2. LC Conditions for separation of O-linked oligosaccharides:
   (a) Gradient: 0–7 min gradient increase from 0 to 5% B.
   (b) 7–46 min gradient increase from 5 to 35% B.
   (c) 46–46.1 min increase from 35 to 80% B.
   (d) 46–54 steady at 80% B.
   (e) 54–54.1 min decrease from 80 to 0% B.

(f) Stop time: 78 min.

(g) Flow rate: 250 μL/min with splitting giving an on-column rate of 5–10 μL/min or with split-less system giving 7 μL/min.

(h) Sample injection volume: 3 μL.

3. Typical MS conditions for oligosaccharide analysis:

(a) Ionization mode: Negative electrospray.

(b) Source voltage (kV): 3.50.

(c) Source current (μA): 11.80.

(d) Capillary voltage (V): –7.97.

(e) Capillary temperature (C): 278.02.

3.10. LC–MS Data Intrepretation

There are a number of different software packages available to interpret LC–MS data. However, the majority of these have been designed for the analysis of peptides. They are still quite useful for the interpretation of sugar data but manual interpretation and verification will still be necessary. Figure 3 shows a typical annotated mass spectrum of oligosaccharides from two different tissue types.

1. Peak picking:

(a) Upload the raw data file from the mass spectrometer. From Thermo instruments these are in the form of .RAW files, whereas Agilent raw files are .YEP files. Both files can be converted to generic mzXML files if appropriate (see Note 15).

(b) Output is a list of "components" as identified by the peak picking software.

2. Component Identification:

(a) Using the peak list as output by the peak picking software, go to the GlyoMod Web site, and enter the list of masses (or alternatively upload file containing one mass per line).

(b) Fill in the relevant options as per sample information as follows (change parameters as necessary)

- Ion Mode and Adducts: $(M–H)^-$
- *O*-linked oligosaccharides/reduced oligosaccharides
- Monosaccharide residues: underivatized
- Ranges of residues (or possible ranges if unsure)

(c) Start GlycoMod. Output is a list of potential components that correspond to the input masses with links to GlycoSuiteDB if available.

3. Structure Verification 1:

(a) Extract the relevant MS^2 data for each of the identified components (*m*/*z*, intensity).

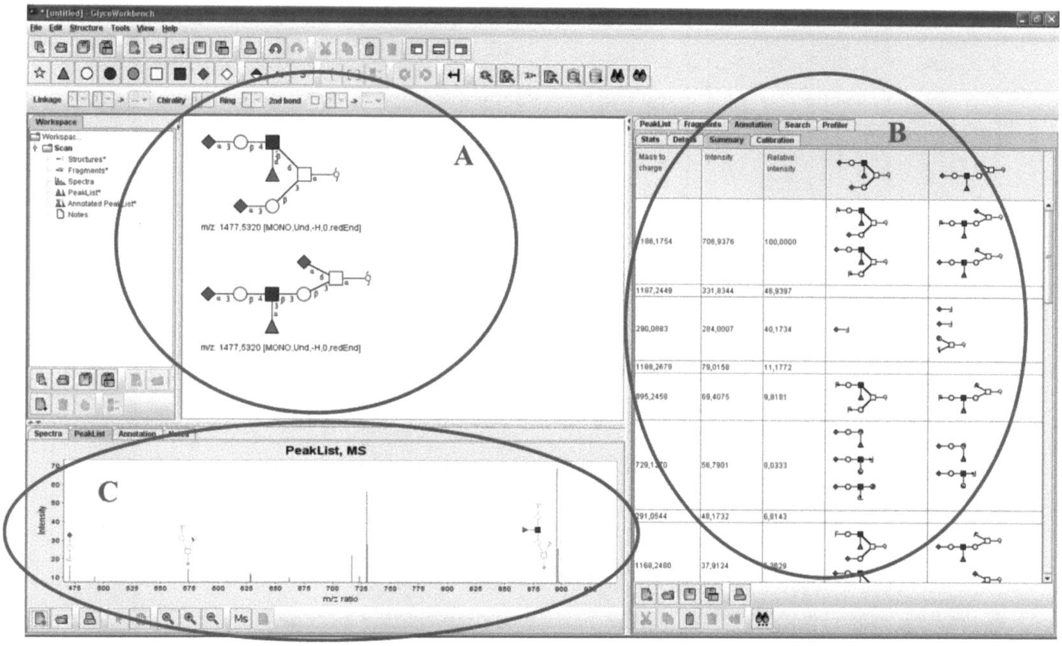

Fig. 4. GlycoWorkBench (http://www.glycoworkbench.org) is used to aid the verification of structures that have been identified using peak picking and peak matching. (**a**) The drawing canvas is used to depict the putative oligosaccharide structure(s) using the consortium for functional glycomics (CFG) notation. The residues are listed at the top on the toolbar. (**b**) Each structure can then be theoretically fragmented. The experimental MS^2 peak list is imported (*m*/*z* values with absolute and/or relative intensities) which can be annotated with the theoretical fragments and the results statistically presented as a table. (**c**) The imported peak list is presented as a spectrum. Manual annotation has been added to aid the readers interpretation of the spectrum.

(b) Using GlycoWorkBench (Fig. 4a), draw the putative structure(s) for the *m*/*z* value that is under investigation (e.g. from suggestions provided by the GlycoMod output from structures previously identified in GlycosuiteDB).

(c) Input the peak list into GlycoWorkBench.

(d) Using the FRAGMENT tool, perform a theoretical fragmentation of the putative structure(s) (Fig. 4b).

(e) Using the ANNOTATE tool, match the theoretical fragments to the MS^2 peak list (Fig. 4c). Repeat for all possible structures. Use the statistical data to identify the most likely structure.

4. Structure Verification 2:

(a) Make a list of possible structures and their corresponding MS^2 spectra.

(b) Go to UniCarb-DB Web site and search for the relevant structures (Fig. 5).

(c) The MS^2 spectra of structures solved by LC–MS are available for download or real-time viewing from the Web site.

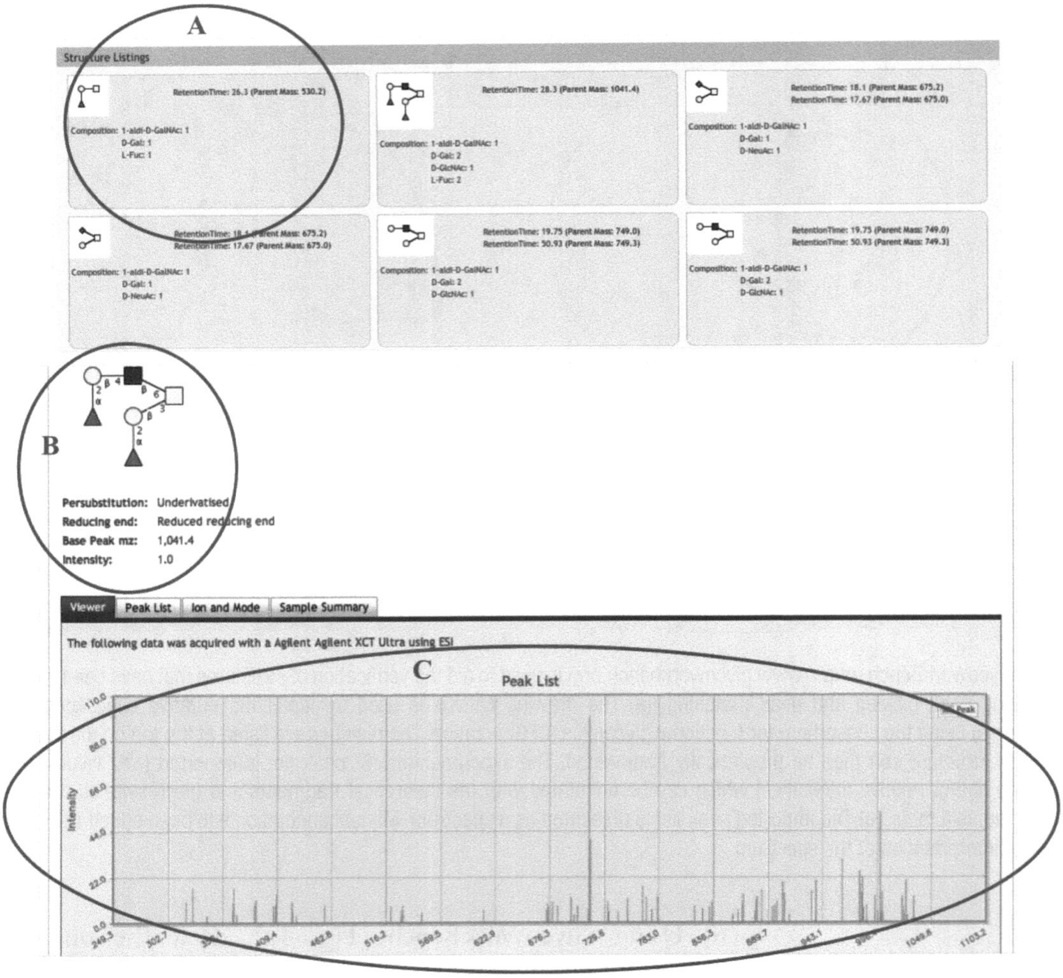

Fig. 5. The UniCarbDB (http://www.unicarb-db.org) is a repository for LC–MS derived MS^2 for oligosaccharides. It can be searched for structures of interest and the peak lists compared to the experimental for structure verification. (**a**) The home page gives a list of all structures held in the database. The structure listing includes *m/z* values, retention times, and residues. (**b**) Clicking on a specific structure brings up more detailed information, including derivatization, reducing or free end and intensity (if available). (**c**) The MS^2 peak list is also available and is presented in a simple viewer. The peak list is available for download. On a tabbed menu can be found more information about sample origin and experimental LC–MS parameters.

Verify that the database MS^2 matches that for the structure under investigation.

(d) Once verification is complete, go to contribute page to upload data for public domain access.

3.11. Statistical Analysis of Data

There are many tools available to do statistical analysis of data sets. Here, the R statistical tool set is presented, as it is freely available.

3.11.1. Conversion of Raw Data

Output from the data interpretation of a sample comprises a list of *m/z* values with corresponding intensities (absolute and/or relative).

These can be further reduced depending on the entire sample set and what the objective of the research is. If, for example it is a longitudinal study that has been undertaken, then it is sufficient to have each of the structures separately. However, if the study is a comparison between disease state and healthy individuals, then it may be more appropriate to reduce the data to monosaccharide compositions.

1. Output the structures and their appropriate relative intensities as a list to an EXCEL file. Convert the intensities to % values.
2. Reduce structures to monosaccharide compositions for the sample, i.e. hexosamine, hexose, fucose, sialic acid, and sulphate.
3. Multiply monosaccharide composition by % intensity for each structure.
4. Sum the monosaccharide compositions for each residue from each structure.
5. The result is five figures representing the total monosaccharide composition for the sample, as exemplified in Fig. 6.

Mass	*Structure*	*HexNAc*	*Hex*	*Fuc*	*NeuAc*	*Sulf*	*Intensity*
588.1	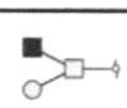	2	1	0	0	0	16.20
750.3	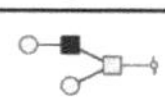	2	2	0	0	0	2.90
791.2	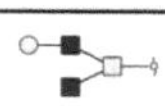	3	1	0	0	0	2.00
896.2	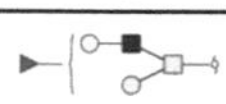	2	2	1	0	0	3.50
953.3	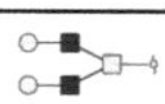	3	2	0	0	0	7.80
1099.3	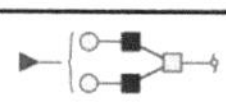	3	2	1	0	0	6.80
1245.4	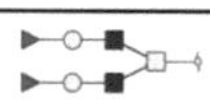	3	2	2	0	0	2.90
879.1	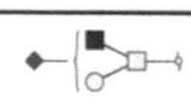	2	1	0	1	0	3.50
1187.4	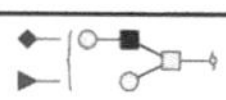	2	2	1	1	0	16.50
1390.4	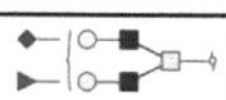	3	2	1	3	0	18.50
2646.4	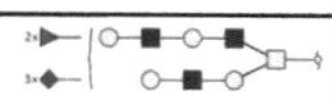	4	4	2	0	1	4.30
1179.2	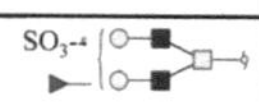	3	2	1	3	1	4.50
2433.8	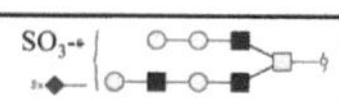	4	4	0	2	1	5.20
2799.9	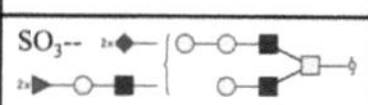	5	5	2	0	0	5.40
Results		2.77	2.13	0.75	0.78	0.15	

Fig. 6. Listed are oligosaccharide structures for one sample. The monosaccharide composition for each structure is also shown as well as the relative intensity as extracted from the raw data. The monosaccharide compositions (results at the end of the table) are arrived at by summing the product of the individual number of residues by the intensity, for each structure.

3.11.2. Input of Data

The R statistical language (http://www.r-project.org/) is a freely available resource for statistical analysis and graphical representations. Alternatively, SPSS, MATLAB, or another statistical software package may be used and apply the statistical techniques that we propose. Here, we give the code as used in R. Table 1 contains a sample data set that can be used to test the R script as given below:

1. Convert data into a simple text file. The first column should consist of sample identifiers. It is also important to assign the state of the samples in a separate column, where 0 denotes "healthy" samples and 1 denotes "disease" samples.
2. Read in data into R, as per following script, substituting your filename for yourdata.txt (see Note 16):

```
SampleData<-read.table('yourdata.txt',header=T,
   row.names=1)
SampleData(1:5,) ## Examine the first five rows
   of the data
```

3.11.3. Exploratory Analysis of Data: Heat Map

1. Heat maps require a specific library in R which is read in as follows:

```
library(gplots)
```

2. For the heat plot, it is only necessary to provide the identifier and the variables by which they will be clustered. From the inputted data set, extract the relevant columns as per the following code:

```
ReducedData<- SampleData(, -c(6, 7))
```

3. The data needs to be scaled to mean 0 and unit variance in order to facilitate graphical interpretation:

```
ScaledData<- scale(ReducedData)
```

4. The data is then converted to a matrix:

```
DataMatrix<- as.matrix(ScaledData)
```

5. Set the colour scale for the heat plot (for a useful guide on heat plot parameters, please see http://www2.warwick.ac.uk/fac/sci/moac/students/peter_cock/r/heatmap/):

```
rc<- rainbow(nrow(x), start=0, end=.3)
cc<- rainbow(ncol(x), start=0, end=.3)
```

6. Run the heat map command:

```
DataHeatMap<- heatmap.2(DataMatrix, scale=
"column", keysize=1, RowSideColors=rc,
margin=c(5, 10), col=bluered, xlab="Specification
Variables", ylab="Patients", tracecol="black",
dendrogram="both")
```

7. Save heat plot as image file, examples of output are shown in Fig. 7.

Table 1
Listed is an example data set for use with the R scripts as provided in the methods section

ID	HexNac	Hex	Fuc	NeuAc	Sulf	State	Status
Sample_01	2.13	1.92	1.94	0.57	0.42	1	State 1
Sample_02	2.05	1.95	1.4	0.72	0.33	1	State 1
Sample_03	1.92	1.81	1.05	0.86	0.18	1	State 1
Sample_04	1.59	1.45	0.88	0.98	0.14	1	State 1
Sample_05	2.95	1.97	0.59	1.23	0.28	1	State 1
Sample_06	1.87	1.84	0.82	0.66	0.35	1	State 1
Sample_07	1.99	1.93	1.5	0.45	0.31	1	State 1
Sample_08	2.11	1.97	1.23	0.54	0.27	1	State 1
Sample_09	1.85	1.84	1.32	0.29	0.31	1	State 1
Sample_10	2.02	1.98	0.66	0.7	0.27	1	State 1
Sample_11	1.97	1.95	0.82	0.48	0.32	1	State 1
Sample_12	1.73	1.65	0.77	1.16	0.11	1	State 1
Sample_13	2.01	1.84	1.2	0.89	0.35	1	State 1
Sample_14	1.75	1.62	1.19	0.77	0.25	1	State 1
Sample_15	2.65	2.28	1.56	0.17	0.82	0	State 2
Sample_16	2.36	2.08	1.55	0.22	0.56	0	State 2
Sample_17	2.52	1.86	1.29	0.08	0.51	0	State 2
Sample_18	2.86	1.96	1.19	0.18	0.74	0	State 2
Sample_19	3.11	1.81	1.32	0.11	0.04	0	State 2
Sample_20	2.82	2.07	1.27	0.06	0.81	0	State 2
Sample_21	2.25	1.98	1.48	0.25	0.1	0	State 2
Sample_22	3.16	1.97	1.33	0.18	0.39	0	State 2
Sample_23	2.39	2.59	1.69	0.29	0.11	0	State 2
Sample_24	2.57	2.43	1.37	0.27	0.68	0	State 2
Sample_25	2.46	2.41	1.15	0.3	0.43	0	State 2
Sample_26	1.68	2.17	1.13	0.36	0.42	0	State 2
Sample_27	2.3	2.08	0.72	0.37	0	0	State 2
Sample_28	2.29	1.99	0.88	1.28	0.27	0	State 2
Sample_29	2.64	2.39	1.17	0.93	0.45	0	State 2
Sample_30	2.58	2.17	0.96	0.76	0.22	0	State 2
Sample_31	2.39	2.14	0.71	0.8	0.22	0	State 2

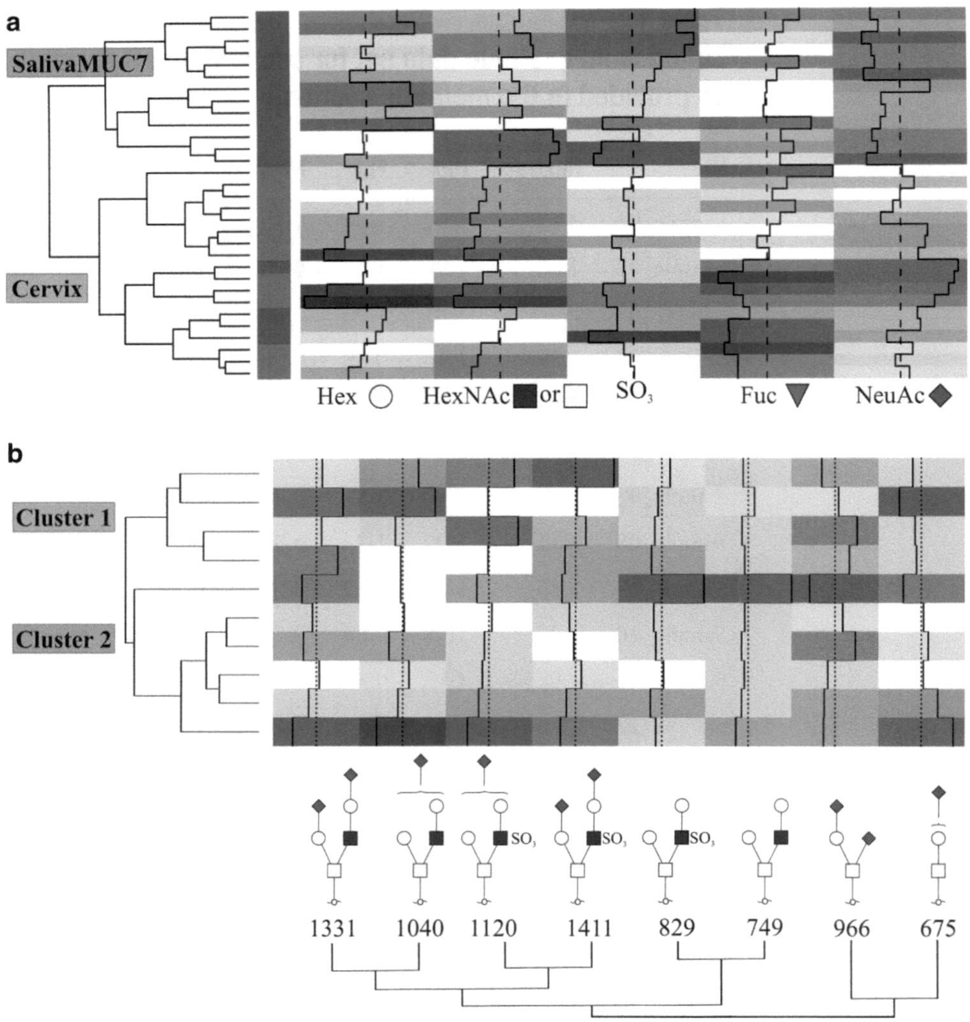

Fig. 7. (**a**) Heat plot shows a typical outcome for statistical analysis of a complete data set. The mass spec information was reduced to a mass spectrometry metric reflecting monosaccharide compositions and these were used to cluster the data. (**b**) Heat plot shows the outcome where a subset of results is used to describe the data. In this example, only certain structures were recorded. These were used as the variables to classify the samples. The variables were clustered to reveal the underlying relationships between the structures.

3.11.4. Statistical Inference

1. Use multivariate logistic regression to model the "disease" status of the samples as a function of the recorded variables (sugar residues):

```
fit <- glm(State~., family='binomial', data=
SampleData)
summary(fit, corr=T)
```

2. Use a model selection procedure to find the most relevant variable(s):

```
summary(step(fit, test="Chisq"))
```

3. Construct a prognostic model for the chosen variable(s), replacing HexNac with the variable(s) relevant to your data, adding a plus sign between two or more variables:

```
plot.new()
plot.window(ylim=c(0,1),   xlim=c(range(Sample
   Data$HexNac)))
   axis(1, lwd=2.5, cex.axis=1.5)
   axis(2, lwd=2.5,cex.axis=1.5)
   points(fit$model$HexNac, fit$fitted)
   title(ylab='Probability of Disease', cex.
   lab=1.5)
   title(xlab='HexNac', cex.lab=1.5)
```

4. Plot the estimated probability substituting your variable for HexNac:

```
fit  <-  glm(State~HexNac,  family='binomial',
   data=SampleData)
summary(fit)
```

5. Check the predictive power of the model. This gives an idea of how the model separates the samples. First create a graphical representation in the form of a box plot:

```
boxplot(fit$fit~fit$model$State,
   col = "grey",
   xlab = "Health Status",
   ylab = "Predicted Probability of being ill")
```

Also examine the mean and standard deviation for the two groups:

```
by(fit$fit, fit$model$State, mean)
by(fit$fit, fit$model$State, sd)
```

6. Summarize the predictive power of the model using the Area Under the Receiver Operating Characteristic (10) Curve (AUC). Read in the relevant library as required by R:

library(ROCR)

(a) Create the ROC curve:

```
pred   <-   prediction(fit$fitted.values,fit
$model$State)
perf <- performance(pred,"tpr","fpr")
plot(perf, main=c('AUC=', paste(round(AUC,3))))
```

(b) Estimate the AUC value and the associated 95% confidence interval:

```
performance(prediction(  fit$fitted.values,
   fit$model$ State),"auc")@y.values((1))
```

A resampling technique (bootstrapping) can be used to estimate the confidence interval. This code can be used with default values as below:

```
AUCboot <- function(formula, data, indices) {
d <- SampleData(indices,)
fit <- glm(formula, family='binomial', data=d)
return(performance(prediction( fit$fitted.values, fit$model$State),"auc")@y.values((1)))}
```

In the following code, *R* denotes the number of resampling iterations that are carried out. Adapt the formula to suit your particular analysis:

```
results <- boot(data=SampleData, statistic=AUCboot, R=1,000, formula= State~HexNac)
results
plot(results)
```

If the histogram of the bootstraped values is symmetric, then the choice of the type of confidence interval has no practical implication. If the histogram of the bootstrapped values is asymmetric, we would prefer the 'bca' (Bias corrected and accelerated) interval:

```
boot.ci(results, type=c("bca", "perc"))
```

4. Notes

1. C refers to the percentage of crosslinked bisacrylamide in relation to the total amount of acryalamide and bisacryamide.
2. All apparatus should be kept at 60°C in an oven.
3. All steps should be completed as quickly as possible from this point onward.
4. Ensure that both chambers are closed before addition of solutions.
5. APS should be freshly made.
6. Leave gels for 24 h before running for best results.
7. If using an equivalent but different system to that described here, volumes of gels may have to be adjusted accordingly.
8. Always use forceps when handling PVDF membrane to reduce contamination.

9. Place a layer of ice on the inside cover of the blotting apparatus. This ensures that the apparatus does not overheat.
10. The staining solution can be stored and reused up to five times.
11. Definition of bands on the blot can be improved by wetting with methanol before imaging.
12. If possible place bands in centre wells, and fill surrounding wells with water to prevent evaporation.
13. Ensure that β-elimination solution is prepared fresh.
14. mEq is equivalent to the amount of exchange resin that would neutralize 1 mM of cations.
15. Many software packages have the capability to accept the majority of proprietary file formats. Otherwise, it is possible to convert the raw data files into a generic file format, such as mzData or mzXML files using a suitable converter such as CompassXport (http://www.ms-utils.org/wiki/pmwiki.php/Main/SoftwareList format conversion).
16. Any line preceded with a hash and a space is treated as a comment by R and therefore not executed.

References

1. Rachagani, S., Torres, M. P., Moniaux, N., Batra, S. K. (2009) Current status of mucins in the diagnosis and therapy of cancer. *Biofactors* **35**, 509–527.
2. Guang, W., Ding, H., Czinn, S. J., Kim, K. C., Blanchard, T. G., Lillehoj, E. P. (2010) Muc1 cell surface mucin attenuates epithelial inflammation in response to a common mucosal pathogen. *J Biol Chem* **285**, 20547–20557.
3. Mejias-Luque, R., Linden, S. K., Garrido, M., Tye, H., Najdovska, M., Jenkins, B. J., Iglesias, M., Ernst, M., de Bolos, C. (2010) Inflammation modulates the expression of the intestinal mucins MUC2 and MUC4 in gastric tumors. *Oncogene* **29**, 1753–1762.
4. Kobayashi, M., Lee, H., Nakayama, J., Fukuda, M. (2009) Roles of gastric mucin-type O-glycans in the pathogenesis of Helicobacter pylori infection. *Glycobiology* **19**, 453–461.
5. Schulz, B. L., Sloane, A. J., Robinson, L. J., Prasad, S. S., Lindner, R. A., Robinson, M., Bye, P. T., Nielson, D. W., Harry, J. L., Packer, N. H., Karlsson, N. G. (2007) Glycosylation of sputum mucins is altered in cystic fibrosis patients. *Glycobiology* **17**, 698–712.
6. Faure, M., Moennoz, D., Montigon, F., Mettraux, C., Mercier, S., Schiffrin, E. J., Obled, C., Breuille, D., Boza, J. (2003) Mucin production and composition is altered in dextran sulfate sodium-induced colitis in rats. *Dig Dis Sci* **48**, 1366–1373.
7. Wada, Y., Tajiri, M., Ohshima, S. (2010) Quantitation of saccharide compositions of O-glycans by mass spectrometry of glycopeptides and its application to rheumatoid arthritis. *J Proteome Res* **9**, 1367–1373.
8. Barthel, S. R., Gavino, J. D., Descheny, L., Dimitroff, C. J. (2007) Targeting selectins and selectin ligands in inflammation and cancer. *Expert Opin Ther Targets* **11**, 1473–1491.
9. Buzas, E. I., Gyorgy, B., Pasztoi, M., Jelinek, I., Falus, A., Gabius, H. J. (2006) Carbohydrate recognition systems in autoimmunity. *Autoimmunity* **39**, 691–704.
10. Crocker, P. R., Paulson, J. C., Varki, A. (2007) Siglecs and their roles in the immune system. *Nat Rev Immunol* **7**, 255–266.
11. Lievin-Le Moal, V., Servin, A. L. (2006) The front line of enteric host defense against unwelcome intrusion of harmful microorganisms: mucins, antimicrobial peptides, and microbiota. *Clin Microbiol Rev* **19**, 315–337.

Chapter 9

O-Glycomics: Profiling and Structural Analysis of Mucin-type O-linked Glycans

Isabelle Breloy

Abstract

The great variability of O-glycan structures makes their analysis a challenging task, which can be solved by the use of several complementary methods. While chromatographic analysis of the fluorescently labeled oligosaccharides shows the quantitative amount of the different glycans in comparison to a standard, mass spectrometry analysis of permethylated oligosaccharides allows identification of new or uncommon glycan structures. In combination with liquid chromatography, all structures present in one sample can be identified. The linkage of the monosaccharides can be analyzed by GC-MS after further derivatization of the permethylated glycans.

Key words: O-glycans, Glycomics, Mucin-type, Hydrazinolysis, 2-Aminobenzamide labeling, HPLC, β-Elimination, Methylation, Methylated glycans, Linkage analysis, Mass spectrometry, LC-MS, GC-MS

1. Introduction

Mammalian mucin-type O-glycans are characterized by a linkage between the anomeric C-atom (C1) of *N*-acetylgalactosamine (GalNAc) and serine or threonine residues within the protein backbone. This core-GalNAc can be elongated with a variety of other monosaccharides. Mucin-type O-glycans show a great variability in structure, composition, and length. These structural variations are not only dependent on the organism, the tissue or specific proteins, but they also show considerable variations in diseases, like cancer. A change in the O-glycosylation pattern can be used as a marker for the development and persistence of carcinomas (1). Therefore, elucidation of the glycan structures is an important and, due to the variability of structures even on one single protein, a challenging task.

Michael A. McGuckin and David J. Thornton (eds.), *Mucins: Methods and Protocols*, Methods in Molecular Biology, vol. 842, DOI 10.1007/978-1-61779-513-8_9, © Springer Science+Business Media, LLC 2012

In this chapter, methods for the characterization of mucin-type O-glycan structures and for the quantification of glycans in complex mixtures are described (Fig. 1). While the chromatographic analysis of fluorescently labeled oligosaccharides is quantitative and sensitive, its disadvantage is the strict dependence on the availability of standard compounds. Novel or isomeric structures can be erroneously identified with isomeric known compounds, unless the chromatographic analysis is extended by mass spectrometric identification. In contrast, the mass spectrometry (MS)-analysis of oligosaccharides is not quantitative (in particular MS based on desorption ionization, like MALDI-MS), but is independent of standard compounds and allows the identification of new structures. With the recently developed method which couples electrospray ionization (ESI)-MS with liquid chromatography (LC-MS), it is even possible to separate isomers to give a comprehensive view on all structures present in a complex sample. The linkage analysis by gas chromatography coupled with MS (GC-MS) reveals insight into the linkage pattern within oligosaccharides, but due to the differing stabilities of desoxy-hexose- (dHex), hexose- (Hex), and *N*-acetylhexosamine- (HexNAc) residues cannot be regarded as a quantitative technique. However, within a group of monosaccharides (for example, GalNAc vs. GlcNAc) reliable estimations of the relative amounts in a sample can be made.

With the methods described here a comprehensive analysis can be performed with less than 100 μg of glycoprotein sample. The methods can be used without adaptation for other types of O-glycosylation, e.g., O-mannosylation or O-fucosylation and an example analysis is shown in Fig. 2.

1.1. Quantitative Analysis of Fluorescently Labeled Glycans by Normal-Phase HPLC

Quantitative O-glycan analysis is done traditionally by HPLC either on a normal- or reversed phase column (2) or by HPAEC (3), which can be coupled with pulsed amperometric detection (PAD). A much higher sensitivity can be obtained by coupling the reducing end of the O-glycans with a fluorophore subsequent to the release from the protein. As there is currently no enzyme available, which cleaves a broad range of different O-glycans (comparable with PNGaseF for *N*-glycans) the nonreductive cleavage must be done chemically. This can be achieved by hydrazinolysis, which was introduced by Patel and Parekh (4). Although handling of anhydrous hydrazine is not easy, there is still no reliable alternative to this method. Labeling of the glycans is most frequently performed with 2-amino-benzamide, which is described in this chapter (5).

1.2. MS Analysis of Methylated Oligosaccharides

A qualitative analysis of glycans by MS can be performed with native or derivatized oligosaccharides and is applicable irrespective of a reducing end. Accordingly, the glycans can be cleaved by reductive β-elimination. As the O-glycosidic linkage is alkali-labile, this can be achieved in dilute alkali and in the presence of sodium

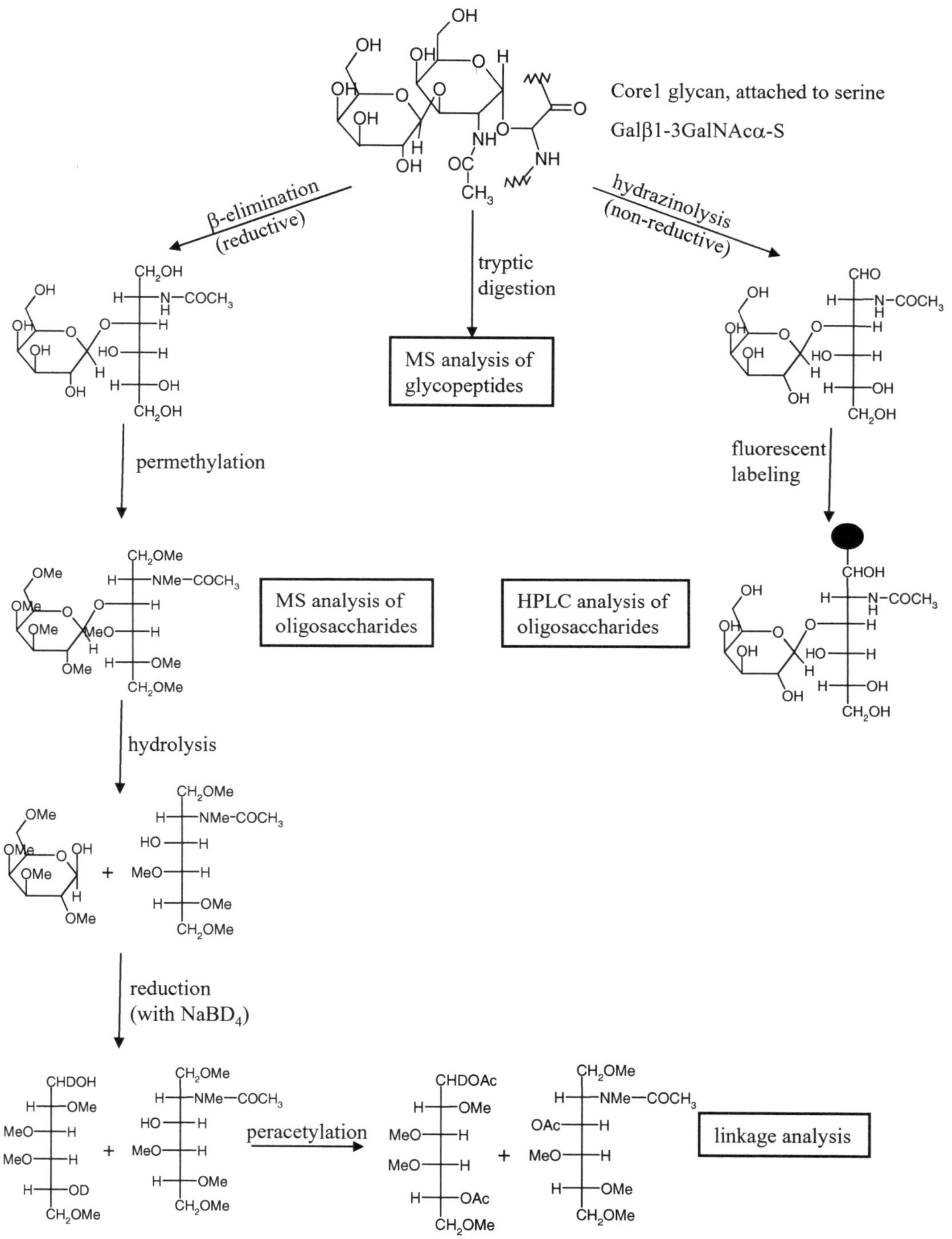

Fig. 1. Alternative routes in the analysis of mucin-type O-glycans. Alternatives in the qualitative and quantitative analysis of mucin-type O-glycans as described in this chapter are shown in an overview. The derivatization process by permethylation is exemplified here in detail with respect to the common core 1 structure (TF-antigen). After reductive β-elimination and permethylation, the oligosaccharides can be analyzed by MS1 (monosaccharide composition and semiquantitative profiling) and MS2 (sequence analysis). Further hydrolysis, reduction and peracetylation allows a linkage analysis by GC-MS of the partially methylated alditol acetates. Glycopeptide analysis by MS is described in another chapter of this issue. The oligosaccharide composition within complex mixtures can be analyzed quantitatively by HPLC after nonreductive cleavage and labeling with a fluorescent dye.

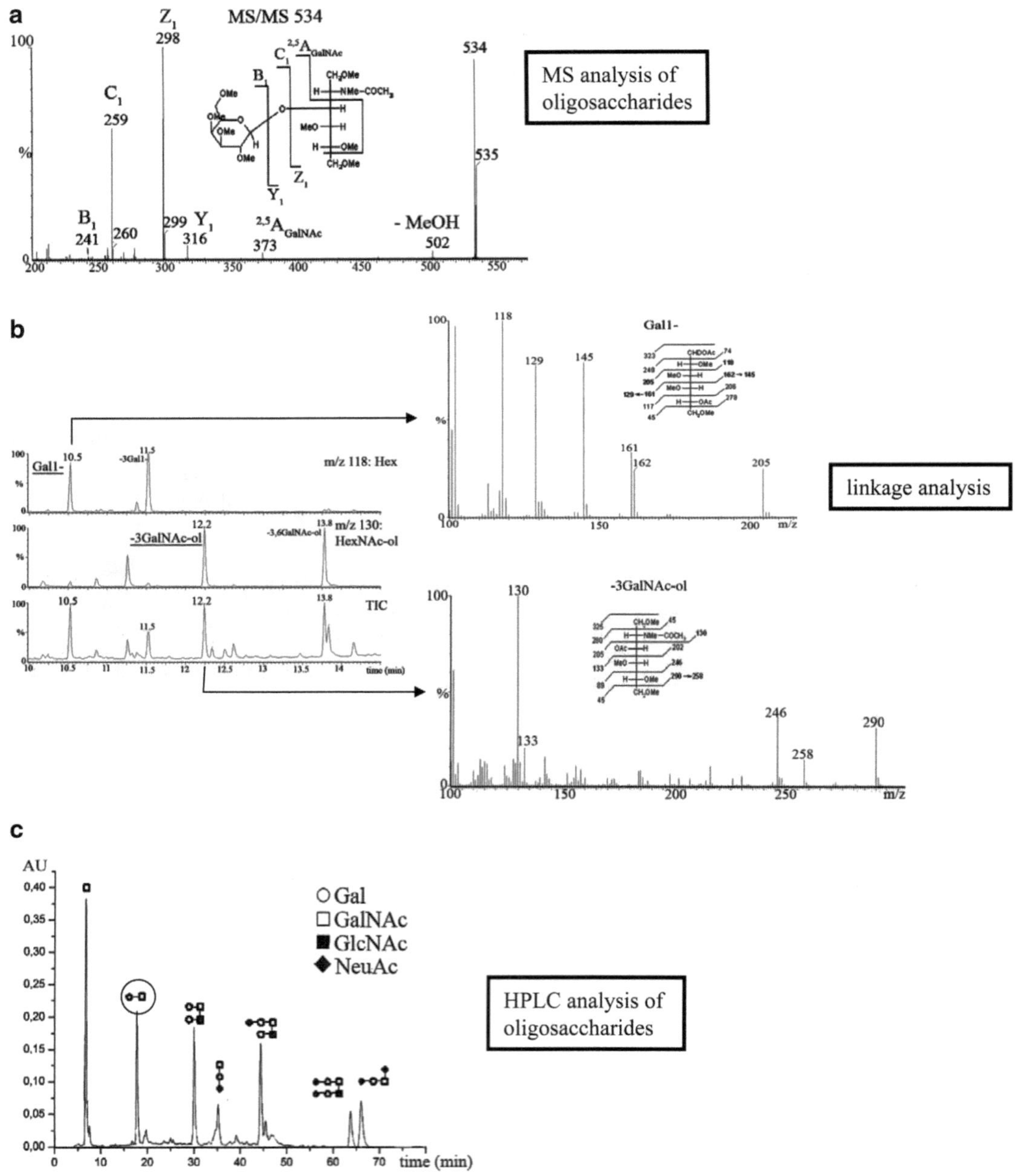

Fig. 2. Analytical data obtained for the disaccharide Galα1-3GalNAc-ol. The alternative analytical routes are exemplified with respect to the analysis of core 1 disaccharide (TF-antigen). (**a**) ESI-Q-Tof-MS/MS spectra of the permethylated alditol as Na^{+}-adduct. The typical C and Z-ions are prominent, but A, B, and Y-ions can also be observed. (**b**) Linkage analysis of mucin-type glycans derived from a recombinant alpha-dystroglycan glycoprotein and expressed in HEK-293 cbells. Besides other structures, the terminal galactose and the subterminal 3-linked GalNAc-ol of the core 1 disaccharide (TF-antigen) could be identified by their retention times and fragmentation patterns in GC-MS. (**c**) HPLC-chromatogram of fluorescently labeled mucin-type glycans derived from a recombinant MUC1 protein and expressed in HEK cells. The different glycan structures are identified by comparison with standard compounds. The TF-antigen is marked with a *circle*. A reliable quantification is possible on the basis of the peak areas.

borohydride as a reducing agent. Reduction of the terminal aldehyde to the respective alditol is necessary to prevent peeling processes (consecutive elimination and rearrangement reactions starting at the reducing terminal sugar). To improve MS detection of the glycans, it is possible to derivatize the hydroxyl groups by substituting the proton by a nonpolar group, like a methyl or a trifluoroacetyl group. The preferred method is still permethylation, which even enables detection of highly acidic sugars like sialic acids by MS (6).

1.3. LC-MS of Methylated Glycans

Permethylated glycans can be analyzed either by directly introducing the methanolic solution into the ion source of ESI instruments using the static nanospray option or by using online nano-LC protocols. We have previously published such a protocol in this series (7), which offers the possibility to separate methylated glycan alditols on reversed phase capillary columns under conditions similar to those of peptide separation. The advantage of the method can be seen in the separation of isomeric compounds prior to MS. O-linked glycans are characterized by a high degree of structural heterogeneity and this is partly due to the formation of isomeric branching isomers. Without prior separation, these isomers would give mixed fragmentation spectra (MS2), which would not allow the unambiguous assignment of glycosidic structures. A further more technical advantage of the described protocol is the prevention of solvent exchange, since the LC of methylated glycan alditols on C18 columns can be performed in aqueous acetonitrile gradients.

1.4. Linkage Analysis of Glycans by Methylation Analysis

Methylation analysis and ^{1}H nuclear magnetic resonance (NMR) spectroscopy are the only available methods to analyze the linkage pattern of a complex glycan. Although NMR gives additional information about the anomeric state of the monosaccharides, methylation analysis is often preferred due to the availability of only limited sample amounts. NMR analysis requires at least 50 nmol of the pure glycan, whereas a methylation analysis can be performed with as little as 5 nmol. It is also possible to analyze a mixture of different glycans by methylation analysis, which avoids a difficult and time-consuming purification to obtain single glycan species. Methylation analysis starts with the permethylated compound, which is hydrolyzed to the partially methylated monosaccharide components, reduced to the respective alditols and finally acetylated to the partially methylated alditol acetates (PMAA). These can be separated by GC and identified by online MS. Hence, each component sugar can be identified via a set of two parameters, the retention time and the mass spectrum. To discriminate between the core-sugar and sugars within the glycan chain, a label can be introduced into C1 by reduction with sodium borodeuteride instead of borohydride. The method has been extensively described elsewhere (8).

2. Materials

2.1. Analysis of Fluorescently Labeled Glycans by Normal-Phase HPLC

1. Dry Argon (see Note 1).
2. Conic glass vial (1 mL) with screw cap and cartridge seal.
3. Water free hydrazine (see Note 2) (toxic, corrosive and unstable, store at 4°C under argon).
4. Re-*N*-acetylation solution (prepare fresh): 2 mM acetic anhydride diluted in a saturated $NaHCO_3$ solution.
5. Suspension of Dowex 50XW8 200–400 mesh can be prepared from the hydrogen form by swelling of the dry beads in 1 M HCl. Aspirate the HCl off with a suction filter and wash the beads with water until the flow-through has pH 5. Store in ddH_2O (1:1) at 4°C.
6. 2-AB labeling solution (prepare fresh): 1 M 2-aminobenzamide in acetic acid and 2 M $NaCNBH_3$ in dry DMSO in a ratio of 2:3.
7. Whatman 3MM cellulose chromatography paper.
8. *n*-butanol/ethanol/water (4:1:1, v/v/v).
9. (Centrifugal) Filter device with a 0.2 μm membrane.
10. Acetonitrile (ACN)/water (3:1, v/v).
11. HPLC running buffer:
    (a) 250 mM NH_4HCO_2 in ddH_2O (pH 4.4).
    (b) Acetonitrile (store in the dark).
12. HPLC apparatus equipped with a normal phase column (amino bonded silica) and a fluorescence detector.

2.2. MS Oligosaccharide Analysis of Methylated Glycans

1. Dry Argon (see Note 1).
2. Nitrogen.
3. 1 M $NaBH_4$ in 50 mM NaOH has to be prepared fresh shortly before usage. $NaBH_4$ reacts with oxygen. Keep under argon.
4. Glacial acetic acid (corrosive).
5. Dowex 50WX8 (see Subheading 3.1, step 4).
6. Conic glass vial (1 mL) with screw cap and cartridge seal.
7. Ethanol of the grade p.a.
8. 0.1% acetic acid in methanol.
9. Water free DMSO, store under argon.

10. DMSO/NaOH suspension prepared after a protocol from Anumula and Taylor (9):
    (a) Make a suspension from 0.1 mL NaOH (50% w/v), 0.2 mL methanol and 6 mL water-free DMSO by vortexing and sonification in a water bath for 3 min (see Note 3).
    (b) Pellet the NaOH by centrifugation at 6,000 × g for 5 min and resuspend in 6 mL DMSO. Repeat this five times.
    (c) Suspend NaOH pellet in 2 mL water-free DMSO and store under argon at −20°C (see Note 4).
11. Methyl iodide (toxic and a carcinogen, store at 4°C).
12. 1 M acetic acid in ddH_2O.
13. $CHCl_3$ p.a.
14. Methanol p.a. (toxic).
15. 2,5-Dihydroxybencoic acid (2,5-DHB) as a MALDI matrix for permethylated glycans can be used saturated in a solution of ACN in 0.1% TFA/ddH_2O (1:2). Dissolved matrix can be stored in a dark place at 4°C for approximately 1 week.

2.3. LC-MS of Methylated Glycans

1. LC-MS data can be acquired on a Q-TOF2 quadrupole-TOF mass spectrometer (Micromass) equipped with a Z spray source. Samples are introduced by an Ultimate Nano-LC system (LC Packings) equipped with a Famos autosampler and a Switchos column-switching module. The column setup comprises a 0.3 × 10 mm trapping column and a 0.075 × 150 mm analytical column, both packed with 3 μm Atlantis dC18 (Waters). The ESI interface comprises a 20 μm internal diameter, 90 μm outer diameter tapered spray emitter (Carbotec) linked to the HPLC flowpath using a 7 nL dead volume stainless steel glass capillary mounted onto the PicoTip holder assembly (New Objective). Of course, the described LC-MS parameters represent only an example that can be replaced by other related instruments and setups.
2. For example, LC-MS data can also be acquired on an HCT ETD II ESI ion trap mass spectrometer (Bruker Daltonics, Bremen, Germany) equipped with a nano ion source. Samples are introduced via an EASY nano LC system (Proxeon, Odense, Denmark) using a vented column setup comprising a 0.1 × 20 mm trapping column and a 0.075 × 100 mm analytical column, both self-packed with ReproSil-Pur C18-AQ, 5 μm (Dr. Maisch, Ammerbuch, Germany).

2.4. Linkage Analysis of Glycans by Methylation Analysis

1. Nitrogen.
2. Oligosaccharide standards for the first experiment (1 μg of each oligosaccharide).
3. 2 M TFA in ddH_2O (corrosive).
4. Ethanol p.a.

5. 10 mg/mL borodeuteride in 2 M NH_4OH (has to be prepared fresh not more than 1 h before usage).
6. Glacial acetic acid (corrosive).
7. 0.1% acetic acid in methanol (toxic).
8. Acetic anhydride (should be stored carefully closed, eventually under argon).
9. $CHCl_3$ p.a.
10. CH_2Cl_2 p.a.
11. GC-MS apparatus (e.g., a Fison MD800 GC/MS (Thermo Fisher, Dreieich, Germany)), equipped with a 15 m RTX5-SILMS column (Restek, Bad Homburg, Germany).

3. Methods

3.1. Analysis of Fluorescently Labeled Glycans by Normal-Phase HPLC

1. Dry the pure protein (see Notes 5 and 6) in a glass vial extensively in a desiccator and keep it under argon during the next steps.
2. Add 50 μL hydrazine (see Note 7) and incubate 5 h at 60°C.
3. Remove the hydrazine in a desiccator by drying the reaction mixture for 1 h. Keep the reaction under argon.
4. Re-*N*-acetylate the glycans with 30 μL re-*N*-acetylation solution during 15 min on ice.
5. Add 60 μL Dowex 50WX8 to the reaction mixture and incubate under gentle agitation for 5 min at room temperature. Pellet the beads during 5 min centrifugation at 16,000 × g. Transfer the supernatant into a new vial. Wash the beads two times with 120 μL ddH_2O and combine the supernatants.
6. Dry the solution in a speedvac.
7. Label the glycans with 5 μL 2-AB labeling solution for 2 h at 60°C (see Note 7).
8. Spot the solution on a whatman filter paper.
9. Remove excessive reagent by paper chromatography with *n*-butanol/ethanol/water (see Note 8).
10. Cut out the application spot and dry it in a desiccator for 2 h.
11. Elute the glycans from the paper with 0.5 mL ddH_2O during 20 h at 4°C under gentle agitation (see Note 9).
12. Remove the paper by filtration through a membrane with 0.2 μm pores (see Note 10).
13. Dry the glycans in a speedvac.

14. Dissolve the glycans in 200 μL ACN/H_2O and store them at −20°C.
15. Couple an HPLC system with a by normal-phase column with a fluorescence detector (see Note 11) using an excitation at 330 nm and an emission at 420 nm.
16. Wash the column with ACN for 30 min with a flow rate of 0.5 mL/min (see Note 12).
17. Wash the column with buffer B for 30 min.
18. Equilibrate the column with buffer A for 30 min.
19. Inject the sample onto the column.
20. Elute the sample with a linear gradient from 80 to 40% ACN in 250 mM NH_4HCO_2, pH 4.4 (see Notes 13 and 14).

3.2. MS Oligosaccharide Analysis of Methylated Glycans

1. Release the glycan chains from the dry glycoproteins or glycopeptides (see Note 15) in a 1.5 mL Eppendorf vial by incubation in 20 μL $NaBH_4$ (see Note 16) in NaOH for 16–18 h at 50°C.
2. Quench the reaction by adding 1 μL glacial acetic acid on ice (see Note 17).
3. Add 50 μL Dowex 50WX8 to the reaction mixture and incubate under gentle agitation for 5 min at room temperature. Pellet the beads during 5 min centrifugation at 16,000 × g. Transfer the supernatant into a 1 mL conic glass vial. Wash the beads with 50 μL ddH_2O, pellet them as above and combine both supernatants.
4. Dry the supernatant under a stream of nitrogen at 40°C and wash two times with 100 μL ethanol. Borate ester should be co-evaporated with 50 μL 0.1% acetic acid in methanol. Repeat this step four times.
5. Dry the glycans extensively in a desiccator and keep them under Argon during the next steps.
6. Resuspend the glycans in 50 μL dry DMSO by vortexing for 1 min.
7. Add 50 μL DMSO/NaOH suspension and incubate the reaction mixture for 30 min at room temperature (see Note 18).
8. Freeze the mixture at −20°C.
9. Add 20 μL methyl iodide to the frozen solution and incubate 30 min at room temperature.
10. Quench the reaction by adding 40 μL 1 M acetic acid.
11. Dissolve the methylated glycans in 300 μL $CHCl_3$ and extract salts by adding 200 μL H_2O, vortexing and taking off the hydrophilic upper phase with the dissolved salts. Repeat this four times (see Notes 19 and 20).

12. Dry the methylated glycans under a stream of nitrogen (see Note 21).
13. Dissolve the glycans in 10 μL methanol.
14. Put 1 μL 2,5-DHB on a MALDI target and mix on the target with 1 μL of the glycan-solution. Let the spot crystallize (see Note 22).
15. Analyze the sample by MALDI-MS in positive ion mode. Most permethylated O-glycans can be detected in the range between 500 and 2,000 Da.

3.3. LC-MS of Methylated Glycans

LC-parameters in LC-QTOF. Inject 1–5 μL permethylated glycans pre-solubilized in ACN and finally dissolved in 10% acetonitrile/0.1% formic acid (FA) onto the analytical column (Atlantis, C18, Waters). Set the flow rate resulting from a 1:1,000 split of the 200 μL/min flow delivered by the system pump at approximately 200 nL/min. The samples are eluted from the analytical column by using a gradient of 0.1% aqueous FA:isopropanol:ACN at 85:5:10 (start conditions) to the same solutions at 80:10:10 (final conditions).

LC-parameters in LC-HCT. Trap the samples initially on C18 (self-packed ReproSil-Pur C18-AQ, Dr. Maisch, Ammerbuch, Germany) before the application of a ternary gradient. Elute the samples onto the analytical column (self-packed ReproSil-Pur C18-AQ) by using a gradient of running buffer A (0.1% FA) and running buffer B (0.1% FA in acetonitrile:isopropanol (1:1)) from 0% B to 20% B (duration 1 min), to 80% B (duration 15 min) followed isocratic conditions at 80% B for 4 min (see Note 23).

MS-parameters (QTOF). Apply +1.7 to +1.9 kV to the stainless steel union in the LC-MS interface. Set the cone voltage to 40 V and the source temperature to 80°C. No drying gas is used in nano LC-MS experiments. Use a collision energy setting of 4 V for the acquisition of survey scans. The argon pressure in the collision cell 0.5–0.6 bar for all experiments. Acquire automatic tandem MS (MS/MS) scans with collision energy settings ranging from 15 V (M+H) to 60 V (M+Na). The data-dependent acquisition of MS and MS/MS spectra is controlled by Masslynx 4.0 software.

Survey scans of 1 s cover the range from *m/z* 400 to *m/z* 1,500. Select singly and doubly charged ions for MS/MS experiments. In MS/MS mode, scan the mass range from *m/z* 40 to *m/z* 1,400 in 1 s, and add up five scans for each experiment.

MS parameters (HCT). Use the NanoFlow-ESI source for the LC-MS experiment. Set the acquisition parameters to standard/enhanced for the mass range mode, the capillary exit to 139.0 V, the accumulation time to 199,853 μs, the ion polarity to positive ion detection, the scan range to 350 to 1,500 *m/z*, the skimmer to 40.0 V, spectra averaging to 6 spectra.

3.4. Linkage Analysis of Glycans by Methylation Analysis

1. Dry the permethylated glycans (as obtained from Subheading 3.2, steps 1 and 2) and dissolve them in 100 µL 2 M TFA. Incubate the solution for 2 h at 121°C (see Note 24) to hydrolyze the glycan chains.
2. Dry the monosaccharides in a stream of nitrogen and wash them by adding 100 µL ethanol and evaporating the ethanol with nitrogen. Repeat this step once.
3. Reduce the monosaccharides by incubation in 100 µL borodeuteride solution for 2 h at room temperature.
4. Quench the reaction by adding 12 µL glacial acetic acid and dry the reaction mixture under a stream of nitrogen.
5. Co-evaporate the borate ester by adding five times 50 µL 0.1% acetic acid in methanol under a stream of nitrogen. Repeat this step four times.
6. Acetylate the monosaccharides by incubation in 100 µL acetic anhydride for 1 h at 100°C (see Note 24).
7. Dry the monosaccharides under a stream of nitrogen and extract them with $CHCl_3/H_2O$ as described under Subheading 3.2, step 11 of this article (see Note 21).
8. Resuspend the monosaccharides in 10 µL CH_2Cl_2.
9. Analyze the monosaccharides by GC-MS (see Note 25). Inject 1–3 µL at an initial temperature of 60°C for 1 min, followed by a gradient of 40°C/min up to 100°C and a gradient of 10°C/min up to 280°C. Ionize the monosaccharides by electron impact at 70 eV and detect positively charged fragments with one scan/second at 400 V (see Note 26).

4. Notes

1. Argon can be dried easily by blowing it through a glass column filled with silica gel.
2. Water free hydrazine is toxic, corrosive, and unstable and has to be stored at 4°C under argon. Any contact with the skin or inhaling of aerosols should be avoided.
3. The suspension has to be fine which can only be achieved by extensive vortexing in every step. If clumps of NaOH can be seen, you have to start over.
4. Prepare a fresh solution when the color of the suspension turns yellow.
5. The protein has to be completely free from salt.
6. 10–30 µg protein, depending on the number of glycan chains linked to the protein.

7. Use a glass syringe or glass pipet.
8. There are several other fluorescent labels (e.g., 2-aminobenzoic acid or 2-aminopyridine) commercially available.
9. The glycans do not migrate under these conditions. Only the mono- and to a lesser extend the disaccharides move slowly with the solvent. For this reason, the area of sample application and up to 1 cm above should be excised.
10. The paper should be cut into small pieces to enhance efficiency of elution.
11. It is faster to use a filter device for a centrifuge, like Ultrafree MC filter devices (Amicon).
12. This method was developed for a Beckman Gold HPLC station equipped with a 5 μm 4.6 × 250 mM Astec NH_2 polymer column and has to be adapted for other equipment.
13. Equilibration times and flow rates may need adaption depending on the column.
14. Using a flow rate of 0.5 mL/min the glycans elute during 80 min under the described conditions.
15. 10–30 μg of the pure glycoprotein or glycopeptide should be used, depending on the amount of glycan chains on the protein.
16. $NaBH_4$ reduces the terminal aldehyde of glycans to the respective alcohol preventing a "peeling reaction" of the sugar (degradation from the reducing end by consecutive elimination and rearrangement reactions) under hot alkaline conditions.
17. By addition of the acid extensive foaming should be observed. If not, the sodium borohydride may have been destroyed to a large extent in addition to an acidic sample which may also inhibit the β-elimination due to a wrong pH.
18. During the incubation time, the mixture should be vortexed if a cloudy sediment of dispersed NaOH is observed.
19. Phase separation can be accelerated by a short centrifugation (1 min at 3,000 × g).
20. The permethylated glycans are in the lower $CHCl_3$-phase.
21. Dry the glycans carefully by heating at no more than 40°C as they are volatile.
22. If no crystals form, the sample most likely contains a larger amount of salt and has to be dried and extracted with $CHCl_3$/H_2O again. Using an alternative matrix like α-*cyano*-4-hydroxycinnamic acid (HCCA) the crystals can be washed with 0.1% TFA with 10 mM ammonium phosphate to remove salt.
23. It is recommended that chloroform-water extracted samples of methylated glycans are further processed by solid-phase

extraction on C18 as described by Dell et al. (10) to avoid clogging of column tips by residual salt. Permethylated glycan alditols elute under the described chromatographic conditions as broad peaks covering a time range of >60 s. Other gradients without isopropanol used conventionally for peptide separation were found to give more sharpened peaks in the range of 10 s, however, under these conditions the effectiveness of glycan elution from the column may be somewhat reduced.

24. Ensure that vials are closed completely and recheck this after 5 min incubation time.
25. This method was developed for a Fison MD800 GC-MS (Thermon Electron) equipped with a 15 m RTX5-SILMS column (Restek) and has to be adapted if other equipment is used.
26. Most fragments are detectable in the range between 100 and 450 Da.

References

1. Brockhausen, I. (1999) Pathways of O-glycan biosynthesis in cancer cells. *Biochim Biophys Acta.* **1473**: 67–95.
2. Hounsell, E. F., Jones, N. J. and Stoll, M. S. (1985) The application of high-performance liquid chromatography to the purification of oligosaccharides containing neutral and cetamido sugars. *Biochem Soc Trans.* **13**: 1061–1064.
3. Campbell, B. J., Davies, M. J., Rhodes, J.M. and Hounsell, E. F. (1993) Separation of neutral oligosaccharide alditols from human meconium using high-pH anion-exchange chromatography. *J Chromatogr.* **622**: 137–146.
4. Patel, T. P. and Parekh, R. B. (1994) Release of oligosaccharides from glycoproteins by hydrazinolysis. *Methods Enzymol.* **230**: 57–66.
5. Bigge, J. C., Patel, T. P., Bruce, J. A., Goulding, P.N., Charles, S. M. and Parekh, R. B. (1995) Nonselective and efficient fluorescent labeling of glycans using 2-amino benzamide and anthranilic acid. *Anal. Biochem.* **230**: 229–238.
6. Ciucanu, I. and Kerek, F. (1984) A simple and rapid method for the permethylation of carbohydrates. *Carbohydr. Res.* **131**: 209–217.
7. Hanisch, F.-G. and Müller, S. (2009) Analysis of methylated O-glycan alditols by reversed-phase nanoLC coupled CAD-ESI mass spectrometry. (Packer, N. H. and Karlsson, N. G. ed.) Humana Press, New York, NY, pp. 107–115.
8. Merkle, R. K. and Poppe, I. (1994) Carbohydrate composition analysis of glycoconjugates by gas-liquid chromatography/mass spectrometry. *Meth. Enzymology* **230**: 1–15.
9. Anumula, K. R. and Taylor, P. B. (1992) A comprehensive procedure for preparation of partially methylated alditol acetates from glycoprotein carbohydrates. *Anal Biochem.* **203**: 101–108.
10. Dell, A., Khoo, K. H., Panico, M., McDowell, R. A., Etienne, A.T., Reason, A.J., Morris, H. R. (1993) FAB-MS and ES-MS of glycoproteins. In: Fukuda, M., Kobata, A., editors. *Glycobiology: A Practical Approach. Oxford: Oxford University Press*: 187–222.

Chapter 10

O-Glycoproteomics: Site-Specific O-Glycoprotein Analysis by CID/ETD Electrospray Ionization Tandem Mass Spectrometry and Top-Down Glycoprotein Sequencing by In-Source Decay MALDI Mass Spectrometry

Franz-Georg Hanisch

Abstract

The sites of mucin-type O-glycosylation are difficult to predict, making structural analysis by mass spectrometry indispensible. This chapter refers to state-of-the-art techniques in the site localization of O-linked glycans and their structural characterization in situ using tandem ESI and MALDI mass spectrometry. Detailed protocols are provided that describe the application of nano-LC-ESI-MS/MS with alternative fragmentation modes (collision-induced dissociation vs. electron-transfer dissociation) for the analysis of O-glycopeptides. Moreover, a top-down sequencing approach by MALDI-MS is presented that is based on the in-source decay of intact glycoproteins or large glycopeptides and allows a ladder sequencing of up to 70 amino acid residues from both termini with unequivocal assignment of modified sites.

Key words: Mucin-type O-glycosylation, O-glycans, Mucin, Glycopeptides, ESI mass spectrometry, Electron-transfer dissociation, MALDI in-source decay

1. Introduction

This chapter cannot review the features of all currently available instrumental and methodological alternatives and the analytical depths achievable with different equipment. The reader should refer to related overviews (1, 2) and consider that each type of mass spectrometer represents an "individual" with respect to ion generation (ESI, MALDI) and ion separation depending on the configuration and type of analyzers (q-tof, tof-tof, q-trap, orbitrap, etc.) each connected with advantages in particular applications while limited with respect to others. Adding to this, the acquisition and evaluation software provided with the instrumentation from

Michael A. McGuckin and David J. Thornton (eds.), *Mucins: Methods and Protocols*, Methods in Molecular Biology, vol. 842, DOI 10.1007/978-1-61779-513-8_10,

different companies has quite distinct features. This introduction focuses on post-source-decay (PSD) MALDI-MS/MS and collision-induced dissociation (CID) or electron-transfer dissociation (ETD) ESI MS/MS applications in the context of site-specific O-glycopeptide analysis. A further topic is the site-specific O-glycopeptide analysis by top-down ladder sequencing of proteins by in-source decay (ISD) MALDI mass spectrometry.

Differing from N-glycosylation, where a consensus sequence controls the enzymatic transfer of the glycan to the protein backbone, the initiation of O-glycosylation by addition of *N*-acetylgalactosamine (GalNAc) to serine or threonine side chains is catalyzed by about 20 isoenzymes with only partially overlapping substrate preferences and without a common sequon (3). This makes O-glycosylation largely unpredictable, although efforts in this direction were made by the creation of a software tool that is based on available structural data and neural network algorithms (Net-O-Glyc 3.1) (http://www.cbs.dtu.dk/services/NetOGlyc/). This situation makes mass spectrometric analysis of O-glycosylation sites indispensable.

In the past, efforts were made to localize alkali-labile-bound O-glycosylation sites by substitution chemistry, the base-catalyzed β-elimination of glycans, and the addition of nucleophilic compounds to yield a labeled product with a defined mass tag. These so-called β-elimination-Michael addition (BEMAD) approaches using ammonia, various alkylamines, or DTT (4–8) show some advantages in MS analysis, but suffer from the disadvantage that information on the glycan structure is lost. Other approaches were based on the partial acid hydrolysis of O-glycopeptides in the gas phase, but were applicable only for small glycan chains lacking labile monosaccharide constituents, like sialic acid or fucose (9). With the advent of soft fragmentation modes in tandem mass spectrometry, the direct mass spectrometric analysis of native O-glycopeptides enables the elucidation of both, the localization of glycan chains in the peptide backbone and the structural features of the glycan in situ.

Sequencing of O-glycopeptides by PSD-MALDI-MS. The process of metastable decomposition of molecular ions during their acceleration or when passing the drift region (PSD) is laser induced and hence dependent on laser power and the density of the matrix cloud formed after the laser shot. As it is a soft fragmentation mode, PSD shows some degree of selectivity with respect to peptide bond cleavage, making it more dependent on the intrinsic chemical properties of the analyte compared to CID-induced fragmentation. On the other hand, it offers the advantage that labile-bound O-glycans remain stable during the fragmentation process. In contrast to ESI, where multiply charged ion species give rise to relatively simple peptide fragmentation patterns with nearly exclusive formation of b- and y-type ions (CID) or c- and z-type ions (ETD),

PSD-MALDI spectra of singly charged molecular ions show a much greater complexity by showing a-, b-, and y-type ions each with −17 (−NH_3) and −18 companions (−H_2O) and finally also internal fragmentation. Glycosylated fragments are accompanied by the respective deglycosylated species, however, at less than 10–15% of the glycosylated precursor indicating a low degree of sugar cleavage during the PSD process (10).

Sequencing of O-glycopeptides by CID/ETD tandem ESI-MS/MS. Initial attempts to localize O-glycosylation sites in glycopeptides by ESI mass spectrometry were based on instruments with a q-tof configuration of mass analyzers and CID for fragment generation (11). A technical limitation of CID for site-specific O-glycopeptide analysis is the preferential fragmentation of the more labile glycosidic bonds. Accordingly, the glycan is lost by fragmentation before the peptide backbone breaks and attachment sites cannot be localized by the respective mass incremental increases of Ser or Thr residues. Some ESI instruments with ion traps or an orbitrap offer ETD as an alternative fragmentation mode (12). ETD is based on the transfer of electrons from a donor compound (like fluoranthen). As the transfer occurs in a random fashion and the radical formed induces breakage of the respective bond, there is no preferential cleavage of labile bonds as observed in CID and fragment intensity shows a more even distribution over the entire peptide. The consequence is that longer stretches of the peptide sequence can be read on the basis of continuous c- or z-ion series and that labile modifications can be localized. However, it is often advantageous to also obtain complementary information from CID experiments and to merge these with ETD fragmentation data (13).

ISD top-down sequencing of O-glycopeptides. ISD MALDI spectra are known to provide sequence information directly from undigested, intact proteins and can be used for partial top-down N- and C-terminal sequencing of proteins. ISD MALDI mass spectrometry was introduced in 1995 by Brown and Lennon (14) and was later applied in the reflectron technology (15, 16). During ISD-MALDI-MS acquisitions, the molecular ions fragment spontaneously within the ion source and yield preferentially c-type ions from the N-terminal and y- or z-type ions from the C-terminal of the protein. The type of fragment ions formed in the MALDI process was claimed to depend to some extent on the matrix used in sample preparation. While dihydroxy benzoic acid (DHB) and, in particular, its mixture with an isomer called "super-DHB" (sDHB) give rise to fragmentation of both the N- and C-terminal of proteins by ion formation of the c- and y- (or z-) series. 1,5-diaminonaphtalene (DAN) has been demonstrated to yield largely c-type ion series (17, 18). Technical developments in MALDI-TOF instrumental features and in sample preparation have led to significant improvements in the performance of ISD-based sequencing of proteins. Currently, spectra can be recorded with mass accuracies in the 10-ppm range and reading lengths of peptide sequences up to 70 amino acid residues can be achieved. Due to a mass cutoff (ions <1,000 Da have to be deflected), ions can be observed typically between 1,000 and 8,000 Da, provided that the sequence contains no disulfide bridges within this region. Another limitation is found in short gaps in the c-ion series caused by proline residues. ISD analysis of peptide ladders can be combined with the analysis of PSD fragments induced by laser-induced dissociation (LID), which yields sequence information on each ladder peptide. Compared to Edman sequencing, the MS-based technique offers advantages by being less expensive and time consuming; also C-terminal sequences can be read and proteins with N-terminal or internal posttranslational modifications are readily sequenced.

Posttranslational protein modifications, even labile-bound phospho-*O*-Ser/Thr and glycans linked O-glycosidically to Ser/Thr, remain bound to the protein during ladder fragmentation in the ion source. This is due to the fact that fragmentation is induced by hydrogen radical transfer from the matrix, a process related to ETD. No energy relocation to labile bonds as in CID occurs on the subnanosecond timescale. The soft fragmentation mode with stochastic, nearly equi-intense ion formation should hence be well-suited for the site-specific analysis of modified proteins (19). Up to now, the ISD top-down sequencing by MALDI MS is restricted in its application to instruments of the Ultraflex series and the UltrafleXtreme (Bruker Daltonics, Bremen, Germany).

2. Materials

2.1. Instrumentation

1. *Nano-LC ESI-MS/MS analysis.* Analyses are performed on the HCT ultra ETD II ESI ion trap mass spectrometer (Bruker Daltonics, Bremen, Germany) equipped with a nanosprayer and nano-ESI source, online liquid chromatography on the EASY-nano-LC (Proxeon, Odense, Denmark). Samples are introduced using a vented column setup comprising a 0.1×20-mm trapping column and a 0.075×100-mm analytical column: both self-packed with ReproSil-Pur C18-AQ resin (Dr. Maisch, Ammerbuch, Germany) (see Note 1).
2. *MALDI-ISD analysis.* UltrafleXtreme MALDI-TOF-TOF mass spectrometer (Bruker Daltonics); MTP Anchor Chip target from Bruker Daltonics (see Note 2).

2.2. Acquisition and Data Evaluation Software

1. *LC-ESI-MS/MS.* Compass 1.3 for esquire/HCT (Bruker Daltonics); HiStar 3.2-SR2 (Bruker Daltonics); BioTools 3.2 (Bruker Daltonics).
2. *MALDI-ISD-MS.* FlexControl 3.3 and FlexAnalysis 3.3 (Bruker Daltonics).

2.3. Chemicals and Labware

1. *Glycopeptides.* A test glycopeptide was synthesized that corresponds to two repeat units of the variable number of tandem repeats domain of MUC1 and contains GalNAc residues at Thr9 and Ser29: HGVTSAPD**T**RPAPGSTAPPAHGVTSAPE**S**RPAPGS TAPPA.
2. *Matrices.* DAN was purchased from ACROS Organics (Belgium) and used without further purification; 4-hydroxy-α-cyano cinnaminic acid (HCCA) was of high quality and ready for use in MALDI MS application.
3. Acetonitrile (HPLC gradient grade); water (for HPLC), and water with 0.1% trifluoroacetic acid (TFA); formic acid 50% (FA).

3. Methods

3.1. PSD-MALDI-MS/MS for Site-Specific O-Glycopeptide Analysis (see Example in Fig. 1)

1. Sample preparation: Solubilize the glycopeptide sample at 10 pmol/μL or lower concentration in 0.1% TFA (v/v) or mixtures of 0.1% TFA and acetonitrile (up to 30% (v/v) acetonitrile for HCCA thin-layer matrix application).
2. Matrix preparation: Prepare a saturated solution of HCCA in acetone containing 3% (v/v) water, 0.003% (v/v) TFA, and 300 μM ammonium phosphate, monobasic (see Note 3).
3. Application of sample onto mass target: Apply a droplet of about 20 μL matrix solution onto a 800-μm anchor plate and move it with the pipette tip over the anchors. Each anchor takes a small amount of the solution, which dries down immediately. Transfer 0.5 μL of the sample solution onto the covered anchor spot. After 3 min, add 0.5 μL 0.1% TFA containing 10 mM ammonium phosphate and remove the complete droplet (see Note 4).
4. Choose a parameter file for MS2 experiments in the LID-LIFT mode (Ultraflex instruments and UltrafleXtreme, Bruker-Daltonics) that uses the following spectrometric parameters: ion source 1 (7.50 kV); ion source 2 (6.70 kV); lens (3.50 kV); reflector 1 (29.50 kV); reflector 2 (13.95 kV); lift 1 (19.00 kV); lift 2 (3.70 kV); positive polarity (see Note 5).
5. Choose appropriate parameters for the laser frequency (1,000 Hz), the digitizer trigger level (1,000 mV), and the gain voltage offset for reflectron mode (2,200) in the Setup page.
6. On the LIFT page, define the parent mass and the precursor ion selector (PCIS) window range. Select the number of laser shots (1,000). Advanced parameters can be defined with respect to the laser power boost for fragment analysis (50%), the detector gain boost (150%), and the PCIS range (0.45%).

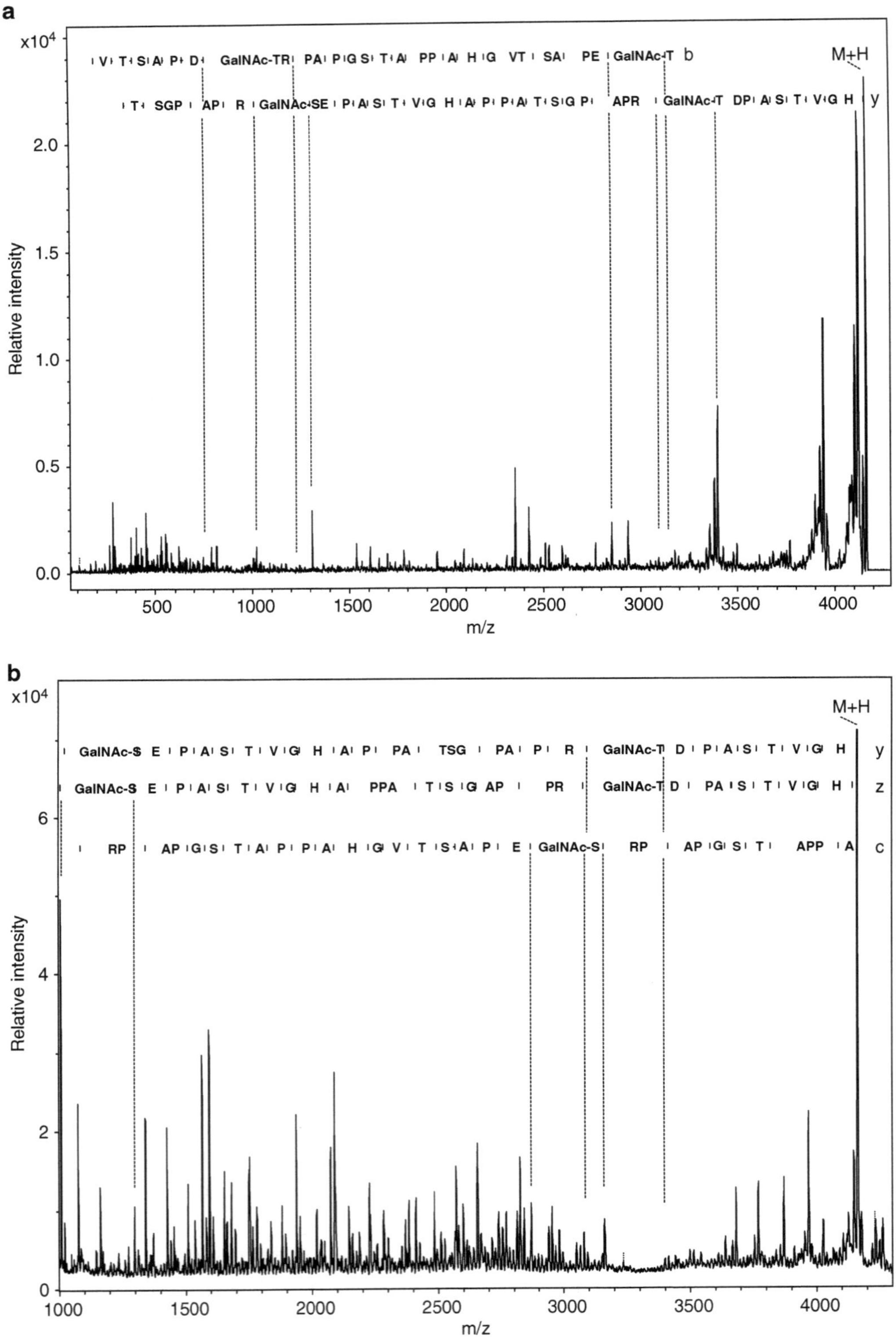

Fig. 1. MALDI-post-source decay and in-source decay sequencing of MUC1-40(GalNAc$_2$). (**a**) PSD-MALDI mass spectrum: A laser-induced dissociation LIFT spectrum of post-source decay fragments was registered on a Bruker UltrafleXtreme instrument using HCCA matrix for sample application (see Subheading 3.1, steps 1–10). (**b**) ISD-MALDI mass spectrum: In-source decay fragments were registered on a Bruker UltrafleXtreme instrument using the DAN matrix for application of the glycopeptide sample as described in the Subheading 3 (see Subheading 3.3, steps 1–8).

7. Measure the parent ion under the above-defined conditions to ensure that the precursor ion is registered with sufficient intensity and the PCIS window was selected appropriately. Accumulate several thousand laser shots.
8. Switch to the "Fragments" mode and increase laser energy manually, if the power boost was not predefined.
9. Start registration of PSD spectra and accumulate 5,000 or more laser shots dependent on sample amount and spectral quality.
10. Save spectrum and open it in FlexAnalysis for mass annotation of ion signals. Optionally, recalibrate the spectrum. Evaluate MS2 spectrum using the manual annotation features of the software (predefined building blocks for mass increments of amino acids, X-Pro diads, and glycosylated Ser/Thr residues to read the peptide sequence and localize the O-glycosylation sites).

3.2. Tandem (CID, ETD) ESI-MS/MS for Site-Specific O-Glycopeptide Analysis

1. Sample preparation: Solubilize the glycopeptide sample in 5% formic acid at <10 pmol/μL (see Note 6).
2. Switch on the ETD source on the Ion/Ion page 1 h before starting the MS experiment.
3. Set instrument parameters for the reactant transfer on the Source page (default settings): ionization chamber (−5.5), gate lens (pass 15.0; block −10.0); transfer (−0.5); focus lens (25.0); partition (−2.0), octopole 1 (−85.0) and octopole 2 (−1.2); octopole RF value for passing/blocking the Cl anions (120.0/0.0); and the Trap Drive value for fluoranthene (25).
4. Select on the Reactant page the reactant-specific accumulation time (1.00 ms) to display the reactant in good conditions in the spectra and the maximal used accumulation time (200 ms). To remove only the reactant from the spectra, activate the checkbox and enter the correct *m/z* value (210). Set the target value for the reactant ion charge control.
5. Method setup (MS parameters): Acquisition of MS1 and MS2 is controlled by the Compass 3.0 software. Editing of the MS part of methods is done with the Bruker esquireControl software. Acquire MS1 scans in the mass range from *m/z* 300 to 2,500 in MS1 with 8,100 amu/s and *m/z* 100–3,000 in MS2 mode. Use "ultrascan" mode for the acquisition of MS2 scans. Select on the MS(n) page after activating under "Auto MS(n) auto MS2," which *m/z* values you would like to include in a list of precursor ions in MS1 (*m/z* value corresponding to the protonated molecular ion of the glycopeptide). Define the number of precursor ions, the absolute and relative thresholds (25,000/5.0%), and on the Ion/Ion page activate "Alternating spectra" in CID and ETD mode. Define under "Auto MS(n)

ETD" the MS/MS stage, the maximal ETD precursor mass, the lower mass cutoff, and the acquisition time for automatic ETD. Save the method.

6. Method setup (LC parameters): The LC method (see below) and the implemented MS method are created in the HiStar software. Aspirate up to 5 μL of the sample into the sample loop and load it onto the trap column using a flow rate of 6 μL/min. Loading pump buffer is 0.1% FA. Elute glycopeptides with a gradient of 0–35% acetonitrile in 0.1% FA over 20 min and a column flow rate of 300 nL/min. Regenerate the column for 2 min in 100% acetonitrile.
7. Data evaluation is performed in the Compass 1.3 Data Analysis software using the deconvolution and manual annotation tools (operating with predefined mass incremental building blocks for amino acids, X-Pro diads, and glycosylated Ser/Thr to determine sequence stretches and glycosylation sites) (see Note 7).

3.3. Top-Down Protein Sequencing by In-Source Decay for Site Localization of O-Linked Glycans (see Example in Fig. 1)

1. Sample preparation: Solublize 100 pmol/μL of the glycopeptide in water/0.1% TFA (see Note 8).
2. Matrix preparation: Prepare a saturated colorless to pale violet solution of DAN in 50% acetonitrile/0.1% TFA (see Note 9).
3. Application of sample onto the MALDI target: Mix 0.5 μL of sample with 1.0 μL of matrix prior to application of 0.5 μL onto the stainless steel MALDI target. Alternatively, MTP Anchor Chip targets can be used. Dry the sample spots at ambient temperature. After introduction of the target into the ion source, inspect the video image of the DAN matrix preparation and ensure homogenous needle-like crystallization (see Note 10).
4. Choose a parameter file for ISD application that uses the following spectrometric parameters: ion source 1 (25.00 kV); ion source 2 (22.25 kV); lens (7.50 kV); reflector 1 (26.50 kV); reflector 2 (13.45 kV); pulsed ion extraction (80 ns); matrix suppression by deflection (900 Da).
5. Choose an appropriate mass range from about 1,000 to 8,000 Da for detection.
6. Accumulate 5,000–10,000 laser shots at 1 kHz acquisition for each sample spot.
7. Choose automatic processing by FlexAnalysis using the SNAP algorithm for monoisotopic peak annotation and internal calibration. During this procedure, the reflector ISD spectrum of the mass calibrant (BSA) is internally calibrated using a mass control list (BSA_ISD mono.mcl), which contains theoretical c-ion masses from 1192.62 to 5784.6 Da. This calibration is automatically applied to the sample spectra.

8. Analyze spectra in FlexAnalysis using the manual annotation tool and predefined mass incremental building blocks for amino acids, X-Pro diads, and glycosylated Ser/Thr to determine sequence stretches and glycosylation sites (see Note 11).

4. Notes

1. Alternative ESI mass spectrometers with ETD option can be used for this type of experiment. Parameters given under methods below are instrument specific and cannot be transferred one to one to other instrumentation.
2. Precursors of the Ultraflextreme MALDI-TOF-TOF, instruments of the Ultraflex series from Bruker, can also be used for TDS by ISD measurements. Instruments from other suppliers have never been reported to allow this type of mass spectrometric application.
3. Optionally, DHB can be used as matrix and applied by the "dried droplet" method. One volume (generally, around 1 μL) of matrix (20 mg/mL in 50% ACN/0.1% aq. TFA) is applied first onto the target and mixed with 0.5 volume of sample by pipetting up and down several times to mix both solutions appropriately. However, the ion intensities of PSD fragments measured with this "cold" matrix are significantly weaker compared to samples applied in DAN matrix.
4. Optionally, the preparation can be washed and recrystallized.
5. Parameters given for PSD-MS/MS measurements in the LIFT mode refer exclusively to the UltrafleXtreme MALDI-TOF-TOF and with some restrictions to the Ultraflex series, but not to similar applications on instruments from other suppliers.
6. The described protocol is formulated for standard glycopeptide compounds. For complex mixtures of peptides/glycopeptides from biological sources, the sample preparation and chromatographic conditions need appropriate adaptation.
7. Since ETD and CID spectra are complementary, it is recommended to use both in localizing glycosylation sites. Under CID conditions, O-linked glycans decompose rapidly with consecutive liberation of sugars and β-elimination from the peptide core with concomitant formation of dehydrated Ser/Thr residues (indicated by mass increments of 69 or 83 Da, respectively). CID spectra are often dominated by the elimination products of partial sugar cleavage, and ions corresponding to peptidic fragments are hardly detectable.
8. An absolute amount of 50 pmol should be placed onto the target to achieve optimal results with respect to ion intensities in the higher mass range.

9. The DAN matrix solution should be always prepared fresh, since the oxidized derivative (indicated by the solution turning dark violet) is not suitable as matrix. A speedy sample application is also recommended to avoid in-solution oxidation of DAN. DAN acts as reducing agent and may cause the partial reduction of disulfide bonds in proteins.
10. If crystallization is heterogenous, add 0.5 μL of pure matrix solution to the spot and wait for recrystallization.
11. It has been claimed that using DAN matrix c-type ions from the N-terminus are formed exclusively. We also observed that C-terminal fragmentation is induced (y and z + 2 ions) and that the preponderance of the latter depends on the glycosylation density of N-terminal peptides. Alternative matrices, like DHB and sDHB, can be used; however, the ISD fragment intensities were generally lower compared to DAN preparations.

Acknowledgments

The structural work was supported in part by the Deutsche Forschungsgemeinschaft Grants HA 2092/15-1 and HA 2092/21-1 (to FGH).

References

1. Geyer, H., Geyer, R. (2006) Strategies for analysis of glycoprotein glycosylation. *Biochim. Biophys. Acta* 1764, 1853–1869
2. Wuhrer, M., Catalina, M.I., Deelder, A.M., Hokke, C.H. (2007) Glycoproteomics based on tandem mass spectrometry of glycopeptides. *J. Chromatogr. B* 849, 115–128
3. Tarp, M.A., Clausen, H. (2008) Mucin-type O-glycosylation and its potential use in drug and vaccine development. *Biochim. Biophys. Acta* 1780, 546–563
4. Rademaker, G.J., Pergantis, S.A., Blok-Tip, L., Langridge, J.I., Kleen, A., Thomas-Oates, J.E. (1998) Mass spectrometric determination of the sites of O-glycan attachment with low picomolar sensitivity. *Anal. Biochem.* 257, 149–160
5. Hanisch, F.-G., Jovanovic, M., Peter-Katalinic, J. (2001) Glycoprotein identification and localization of O-glycosylation sites by mass spectrometric analysis of deglycosylated/alkylaminylated peptide fragments. *Anal. Biochem.* 290, 47–59
6. Mirgorodskaya, E., Hassan, H., Clausen, H., Roepstorff, P. (2001) *Anal. Chem.* 73, 1263–1269
7. Zheng, Y., Guo, Z., Cai, Z. (2009) Combination of beta-elimination and liquid chromatography/quadrupole time-of-flight mass spectrometry for the determination of O-glycosylation sites. *Talanta* 78, 358–363
8. Wells, L., Vosseller, K., Cole, R.N., Cronshaw, J.M., Matunis, M.J., Hart, G.W. (2002) Mapping sites of O-GlcNAc modification using affinity tags for serine and threonine post-translational modifications. *Mol. Cell Proteomics* 1, 791–804
9. Mirgorodskaya, E., Hassan, H., Wandall, H.H., Clausen, H., Roepstorff, P. (1999) Partial vapor-phase hydrolysis of peptide bonds: A method for mass spectrometric determination of O-glycosylated sites in glycopeptides. *Anal. Biochem.* 269, 54–65
10. Goletz, S., Thiede, B., Hanisch, F.-G., Schultz, M., Peter-Katalinic, J., Müller, S., Seitz, O., Karsten, U. (1997) A sequencing strategy for

the localization of O-glycosylation sites of MUC1 tandem repeats by PSD MALDI mass spectrometry. *Glycobiology* 7, 881–896

11. Hanisch, F.-G., Green, B.N., Bateman, R., Peter-Katalinic, J. (1998) Localization of O-glycosylation sites of MUC1 tandem repeats by QTOF mass spectrometry. *J. Mass Spectrom.* 33, 358–362
12. Hogan, J.M., Pitteri, S.J., Chrisman, P.A., McLuckey, S.A. (2005) Complementary structural information from a tryptic N-linked glycopeptide via electron transfer ion/ion reactions and collision-induced dissociation. *J. Proteome Res.* 4, 628–632
13. Darula, Z., Chalkley, R.J., Baker, P., Burlingame, A.L., Medzihradszky, K.F. (2010) Mass spectrometric analysis, automated identification and complete annotation of O-linked glycopeptides. *Eur. J. Mass Spectrom.* 16, 421–428
14. Brown, R., Lennon, J.J. (1995) Sequence-specific fragmentation of matrix-assisted laser-desorbed protein/peptide ions. *Anal. Chem.* 67, 3990–3999
15. Suckau, D., Cornett, D.S. (1998) Protein sequencing by ISD and PSD MALDI-TOF MS Anal. Mag. 26, 18–21
16. Suckau, D., Resemann, A. (2003) T3-sequencing: Targeted characterization of the N- and C-termini of undigested proteins by mass spectrometry. *Anal. Chem.* 75, 5817–5824
17. Resemann, A., Wunderlich, D., Rothbauer, U., Warscheid, B., Leonhardt, H., Fuchser, J., Kuhlmann, K., Suckau, D. (2010) Top-down *de novo* protein sequencing of a 13.6 kDa camelid single heavy chain antibody by matrix-assisted laser desorption ionization-time-of-flight/time-of-flight mass spectrometry. *Anal. Chem.* 82, 3283–3292
18. Resemann, A., Suckau, D. (2009) Automated acquisition of MALDI-ISD spectra for the N- and C-terminal sequence determination of intact proteins, Bruker Daltonics, Technical Note #TN-36
19. Hanisch, F.-G. (2011) Top-down sequencing of O-glycoproteins by in-source-decay matrix-assisted laser desorption ionization mass spectrometry for glycosylation site analysis. *Anal. Chem.* 83, 4829–4837

Chapter 11

Analysing Mucin Degradation

Stephen D. Carrington, Jane A. Irwin, Li Liu, Pauline M. Rudd, Elizabeth Matthews, and Anthony P. Corfield

Abstract

Turnover of mucins in supramucosal gels is essential for the removal of surface contaminants, and the maintenance of normal mucosal barrier function. In addition to the well-known processes promoting the physical turnover of mucus gels, extracellular mucin degradation also requires the coordinated action of a range of enzyme activities including glycosidases and proteases. These are collectively termed "mucinase". Derangements of mucinase activity lead to downstream barrier defects and mucosal disease. This chapter is focussed on methods that can be used to assess the degradation of whole mucins and isolated mucin glycans. A range of approaches is described using labelled or unlabelled substrates utilised in assays based on 96-well plates, size exclusion chromatography, and NP-HPLC. These are suitable for defining the extent and progress of mucin degradation in different mucosal systems, and identifying abnormalities and critical control points.

Key words: Mucin, Mucinase, Glycosidase, *O*-linked glycans, Degradation, Hydrolase, Glycoprotein

1. Introduction

Supramucosal defensive mucus gels provide for the hydration and mechanical protection of moist mucosal surfaces. A property of such gels, which is fundamental to normal barrier function, is their capacity to turn over. In this way, adherent microbes, toxins, and other environmental contaminants are removed. Such turnover relates to both secreted and cell surface mucins. However, this chapter focuses on secreted mucins, and information on the turnover of cell surface mucins can be found in Chapter 7 of this volume.

Well understood physical mechanisms, such as peristalsis in the gut and mucociliary clearance in the lung, promote mucus turnover at different locations. However, for these mechanisms to function optimally, the biophysical properties of the mucus must

Michael A. McGuckin and David J. Thornton (eds.), *Mucins: Methods and Protocols*, Methods in Molecular Biology, vol. 842, DOI 10.1007/978-1-61779-513-8_11, © Springer Science+Business Media, LLC 2012

be physiologically "tuned". This is thought to be achieved largely through changes in the hydrated mucin glycoproteins that form the molecular scaffold of mucus. It provides for the maintenance of a competent defensive barrier that can integrate the requirements for physical protection with the optimal retention of secreted innate and adaptive immune effector molecules, and in some cases the presence of a beneficial microbial biofilm. It is now well recognised that a key component of this homeostatic balance is the enzymic degradation of mucins. Indeed, dysregulation of such degradation is clearly implicated in many disorders of mucosal barrier function, where mucus is either excessively degraded (1, 2) or accumulates abnormally and fails to turn over (3). The degradation of supramucosal mucus gels is the product of many factors that are location-specific. Clearly, the profile of degradative enzymes present is decisive. However, other substrate-related factors include: the profile of expressed mucin proteins, their glycosylation, and the nature and abundance of the mucin cross-linkages in the biopolymeric matrix.

Mucin-degrading enzymes may potentially include proteinases, peptidases, glycosidases, sulphatases, glycosulphatases, and esterases of prokaryotic or eukaryotic origin (4–7). In addition, the presence and action of other hydrolases, including phosphatases, esterases, and lipases should also be considered, since mucins may also be modified by phosphorylation and the addition of fatty acids (8, 9). The balance between endogenous (secreted) and exogenous (microbial) sources of such activities varies at different epithelial locations. Mucosal surfaces supporting a prominent biofilm (such as the colon and reproductive tract) tend towards microbial sources of mucinase activity, and poorly colonised surfaces (such as the eye and the lung) are probably more reliant on host-derived enzymes (10). Beyond regulation of the biophysical properties of mucus gels, mucinase activity also has a potential role in generating the ligands and metabolic feedstocks required for the establishment of stable microbial biofilms at mucosal epithelial surfaces in health and disease states. It may also be required for maintaining optimal levels of key carbohydrate ligands with the capacity to block the adhesion and pathogenicity of certain microorganisms (7, 8, 11, 12).

While proteases can depolymerise mucins, and may promote the access of other degradative enzymes into mucus gels, the comprehensive degradation of mucin monomers requires the systematic removal of their glycans before extensive proteolysis can proceed. The proteolytic degradation of mucins has been extensively considered in a previous edition of this volume (13).

Mucin glycans are closely packed along repeating peptide sequences of the apomucin core. This may sterically hinder the access of endoglycosidases with the capacity to remove such glycans in their entirety. Thus, the degradation of most extended *O*-glycans is usually a progressive process involving the collaboration of exoenzymes which initially degrade the peripheral capping structures of the glycans, followed by their polylactosamine backbones, and subsequently their variable cores (to which access may only be gained after more peripheral structures have been removed) (8, 12, 14). It is currently unknown whether a specific propensity to degradation is biosynthetically programmed into the glycosylation patterns of mucins at given mucosal sites.

Mucinase activity is best assessed using a whole mucin substrate and analysing its progressive degradation to low-molecular-weight fragments. The presence of individual enzyme activities resulting in the release of single components from mucin substrates does not give a complete analysis of the overall mucin-degrading potential of the sample under study. Furthermore, the glycosylation phenotypes of mucin substrates, and their specific interaction with the profile of degradative activities at particular mucosal surfaces, are both determinants of the outcome of mucin degradation. Thus, if fragmentation patterns are to be used as a read-out for mucinase activity, the structural variability between apomucin peptides and their individual patterns of glycosylation must be considered. With increasing knowledge of the structure and organisation of mucin (4, 8, 9), it is now becoming possible to match fragmentation patterns with known sequence information for both peptides and oligosaccharides. However, the analysis of size-fractionated products for carbohydrate and/or amino acid composition is time-consuming and requires large amounts of relatively homogeneous substrate. In this context, the use of recombinant mucins as standardised substrates would be desirable, but these products are not yet widely available. Other approaches such as plate-based assays are required for high-throughput screening. Nevertheless, a careful examination of the fragments derived from incubations of specific mucins with mucinase activity from different sources has the potential to yield information on successive degradative steps.

The progressive deglycosylation of mucins and their glycopeptide fragments can be tracked on slot blots of fractions using panels of lectins or lectin arrays (15). This can be particularly useful when examining the effects of inhibiting specific glycosidase activities on the overall outcome of mucin fragmentation. In particular, this approach has the potential to define key control points in mucin degradation related to the initial removal of capping structures on mucin glycans, which have the potential to block downstream degradation of mucin glycopeptides. A caveat here is the need to compare the profiles generated using lectins, with the same samples detected using generic methods for staining glycoproteins. This is because the comprehensive removal of a single sugar from glycopeptides, when they are detected by lectin staining alone, can give the *impression* of complete degradation of a major glycopeptide when this is not the case.

Screening for mucinase activity is possible using radiolabelled mucins prepared from organ or cell cultures. Mucins can be labelled with suitable precursors; for example, cultures of colonic mucosa can be labelled with [^{35}S]sulphate, [^{3}H]glucosamine, and [^{3}H] threonine. The purified mucins are subsequently incubated with a source of mucinase activity, and the degradation is assessed by gel filtration (4, 12, 16, 17). While extremely useful, this approach has the drawback that the results may be batch dependent, and the preparation and storage of large amounts of radiolabelled mucins from cell, tissue, or organ culture is generally impractical.

Solid phase assays are an alternative to the radioactive methods. They utilise biotinylated mucin substrates tagged through either their peptide or carbohydrate moieties (18, 19) and have been adapted from similar assays for protein substrates (20). They require small amounts of pure substrate, have high sensitivity, are rapid, and can be applied to large numbers of samples. Furthermore, they are suitable for the rapid screening of "total" mucinase activity from enzymatically heterogeneous samples. Such assays are also potentially amenable to further miniaturisation. Tagging mucins through either their protein or carbohydrate residues has implications for assay interpretation, since the release of bound mucin from a plastic surface may be variably dependent on (1) substrate degradation through mucinase action over the whole domain structure of the mucin or (2) cleavage of regions of the mucin molecules specifically responsible for immobilisation to the surface of the plate. Paradoxical results may arise where mucins bind to the plastic surface of microtitre plates through hydrophobic peptide sequences. Here, mucins labelled through their carbohydrate may *appear* to be fully digested when exposed only to peptidases, despite any released glycopeptides being protected from digestion by their glycan "coat". Conversely, where immobilisation is mainly related to *ionic* interactions between charged carbohydrates and the plate surface, sialidase or sulphatase activity alone may release the majority of bound mucin, while causing only *limited* overall mucin degradation. In cases where the mucin has been labelled through amine groups this gives the false impression of proteolytic activity. In practice, mucin binding to microtitre plates probably results from multiple and diverse binding interactions: mitigating such apparently paradoxical results. However, it should be emphasised that the composition of mixed enzyme preparations cannot be deduced directly from plate assays. Other approaches, such as size fractionation chromatography or electrophoresis, are more suited to determining this, since they allow for the comparison of profiles obtained through both methods of biotinylation and their correlation with results obtained using commercially available peptidases and glycosidases.

Beyond depolymerisation and partial degradation of mucins by proteases, their further degradation requires the presence of a suitable range of glycosidases, and potentially esterases and sulphatases, which act in concert. There is great potential diversity of the *O*-linked structures upon which such activities might act, and few standards are routinely available commercially. While there is a need for new and specific synthetic glycosidase substrates, the profile of glycans present on mucins from different locations represents a diverse "bank" of natural substrates. These can be released and profiled by HPLC-based methods (21), which have been developed for the purposes of glycoanalysis and enzyme discovery. While, these methods can specifically track the degradation of specific glycan peaks, they can also be harnessed to examine overall mucinase activity that specifically relates to mucin deglycosylation. This can provide a proportionate readout of overall glycan degradation by comparing integrated peak areas before and after digestion between different enzyme sources. Furthermore, if glycan structures can be assigned from existing databases, detailed information can be obtained about specific glycosidases through the cleavage of known monosaccharide linkages.

No single method of analysis can yet provide complete information on the nature and the dynamics of mucinase activity. Therefore, the techniques listed in this chapter are not a comprehensive list. Other techniques useful for the study of mucin fragmentation, such as rate zonal centrifugation and agarose gel electrophoresis, are described in Chapter 2 of this volume. In addition, rheological readouts are useful on whole mucus samples when examining the dynamic effects of mucinase already present within samples of mucus (22). Finally, a previous edition of this volume specifically describes additional methods of surveying the glycosidase profile of different sources of mucinase activity (23).

2. Materials

2.1. Enzyme Sources

The nature of possible samples to be screened for mucinase activity is diverse and the sources given here serve only as examples.

1. Commercially prepared enzymes: proteases, e.g. trypsin, pronase E, pepsin; glycosidases, e.g. α-sialidase, β-galactosidase, β-*N*-acetylhexosaminidase, α-fucosidase, and *O*-glycanase.
2. Bacterial culture supernatants and cell suspensions or cell extracts.
3. Animal/human secretions and excretions that contain enzymes can also be used, e.g. plasma, urine, tears, faecal extracts, sputum, mucosal washings.
4. Animal and insect cell culture supernatants, cell suspensions, or cell extracts.

2.2. Lectins, and Other Reagents

1. Lectins. The following examples are available as either biotin-tagged or peroxidase conjugated reagents. Sigma-Aldrich Company Ltd. Dorset (peroxidise conjugated): *Triticum vulgaris* (GlcNAc, NeuNAc); *Dolichos biflorus* (α—GalNAc); *Ulex europaeus* (α1-2 fucose). Vector, Peterborough, UK (biotin tagged): *Sambucus nigra* (α2,6 sialic acid); *Maackia amurensis* (α 2,3 sialic acid). Vector reagents are detected using the biotin–avidin system from the same manufacturer.
2. Glycosidase substrates. Synthetic (4-nitrophenyl-glycosides) and fluorimetric (4-methylumbelliferyl glycosides) glycosidase substrates and appropriate buffers for initial profiling of activities present in the enzyme source are described in a previous edition of this book (23). They are available through several suppliers, including Sigma-Aldrich; Oxford Glycosciences, Abingdon, UK; Dextra, Reading, UK; Boehringer Mannheim, Lewes, UK; and Chemica Alta, Edmonton, Canada.
3. Glycosidase inhibitors. The following examples, which are relevant to studies of mucin degradation, are available from Sigma-Aldrich. Deoxyfuconojirimcin (fucosidase inhibitor); siastatin B (sialidase inhibitor); *N*-acetyl-2,3-dehydro-2-deoxyneuraminic (sialidase inhibitor); 1-10-phenanthromonohydrate (β-galactosidase inhibitor).

2.3. Whole Mucin Substrates

2.3.1. Unlabelled Mucins

1. Mucins prepared from mucosal tissue or mucosal scrapings. Commercial sources of mucins such as porcine stomach mucin (PSM), and bovine submaxillary mucin (BSM) are also useful (see Note 1) but may require further purification.
2. Prepare mucin using density gradient centrifugation and gel filtration and characterise for the presence of noncovalently associated contaminants as described in Chapter 2.

2.3.2. Radiolabelled Mucins

1. Radiolabelled mucin singly or dual labelled with [^{3}H]glucosamine or [^{3}H]threonine combined with either [^{35}S]sulphate or [^{14}C] threonine. This mucin must be purified and should elute as a single high-molecular-weight peak at the void volume of a Sepharose CL-2B column. The preparation of these substrates from tissue explants in organ culture is detailed in a previous edition of this volume (16).
2. Gel filtration buffer: 10 mM Tris–HCl, pH 8.0.
3. Sepharose CL-2B (Sigma-Aldrich Company Ltd. Dorset, UK).

2.3.3. Biotinylated Mucins

1. Substrates for biotinylation are described in Subheading 2.3.1 above.
2. Sephadex G25 (Sigma-Aldrich Company Ltd. Dorset, UK). Sephadex column preparation: hydrate Sephadex G25 in phosphate-buffered saline (PBS) (see Subheading 2.3.3, step 8) and pack

in an all-glass column approx 1 × 15 cm. Thoroughly equilibrate the packed column in PBS.

3. Protein biotinylation reagent: *N* -biotinyl-6-aminocaproic-*N*-hydroxysuccinimide ester (BNHS) (Sigma-Aldrich Company Ltd. Dorset, UK).
4. Carbohydrate biotinylation reagent: *N*-biotinyl-6-aminocaproic-acid hydrazide (BACH) (Sigma-Aldrich Company Ltd. Dorset, UK).
5. Dimethylformamide (DMF).
6. Dimethylsulphoxide (DMSO).
7. Sodium periodate.
8. PBS: 0.375 g of sodium dihydrogen phosphate dihydrate, 1.155 g of disodium hydrogen phosphate, and 8.765 g of sodium chloride in 1,000 mL water, pH 7.4.

2.4. Other O-Glycosylated Substrates and Mucin Glycans

1. Where substrates are to be used in assays focussed on exoglycosidase or endoglycosidase activities targeted at *O*-glycans, other *O*-glycosylated nonmucin substrates, such as fetuin and IgA, or synthetic glycopeptides synthesised by chemical and enzymatic methods (e.g. Sussex Research Ottawa, Canada) may also be used.
2. Glycans isolated from these glycoproteins or from whole mucins are used specifically in assays focussed on the role of exoglycosidases in mucin degradation.

2.5. Mucinase Assays

2.5.1. Size Exclusion Assays with Unlabelled Mucin

1. Mucin substrates are described in Subheading 2.3.1 above.
2. Microcon Ultracel YM-30 filters (Millipore (UK) Limited Watford, UK) or equivalent.
3. Synthetic glycosidase substrates and appropriate buffers are described in Subheading 2.2 above.
4. Sepharose CL-2B (Sigma-Aldrich Company Ltd. Dorset, UK) packed into a 1 × 30 cm column equilibrated in 80 mM Tris–HCl buffer pH 7.6.
5. Periodic acid Schiff's stain.
6. Slot blotting manifold and blotting membrane, Immobilon PVDF.
7. Biotin-labelled lectins and detection reagents (see Subheading 2.2, step 1).

2.5.2. Size Exclusion Assays with Radiolabelled Mucin

1. Sepharose CL-2B equilibrated in gel filtration buffer.
2. Gel filtration buffer 10 mM Tris–HCl, pH 8.0, or 4 M guanidine hydrochloride/PBS buffer.
3. Scintillation fluid, Optiphase HiSafe 3, (PerkinElmer, Buckinghamshire, UK).

2.5.3. Plate Assays with Biotinylated Mucin: Coating of Plates

1. Microtitre plates, 96-well, Nunc-Immuno™ plates, MaxiSorp Surface™ (Thermo Fisher Scientific, Roskilde, Denmark) (see Note 2).
2. Coating buffers: 0.1 M sodium acetate buffer, pH 5.0, for carbohydrate-labelled mucin and 0.1 M sodium phosphate buffer, pH 7.0, for protein-labelled mucin (see Notes 3 and 4).

2.5.4. Plate Assays with Biotinylated Mucin: Detection

1. PBS (see Subheading 2.3.3, item 8).
2. PBST: Add Tween-20 to PBS to give a final concentration of 0.2% (v/v).
3. Blocking buffer: 1% (w/v) bovine serum albumin in PBST. Use enough to fill the well, 200–300 μL.
4. Streptavidin-horseradish peroxidase (HRP) solution: Streptavidin-HRP conjugate (Vector Laboratories, Ltd., Peterborough, UK) at 1 mg/mL is to 1:1,500 in blocking solution (75 μL/well).
5. OPD Solution: 1,2-phenylenediamine dihydrochloride (Dako UK Ltd., Cambridgeshire UK). Dissolve four (2 mg) tablets in 12 mL of distilled water and add 5 μL of 30% hydrogen peroxide immediately prior to use.
6. Stop solution: 0.5 M sulfuric acid (28 mL of 95–97% acid in 1,000 mL of distilled water).

2.5.5. HPLC Assays Using Mucin Glycans

1. The high performance liquid chromatography equipment used to develop the methods for glycan profiling was the following: Normal phase (NP) HPLC: TSK-Gel Amide-80 4.6 × 250 mm column (Anachem, Luton, UK) mounted on a 2695 Alliance separation module equipped with a Waters temperature control module and a Waters fluorescent detector 2475 (Waters, Milford, MA, USA). A Waters Empower chromatography workstation software build 1154 is used for system control and data processing. Similar HPLC set-ups can be used.
2. Amicon centrifugal filters: 10 kDa molecular mass cut-off (Millipore).
3. 2-Aminobenzamide (2-AB) kit from Ludger Ltd (The Oxford BioBusiness Center, Oxford, UK).
4. 28% (w/v) NH_3 H_2O (Sigma-Aldrich) saturated with $(NH_4)_2CO_3$ (Fluka, for HPLC use, with a precipitate of white salt). The solution can last for a few months if the bottle is properly sealed.
5. Freshly made 1% (v/v) formic acid (Sigma-Aldrich, 98–100%) in H_2O (Millipore grade).
6. Whatman 3MM chromatography paper (15.0 cm × 100 m, Whatman Schleicher Schuell).

7. Phosphate buffer: 50 mM, pH7 (0.292% (w/v) NaH_2PO_4 + 0.773% (w/v) Na_2HPO_4).
8. HPLC buffer A: 50 mM formic acid adjusted to pH 4.4 with ammonia solution.
9. HPLC buffer B: HPLC grade acetonitrile (Romil-SpS™).
10. 2AB-labelled dextran ladder (2AB-glucose homopolymer, Ludger Ltd).

3. Methods

3.1. Mucinase Assay Using Unlabelled Substrates and Detection by Lectins

This approach is particularly useful for the identification of glycosidases with importance for overall mucin degradation in particular epithelial systems.

1. Re-suspend lyophilised mucins to a final concentration of 1 mg/mL (see Note 5) in an appropriate incubation buffer (see Note 6).
2. Add 1–3.5 mL of enzyme source (see Note 7) to a 300 kDa cut-off centrifugal filter (PALL Corporation) and centrifuge at 7,500 × *g* for 15 min at 4°C. The filtrate is used, as it contains material from which high-molecular-weight glycoproteins have been removed.
3. Establish baseline glycosidase activities in the enzyme sample using assays described in a previous edition of this book (23). Dilute the enzyme preparation in the same buffer as the substrate if required (see Note 8).
4. Mix 50 μL of re-suspended mucin with 10 μL of enzyme and 40 μL of an appropriate buffer in triplicate in a 96-well microtitre plate (see Note 9).
5. A number of both positive and negative controls should be carefully chosen and run alongside these incubations (see Note 10).
6. Incubate at 37°C for 24 h (see Note 11).
7. Dilute each sample with 400 μL of incubation buffer and immediately freeze at −20°C until ready to start gel filtration.
8. Load each 0.5 mL sample onto a 1 × 30 cm Sepharose CL-2B column equilibrated in 80 mM Tris–HCl buffer pH 7.6 and elute with the same buffer. Collect 60 × 0.5 mL fractions.
9. Test 5–100 μL of each fraction to obtain a baseline molecular size profile for all mucin fragments in each incubation using a slot blot assay with a Periodic acid Schiff's stain (Chapter 3) (see Note 12).

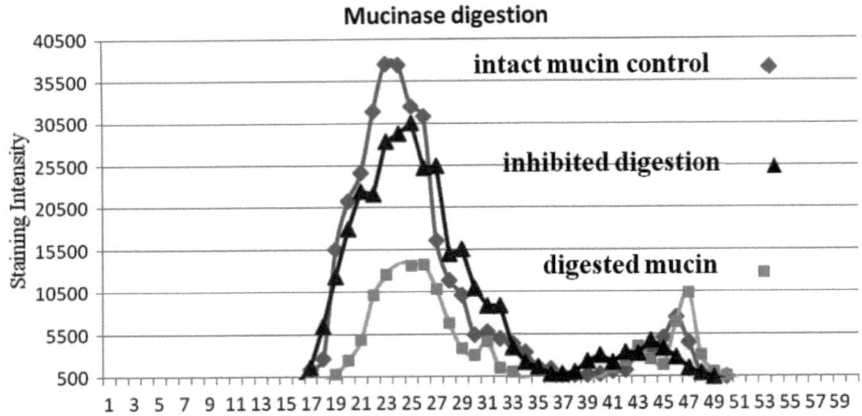

Fig. 1. Profile of semipurified ocular mucins incubated with mucinase from tears. Aliquots were run on Sepharose CL-2B and fractions were slot blotted and stained with PAS: mucin + buffer only (*filled diamond*—baseline control); mucin + mucinase + glycosidase inhihibitor (*filled triangle*—In this case showing that inhibition of a single specific glycosidase substantially blocks digestion); mucin + mucinase (*filled square*—showing substantial degradation manifest as a reduction in the size of the high-molecular-weight peak between fractions 17 and 35). All incubations were carried out at pH 7 for 24 h at 37°C.

10. Probe identical slot blots in triplicate with a selected panel of HRP-tagged lectins and develop using an appropriate HRP substrate (see Note 13). Quantify staining responses densitometrically.
11. Scrutiny of mean sample profiles (for example see Fig. 1) will enable the designation of a specific fraction number that denotes an arbitrary cut-off between high-molecular-weight material present before digestion (usually defined as the void volume excluded peak from the fraction profile) and lower-molecular-weight mucin fragments, mostly representing degraded mucin fragments. The loss of integrated peak area from the high-molecular-weight peak between the mucinase digested and control (buffer only) profiles can then be expressed as a percentage which represents the loss of high-molecular-weight lectin-stainable material attributable to digestion (see Notes 14 and 15).

3.2. Mucinase Assay Using Radioactive Substrates

1. Mix 20–500 μL of enzyme extract or commercially available enzyme (see Subheading 2.3.2, step 1) with 5,000–50,000 cpm of radiolabelled mucin in incubation buffer in a final volume of 1 mL, maximum (see Notes 16 and 17).
2. Incubate at 37°C for periods up to 24 h (usually 6, 12, or 24 h, but start with 24 h if the activity is unknown).
3. Prepare a blank and incubate under the same conditions as step 2 above (see Note 18).

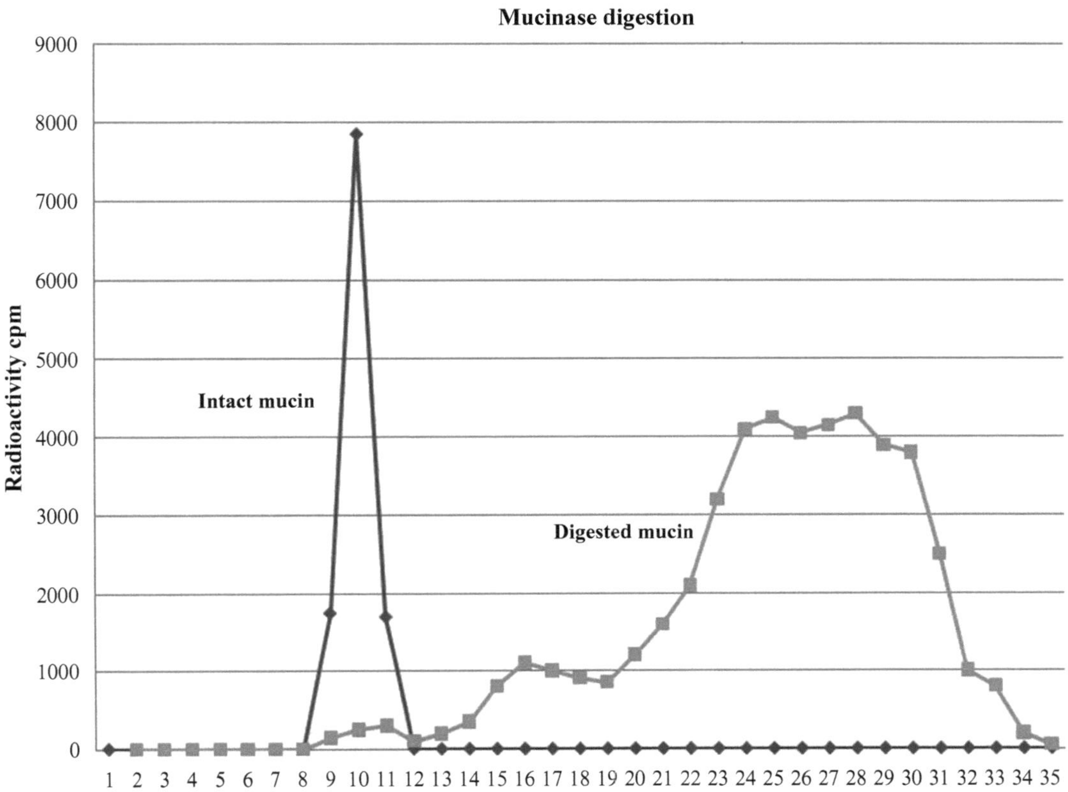

Fig. 2. Identification of the products of mucinase activity by Sepharose CL-2B chromatography. Intact human colonic mucin, derived from colonic explants in culture, labelled with [^{3}H]-D-glucosamine (*filled diamond*). The products of incubation with human faecal extract are shown (*filled square*) and the fractions identified as digested mucin.

4. After incubation, either load onto Sepharose CL-2B column immediately, or freeze at–20°C until you are ready to start gel filtration.
5. Load as a 1 mL sample onto a 1 × 30 cm Sepharose CL-2B column equilibrated in 10 mM Tris–HCl, pH 8.0, and elute with the same buffer.
6. Collect 30 × 1 mL fractions
7. Mix the entire fraction (or 0.5 mL where >50,000 cpm are present) with scintillation fluid and measure the radioactivity.
8. Compare the profiles of the test sample with the blank incubations. Identify the region of low-molecular-weight product that represents degraded mucin. Figure 2 gives an example profile.
9. Subtract the blank incubation (background) from the test incubation, and assess the proportion of low-molecular-weight product formed from the high-molecular-weight substrate.

3.3. Biotinylation of Mucin Substrates

3.3.1. Biotinylation Through the Protein Moiety

1. Dissolve BNHS in DMF to give a final concentration of 20 mg/mL.
2. Dissolve 1 mg of mucin in 0.9 mL of PBS (larger batches can be prepared at the same mucin buffer ratio).
3. Add 0.1 mL of BNHS solution and incubate at 4°C overnight or at room temperature for 4 h.
4. Load the biotin/mucin incubation (1 mL) onto a Sephadex G25 column and run in PBS buffer, collecting 1 mL fractions up to 30 mL total volume. The Sephadex column is discarded (see Note 19).
5. Test fractions from the column for mucin using the slot blot assay with the periodic acid Schiff's stain (see Chapter 3).
6. Test fractions for biotin labelling by adherence to 96-well microtitre plate, followed by streptavidin labelling (see Subheading 3.4).
7. Pool the labelled fractions (approx 5 × 1 mL) to give a concentration of 20 g/mL mucin and aliquot as 0.2 mL samples. Store at 4°C until used.

3.3.2. Biotinylation Through the Carbohydrate Moiety

Carbohydrate labelling requires the periodate oxidation of the carbohydrate moieties of the mucin before biotinylation of these oxidised residues. In the case of colonic mucins, the presence of *O*-acetyl esters will block this oxidation and a saponification step must be undertaken first.

1. Dissolve 1 mg of mucin in 0.5 mL of 0.1 M sodium hydroxide and incubate for 45 min at room temperature. Neutralise to approx pH 7.0 with 0.05 mL of 1 M HCl. Check the pH. This is the saponification step.
2. Adjust to pH 5.5 with 0.1 M acetate buffer. Add sodium periodate so that the final concentration is 1 mM.
3. Incubate for 20–60 min at room temperature
4. Apply the oxidised mucin to a Sephadex G25 column as for the biotinylation of protein- labelled mucin (see Note 20) and collect fractions.
5. Detect mucin-containing fractions using the slot-blot assay with the periodic acid Schiff's stain (see Chapter 3) and pool these fractions (approx 5 × 1 mL).
6. Add BACH in DMSO to a final concentration of 1 mM.
7. Incubate at room temperature for 2 h or overnight at 4°C.
8. Separate the biotinylated mucin on a Sephadex G25 column as for protein-labelled mucin.
9. Collect fractions and detect mucin using the slot blot assay with the periodic acid Schiff's stain and with the 96-well microtitre plate assay (see Subheading 3.4).

10. Pool the labelled fractions (approx 8 × 1 mL) to give a concentration of 12.5 μg/mL mucin and aliquot as 0.2 mL samples. Store at 4°C until used.

3.4. Mucinase Plate Assay Using Biotinylated Mucin Substrates

3.4.1. Coating of Microtitre Plates

1. Dilute the biotinylated mucin in coating buffer (see Note 4).
2. Carefully place the chosen volume in the bottom of the microtitre plate well (see Note 3). Typically, about 50 μL is used.
3. Incubate the plates overnight at 4°C for the mucin to adsorb onto the plate.
4. Empty the plates carefully.
5. Wash once with incubation buffer 1 × 50 μL followed by 3 × 200 μL. The buffer used is the buffer that is to be used in the assay (see Notes 16 and 17).

3.4.2. Digestion of Coated Plates

1. Prepare a suitable dilution of the enzyme sample in incubation buffer (see Notes 16, 17, and 21).
2. Place 60–100 μL of enzyme preparation in each well (i.e. more than the volume of mucin solution which was used to coat the plate).
3. Incubate at 37°C typically for 1–2 h.
4. Carefully remove the digestion media.
5. Wash four times with 200 μL of PBS.

3.4.3. Detection of Labelled Mucins

1. Block non-specific binding with 150–300 μL of blocking buffer for 1 h at room temperature or overnight at 4°C.
2. Empty the plates and wash twice with 200 μL PBS per well.
3. Incubate with streptavidin-HRP solution (75 μL) for 60 min at room temperature.
4. Empty the plates and wash with PBS 1 × 100 μL and 3 × 200 μL.
5. Place 100 μL of OPD solution in wells and incubate in the dark for up to 60 min.
6. Stop the reaction and develop the colour by the addition of 100 μL of stop solution (0.5 M H_2SO_4).
7. Measure the absorbance at 490 nm (see Note 22).

3.5. HPLC Assays Tracking Mucin Glycans and Their Degradation

3.5.1. Chemical Release of Glycan Substrates from Glycoproteins

1. Take 100 μg of glycoprotein and mix with 100 μL of 28% NH_3 H_2O saturated with $(NH_4)_2CO_3$.
2. Add a further 10 mg of NH_4HCO_3 to the solution.
3. Incubate the mixture at 60°C for 40 h (see Note 23).
4. Evaporate the solution in a Speedvac under vacuum at 35°C to remove the NH_4HCO_3 (desalting step).
5. Repeatedly re-dissolve the sample in H_2O and re-dry until no white solid is visible (confirmation of desalting).

6. When the salt has been removed, treat with 20 μL 1% formic acid at room temperature for 40 min.
7. Prepare working solution of 2-AB using the Ludger Tag™ kit as follows: Take 150 μL of acetic acid and mix with the entire bottle of DMSO provided. Add 100 μL of the mixture to the bottle of 2AB provided and mix. Add 100 μL of this mixture to the bottle of sodium cyanoborohydride powder provided and mix until the solid is dissolved. The final mixture is ready for sugar labelling.
8. Dry the sample under vacuum as described in step 4 and incubate with 5 μL of 2-AB solution at 65°C for 30 min and then agitate. Continue the incubation for a further 90 min.
9. Remove the excess 2-AB by ascending paper chromatography in acetonitrile on Whatman 3MM paper strips (~3 × 12.5 cm): Spot the labelled sugar (~5 μL, as indicated in the previous step) onto the paper 1 cm from one end. Stand the strip vertically in acetonitrile with the sample facing down. Ensure that there is sufficient acetonitrile to barely submerge the end of the paper leaving the sample spot just above the level of the solvent. The excess 2AB will migrate upwards with acetonitrile along the paper strip, while the labelled glycan sample remains at the original position. The sample spot is then cut out of the paper, and placed in a syringe. 250 μL of H_2O is then drawn into the syringe and the sample is left to soak for 10 min. The water (with dissolved sample) is then recovered and placed in an Eppendorf. This process is repeated a further three times on the same sample spot, yielding the labelled sample in 2 mL of water (24).
10. Load 10% of the final glycan sample on NP-HPLC to obtain the glycan profile (designated P1) (see Fig. 3). Set the gradient conditions of HPLC for a linear gradient of 26–52% HPLC buffer A (50 mM formic acid adjusted to pH 4.4 with ammonia), over 104 min at a flow rate of 0.4 mL. Samples (20 μL) are mixed with 80 μL of acetonitrile before injection into the HPLC column. The fluorescence is measured at 420 nm with excitation at 330 nm.

3.5.2. Preparation of Enzyme Source

1. Take 100 μL bacterial culture supernatant (or other appropriate enzyme source—see Subheading 2.1) and concentrate it by centrifugal filtration at 4,000 × *g* for 20 min through an Amicon centrifugal filter (15 mL, 10 kDa molecular mass cut-off).
2. Discard the filtrate and retain the retentate.
3. Add 50 mM phosphate buffer to the retentate to reconstitute the original volume.
4. Repeat steps 1–3 three times.
5. The final retentate is ready for use, as low-molecular-weight compounds, including glycans, have been removed (see Note 25).

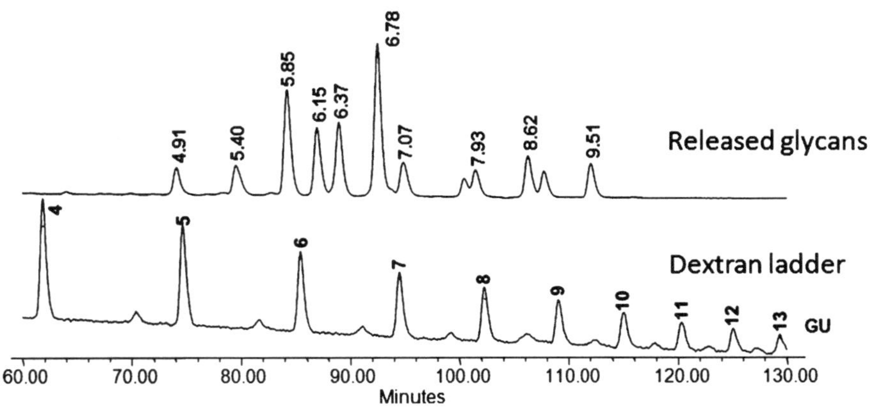

Fig. 3. Integration of glycan HPLC profiles. The upper trace is a standard dextran ladder with varying chain length (from 10 to 150 kDa), which is used for calibration. The numbers relate to the number of glucose units (GU) in the dextran chain. The lower trace shows glycans released from bovine fetuin (an example of P1- a profile of chemically released and 2-AB tagged glycans unmodified by glycosidase digestion). The sugars in the mixture are eluted at different times in ascending order based on the molecular weight. The numbers adjacent to each peak are GU values derived with reference to the dextran ladder (see Note 24).

3.5.3. Digestions to Identify Exoglycosidase Activity

1. Add 10% of the labelled glycan substrate (prepared as described under Subheading 3.5.1) to 100 μL ultrafiltered enzyme in an appropriate incubation buffer, e.g. PBS or 0.1 M citrate-phosphate buffer, pH 5 (see Subheading 3.5.2) and incubate at 37°C overnight.
2. Remove the proteins from the sample by filtration through Millipore centrifugal filters (10 kDa cut-off). The filtrate contains the glycans.
3. Dry the filtrate containing glycans under vacuum and re-suspend in 20 μL of H_2O.
4. Add 80 μL of acetonitrile and load the sample (now in 80% acetonitrile) onto the HPLC to obtain the profile of glycans digested by the culture (designated P2).
5. A mixture of glycan and bacterial culture (without incubation at 37°C overnight) is also profiled by HPLC, and used as a control.
6. Compare P1 (i.e. the peaks before digestion- see Subheading 3.5.1, step 10) and P2 (see step 4 above). The shift of any peak in P2 indicates the presence of an exoglycosidase active against the terminal sugar present on the sugar represented by the peak. Figure 4 gives an example based on a commercially available enzyme (NAN1) used to digest glycans derived from a standard glycoprotein (fetuin).
7. When coupled with the analysis of glycan structures (see Note 26), this is also a powerful method for detecting and characterising unknown exoglycosidase activities (see Note 27), and mapping them to the degradation of specific glycan structures that are part of the natural repertoire of specific mucins.

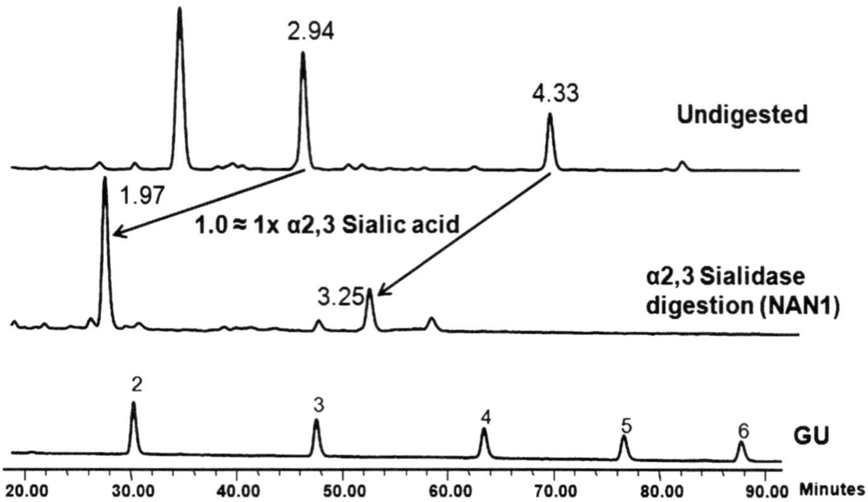

Fig. 4. Analysis of glycan structure based on exoglycosidase digestion coupled with HPLC profile. Different exoglycosidases have different substrate specificities. The digestion of a given terminal sugar from a glycan structure by the corresponding exoglycosidase causes a peak shift. The difference in Glucose Unit (GU) values of the peaks before and after digestion indicates the cleavage of a specific sugar and linkage. Sequential digestion of a glycan by a suite of commercially available exoglycosidases can thus give detailed information about the structure of the entire glycan (see Note 26).

3.5.4. Digestions to Identify Endoglycosidase Activity (see Note 28)

1. Add 100 μg of unlabelled mucin (see Subheading 2.3.1) or other glycoprotein (see Subheading 2.4, step 1) to 100 μL ultrafiltered enzyme source (see Subheading 3.5.2) and incubate overnight at 37°C.
2. Remove the proteins from the sample by filtration using Millipore centrifugal filters (Mr cut-off 10 kDa). The flow-through fraction contains the glycans removed from the glycoprotein by endoglycosidase activity.
3. Dry the filtrate under vacuum and incubate the dried sample with 20 μL of 1% (w/v) formic acid at room temperature for 40 min.
4. Add 5 μL of 2-AB (made up according to the manufacturer's instructions) to the dried sample and incubate at 65°C for 2 h.
5. Remove the excess 2-AB by chromatography using 3MM Whatman paper and elute the glycans with H_2O from the chromatographic paper (see Subheading 3.5.1, step 9).
6. Dry the eluted sample in a speedvac and re-suspend in 20 μL of H_2O.
7. Load the glycan sample onto HPLC to generate a glycan profile (designated P3).
8. A mixture of glycan and bacterial culture is used as a control (see Subheading 3.5.3, step 5).
9. A comparison of P3 (the profile of glycoproteins released from the mucin by the enzyme source) with P2 (the profile of

chemically released glycans from the same glycoprotein digested by the same enzyme source—see step 4, above), indicates if there is an endo-glycosidase activity present that can release the glycans (evident on P3) from the mucin. Further comparison between P2 and P1 (the undigested profile of chemically released glycans) will indicate whether the released structures have also been subjected to exoglycosidase activity present within the same enzyme source (see Notes 29 and 30). Conversely, the absence of a known peak, present in P2, from P3 tends to indicate the absence of the corresponding endoglycosidase.

3.5.5. Comparison of Overall Mucinase Activity Between Enzyme Sources Using HPLC

The above approach is principally of use in detecting specific glycosidases with activities against individual structures present within HPLC glycoprofiles. However, it can also be used to provide proportionate comparisons of overall glycosidase activity between different enzyme sources, assessed against a particular glycan pool. This is possible because the area under each peak is representative of the quantity of the glycan within the peak (see Note 31).

1. Digestions are undertaken as described in Subheading 3.5.3. Care is required to use equal quantities of glycans and buffers, and vary only the enzyme source.
2. Instead of tracking changes in GU value of individual peaks, the overall reduction in the integrated peak area between P1 and P2 is compared between enzyme sources to obtain a percentage difference, using a similar approach to that described for whole mucins (see Subheading 3.1, steps 11 and 12). This is representative of differences in overall exoglycosidase activity between enzyme samples and requires the selection of a suitable GU cut-off, below which peaks are excluded from the comparison (see Note 30).
3. Further digestions may also be undertaken as described in Subheading 3.5.4. Again, all conditions should be identical for the profiles to be compared, other than the enzyme source.
4. Instead of tracking changes in GU value of individual peaks, the total peak area of P3 can be compared between enzyme sources to obtain a percentage difference which is indicative of overall differences in endoglycosidase activity. Again, it is advisable to select a suitable GU value cut-off, below which peaks are excluded from the comparison (see Note 31). The presence of only monosaccharides in P3 precludes the ability to differentiate endo- and exoglycosidase activity (see Note 30). However, it can be useful for tracking the release of glycans from whole mucins.

4. Notes

1. The choice of the mucin source for mucinase assays is of great importance. If possible, it is always best to prepare a mucin substrate from the mucosa that is the target for mucinase study. In cases where it is not possible to obtain suitable samples, e.g. owing to ethical considerations, or to the low abundance of material from minor mucosal surfaces, it may be possible to obtain equivalent material from normal animals. If "nontypical" mucins are chosen and used to probe the presence of mucinase activity in normal and/or disease situations, the interpretation of the physiological significance of the results needs careful consideration. Such mucins can of course be used simply to detect the presence of any mucinase activity.
2. The choice of plate type and buffer is critical. The amount of mucin that binds to the plate is dependent on both the type of plastic that the plate is composed of and any treatment it has undergone (such treatments affect the charge of the plastic). The pH affects the charge of the mucin and therefore the adherence to the plate. Preliminary experiments are crucial to assess the effect of both the plate type and the pH of the buffer used.
3. When preparing the plates, it is essential to dilute the mucin in the desired coating buffer before addition to the plate. When placing the diluted mucin in the plate, great care must be taken to not to touch the sides of the well where the mucin will not be required to coat. Also when transporting the plate, it is essential not to distribute the mucin outside the normal coating area. The mucins stick rapidly and disturbance will increase the coated surface area.
4. The chemical composition and pH of the coating buffer are critical. Avoid using buffers containing amino groups (Tris, HEPES, glycine), particularly with protein-labelled mucins. Instead, use buffers such as acetate or citrate.
5. Care should be taken to standardise conditions of mucin purification if assay results are to be compared. To determine a dry weight for the mucin samples, they should be completely desalted using Sephadex G25 media, (bead size 50–150 μm and fractionation range 1,000–5,000 Da) and a dry weight obtained after complete freeze drying. While weighing small amounts of substrate accurately can be problematic, a range of protein assays have proved unreliable for mucin quantification. It should be noted that these assays require the amount of substrate used to be visualised on blots detected by lectins. Thus, an approximate quantification of the amount of mucin

used is generally sufficient. This can be achieved empirically by slot blotting 1 μL of the diluted substrate to be used in the assay, and staining the blot with Periodic acid Schiff's reagent. A clearly visible staining response normally indicates that sufficient substrate is present for subsequent lectin detection.

6. The choice of suitable incubation buffer should cover the range of potential glycosidase activities identified in the glycosidase assay. Sodium citrate and acetate buffers can be used at lower pH. Phosphate and Tris buffers can be used at higher pH. The assay is designed to be run at the physiological pH of the mucosal surface under study.
7. If the enzyme sample is a mucus sample such as sputum it will be necessary to spin down the sample to pellet contained mucins or other biopolymers. The supernatant is then used as an enzyme source.
8. The glycosidase activities detected can be used to select a suitable panel of lectins for staining a variety of specific sugars and linkages that may be cleaved by the activities in the sample. Quantifying these activities enables the standardisation of glycosidase activities across biological replicates by dilution where this is appropriate to the specific objective of the assay.
9. It is most important that the sample is mixed thoroughly. This can be achieved by gently pipetting the sample 5–10 times. The use of the microplate is to facilitate reading of multiple assays and replicates. The concentration of mucin is selected to exceed the saturated binding capacity of the multiwell plate, since the incubated volume is recovered and subsequently loaded onto a size exclusion column. The selection of the type of plate is not critical but variants with lower binding capacity are preferred.
10. It has been shown that there are structural sites contained within mucins that are preferential locations for spontaneous cleavage. Also, native proteolytic activity and bacterial glycosidases are both associated with the mucin breakdown so control incubations should be carefully chosen. A mucin only buffer control should be run to provide a baseline for such background degradation. Broad range protease inhibition should be used inhibit potential mucin degrading proteolytic activity and allow the specific action of glycosidases to be resolved. Specific glycosidase inhibitors can be used to identify important mucin-degrading glycosidases and highly purified commercial enzymes can be used to validate significant observations.
11. It may not be necessary to incubate for this length of time to obtain stable intermediate glycopeptides, so preliminary incubations at 6, 12, and 18 h can be run to determine if this is the case.

12. Thorough mixing of the sample before pipetting is essential for reproducible results. This should initially be assessed on triplicate runs of each sample.
13. A useful preliminary panel of lectins could include those from *Triticum vulgaris*, *Dolichos biflorus*, *Ulex europaeus*, *Maackia amurensis*, and *Sambucus nigra*. 1 mL of 10 mg/mL Diaminobenzidine in 50 mM Tris–HCl pH 7.6 added to 13.5 μL of hydrogen peroxide is routinely used as a peroxidase substrate to develop blots, but there are also many other suitable alternatives which are commercially available. Where you intend to make direct comparisons of the profiles, it is essential to detect and develop blots under identical conditions.
14. In order to make legitimate comparisons and pool data to generate mean lectin staining profiles, it must first be established that the variability observed between repetitions of incubation and slot blots is not significantly different. This can be achieved by comparing the variation observed in the replication of a single incubation with the variation observed across different biological samples (25). Once it is established that the changes in the staining patterns observed are not significantly different between each replicate incubations and replicate slot blots, it is then legitimate to plot profiles from the mean of data from each fraction number.
15. Lectin staining varies with the availability of sugar structures for binding and is subject to changes in the tertiary and quaternary status of peptides, and steric factors related to the density of mucin glycans. Thus, it is important to note that changes in the profile observed in any one assay cannot be assessed in terms of absolute changes in glycopeptide size. Instead, observations are made in terms of the percentage loss of stainable material. The percentage loss of staining from the high-molecular-weight mucin peak will vary for each lectin, and in our hands is highly sensitive to specific glycosidase inhibition, rendering this approach particularly sensitive in defining critical control points in mucin degradation attributable to the cleavage of specific sugar residues.
16. Mucinase activity usually shows a broad pH range as the complete activity is a composite of enzymes acting together. Suitable incubation buffers should be chosen to cover the potential range of activity expected. Typically, acetate buffers are used at lower pH and Tris at higher pH values. Highly purified enzymes available for commercial sources can be used to validate plate assays, and to generate specimen fragmentation profiles.
17. The use of phosphate buffers should be avoided if possible because phosphate may inhibit sulphatase activities, and thus affect overall mucinase activity.

18. Blank incubations can be prepared using enzyme extract that has been heat treated for 5 min at 95°C and centrifuged. Alternatively, the enzyme can be substituted by the same volume of incubation buffer. The suitability of either of these blanks should be tested for interference in each assay system.
19. The Sephadex G25 step removes free biotin. It is necessary to discard the column because irreversible adsorption of biotin occurs.
20. It is necessary to remove any unreacted periodate from the mucin before the biotinylation reaction is started because it interferes with the reaction. The use of a Sephadex G25 column is an easy method to do this and provides a high recovery.
21. The dilution of enzyme samples should be made in incubation buffer suitable for the enzyme itself and for the mucinase plate assay. This is easiest with commercial enzymes in which the amount of enzyme is known. In cases in which crude preparations are being used, such as faecal extracts, urine, serum, bacterial, or cell cultures as cell suspensions, or cell supernatants, the samples should be tested directly and as dilutions in incubation buffer. It is wise to test at high, low, and intermediate pH ranges.
22. If a 490-nm filter is not available, use the nearest available wavelength. If the plate reader uses a reference filter for standardisation, wavelengths between 630 and 650 nm (or nearest available) should be used.
23. The working conditions for *O*-glycan release by ammonia based non-reductive β-elimination used here are those originally reported by Huang et al. in 2001 (26). Optimal conditions for maximal glycan recovery and minimal glycan degradation can vary with the profile of glycan structures and may require further optimisation if yields are not sufficient.
24. GU values are proxy "molecular weights" determined as follows: (a) Take the retention times of the peaks of your dextran ladder (e.g. the peak marked GU4 in Fig. 3 = Dextran 4 with a molecular weight of 40,000 Da, GU5 = Dextran 5 with a molecular weight of 50,000 Da and so on; (b) Calculate the Log 10 of the proxy molecular weight values; (c) Generate a plot of Log 10 molecular weight versus retention time; (d) For each of the analyte peaks determine the Log 10 molecular weight from the calibration curve; (e) To calculate the GU value, calculate the antilog of the experimentally determined Log 10 molecular weight value and then divide the result by 10,000. It is noteworthy that the GU value of a given structure is highly dependent on the HPLC equipment used (especially the column) and the gradient applied. In order to compare results from different batches, the HPLC conditions used need to be standardised.

25. Ultrafiltration using Amicon centrifugal filters is an efficient method for the removal of background sugar from the bacterial supernatant (or other samples). However, the conditions given here are for the 15 mL filters. Conditions can vary with different filter size and also depend on the concentration and composition of the samples. Generally, the indicator of efficient removal of small molecules from the sample is the colour which becomes much lighter after filtration.

26. The analysis of *O*-glycans can be accomplished by assigning structures and linkages using a panel of commercially available glycosidases and detecting peak shifts. The contribution of each individual monosaccharide to the overall GU value of a given glycan can be calculated and the incremental values can therefore be used to predict the structure of an unknown sugar from its GU value. Owing to the fact that each exoglycosidase has precise specificity, the result of the digestion is able to define the position of the target monosaccharide as well as the linkage through which the terminal sugar attaches to the glycan. Mass spectrometry is commonly also required as part of the analytical strategy in cases where digestion is blocked (e.g. by the presence of sulphated sugars). Known glycan structures and their GU values can be obtained from publicly available databases (see Note 27). Further details on this strategy are beyond the scope of this chapter, but can be found elsewhere (21, 27). It is noteworthy that the shift in GU value caused by the removal of any particular monosaccharide has certain variability (standard deviations are available from the databases specified in Note 27). Furthermore the GU value of sialic acid is slightly less stable than other monosaccharides and is influenced by the structure that it is attached to.

27. The identification of exoglycosidase activities from uncharacterised biological sources generates similar digested vs. undigested profiles from which information on the glycosidase content of the sample can be logically inferred provided structural information is obtained on the target glycan. At present, databases of *N*-linked glycans (e.g. http://glycobase.nibrt.ie:8080/database/show_glycobase.action) are more comprehensively populated than those containing *O*-linked glycans. Furthermore, it should be noted that the incremental GU value for addition of known monosaccharides to *N*-linked glycans may be different to that for *O*-linked glycans. A database of *O*-linked glycans is under construction and some structures are available on the web site http://glycobase.nibrt.ie:8080/database/show_olinks.action.

28. The presence of such activities in samples is expected to be uncommon. However, if detected they would be of great potential relevance to the development of improved methods for

O-glycan release for analytical purposes. The common chitobiose core structure of N-linked glycans can be removed by a commercially available universal *N*-glycanase, whereas the variable core structures of *O*-glycans currently require chemical release engendering variable levels of glycan degradation or "peeling" (28).

29. It is not unlikely that exoglycosidases act on particular glycans to truncate them *before* the action of endoglycosidases, since currently known *O*-glycanases are active against short oligosaccharide structures. Nevertheless, the presence of endoglycosidases active against longer structures cannot be ruled out, and their discovery would indeed provide useful analytical reagents. It is also possible that multiple endoglycosidases are present that are capable of removing multiple glycans from the mucin. These might be expected to give rise to distinct peaks in P3. However, they could also be cleaved back by exoglycosidases to generate shorter GalNAc terminated oligosaccharides on P3. This possibility can be identified by comparing peak areas between P1, P2, and P3, but a full understanding of the sequence of degradation and the enzymes involved is likely to require a detailed understanding of the structures present. In interpreting P3, it is important to remember that the glycans are labelled *after* digestion. Thus, monosaccharides other than the GalNAc residue that is normally labelled in P1 and P2 (where 2-AB labelling is conducted *after* chemical release but *before* digestion), will also appear in the profile if exoglycosidases are present in addition to endoglycosidases.

30. In the unlikely event that enzyme sources contain the entire suite of glycosidases capable of completely digesting the glycans, leaving only monosaccharides, this method cannot distinguish between the presence of endo- and exoglycosidases. This is because the glycans attached to the protein, and released glycans, can both be sequentially digested to generate monosaccharide peaks. Thus, no difference between P3 and P2 would be detected.

31. Upon digestion, the peak areas attributable to a digested peak or peaks are shifted towards the left of the profile (a lower GU value). Thus, comparisons of profiles should be made on the basis of selecting an appropriate GU cut-off that excludes the lower GU peaks representative of monsaccharides and smaller degraded saccharides (normally disaccharides). The smallest GalNAc initiated disaccharide (core 1 structure) bears the GU value of about 1.79. Thus, GU values lower than 1.5 should not be selected. It cannot always be assumed that each peak in glycan profiles contains only a single structure, so care is required in data interpretation. Likewise, in some cases partially degraded structures may not move below a GU of 1.79 (e.g. where a

restricted combination of glycosidases are present), so there will always be a requirement for preliminary examination of the profiles before the selection of an appropriate GU value cut-off.

Acknowledgements

This work was supported by grants 046530/Z/96 and 051586 /Z/97 from the Wellcome Trust (APC). Research in the labs of SDC and PMR is currently supported in part through Strategic Research Cluster Grants from Science Foundation Ireland (award codes 07/SRC/B1156 and 08/SRC/B1393).

References

1. Wiggins, R., Hicks, S. J., Soothill, P. W., Millar, M. R., and Corfield, A. P. (2001) Mucinases and sialidases: their role in the pathogenesis of sexually transmitted infections in the female genital tract. *Sex. Transm. Infect.* 77, 402–408.
2. Dwarakanath, A. D., Campbell, B. J., Tsai, H. H., Sunderland, D., Hart, C. A., and Rhodes, J. M. (1995) Faecal mucinase activity assessed in inflammatory bowel disease using 14 C threonine labelled mucin substrate. *Gut* **37**, 58–62.
3. Evans, C. M., and Koo, J. S. (2009) Airway mucus: the good, the bad, the sticky. *Pharmacol. Ther.* **121**, 332–348.
4. Linden, S. K., Sutton, P., Karlsson, N. G., Korolik, V., and McGuckin, M. A. (2008) Mucins in the mucosal barrier to infection. *Mucosal Immunol.* **1**, 183–197.
5. Malloy, B., Hart, G. W., and Stanley, P. (2009) Structural Analysis of Glycans., in *In Essentials Of Glycobiology.* (Varki, A., Cummings, R. D., Esko, J. D., Freeze, H. H., Stanley, P., Bertozzi, C. R., Hart, G. W., and Etzler, M. E., Eds.) 2nd ed., pp 661–678, Cold Spring Harbor Laboratory Press, New York.
6. Rho, J. H., Wright, D. P., Christie, D. L., Clinch, K., Furneaux, R. H., and Roberton, A. M. (2005) A novel mechanism for desulfation of mucin: identification and cloning of a mucin-desulfating glycosidase (sulfoglycosidase) from Prevotella strain RS2. *J. Bacteriol.* **187**, 1543–1551.
7. Carrington, S. D., Clyne, M., Reid, C. J., FitzPatrick, E., and Corfield, A. P. (2009) Microbial interaction with mucus and mucins. In: *Microbial Glycobiology* (Moran, A., Holst, O., Brennan, P., and von Itzstein, M., Eds.), pp 655–671, Academic Press.
8. Corfield, A. P., Carroll, D., Myerscough, N., and Probert, C. S. (2001) Mucins in the gastrointestinal tract in health and disease. *Front. Biosci.* **6**, D1321–D1357.
9. Thornton, D. J., Rousseau, K., and McGuckin, M. A. (2008) Structure and function of the polymeric mucins in airways mucus. *Annu. Rev. Physiol.* **70**, 459–486.
10. Patsos, G., and Corfield, A. P. (2009) O-Glycosylation: Structural Diversity and Functions. In: *Fundamentals of Glycosciences. The Sugar Code.* (Gabius, H.-J., Ed.), pp 111–137, Wiley/VCH.
11. Corfield, A. P., Wiggins, R., Edwards, C., Myerscough, N., Warren, B. F., Soothill, P., Millar, M. R., and Horner, P. (2003) A sweet coating--how bacteria deal with sugars. *Advances in Experimental Medicine and Biology* **535**, 3–15.
12. Roberton, A. M., and Corfield, A. P. (1998) Mucin degradation and its significance in inflammatory conditions of the gastrointestinal tract. In: *Medical Importance of the Normal Microflora* (Tannock, G. W., Ed.), pp 222–261, Kluwer Academic Publishers, Dordrecht.
13. Hutton, D. A., Allen, A., and Pearson, J. P. (2000) Proteinase activity. *Methods in Molecular Biology (Clifton, N.J)* **125**, 393–401.
14. Corfield, A. P. (2007) The Glycobiology of Mucins in the Human Gastrointestinal Tract. In: *Glycobiology* (Sansom, C., and Markman, O., Eds.), pp 248–260, Scion Publishing Limited.
15. Tateno, H., Uchiyama, N., Kuno, A., Togayachi, A., Sato, T., Narimatsu, H., and Hirabayashi, J. (2007) A novel strategy for mammalian cell surface glycome profiling using lectin microarray. *Glycobiology* **17**, 1138–1146.

16. Corfield, A. P., Myerscough, N., Van Klinken, B. J., Einerhand, A. W., and Dekker, J. (2000) Metabolic labeling methods for the preparation and biosynthetic study of mucin. *Methods in Molecular Biology (Clifton, N.J)* **125**, 227–237.
17. Corfield, A. P., Wagner, S. A., O'Donnell, L. J., Durdey, P., Mountford, R. A., and Clamp, J. R. (1993) The roles of enteric bacterial sialidase, sialate O-acetyl esterase and glycosulfatase in the degradation of human colonic mucin, *Glycoconj. J.* **10**, 72–81.
18. Colina, A. R., Aumont, F., Belhumeur, P., and de Repentigny, L. (1996) Development of a method to detect secretory mucinolytic activity from Candida albicans, *J. Med. Vet. Mycol.* **34**, 401–406.
19. Colina, A. R., Aumont, F., Deslauriers, N., Belhumeur, P., and de Repentigny, L. (1996) Evidence for degradation of gastrointestinal mucin by Candida albicans secretory aspartyl proteinase. *Infect. Immun.* **64**, 4514–4519.
20. Koritsas, V. M., and Atkinson, H. J. (1995) An assay for detecting nanogram levels of proteolytic enzymes. *Anal. Biochem.* **227**, 22–26.
21. Royle, L., Matthews, E., Corfield, A., Berry, M., Rudd, P. M., Dwek, R. A., and Carrington, S. D. (2008) Glycan structures of ocular surface mucins in man, rabbit and dog display species differences, *Glycoconj. J.* **25**, 763–773.
22. Innes, A. L., Carrington, S. D., Thornton, D. J., Kirkham, S., Rousseau, K., Dougherty, R. H., Raymond, W. W., Caughey, G. H., Muller, S. J., and Fahy, J. V. (2009) Ex vivo sputum analysis reveals impairment of protease-dependent mucus degradation by plasma proteins in acute asthma. *Am. J. Respir. Crit. Care Med.* **180**, 203–210.
23. Corfield, A. P., and Myerscough, N. (2000) Glycosidase activity, *Methods in Molecular Biology (Clifton, N.J)*. **125**, 403–416.
24. Royle, L., Radcliffe, C. M., Dwek, R. A., and Rudd, P. M. (2006) Detailed structural analysis of N-glycans released from glycoproteins in SDS-PAGE gel bands using HPLC combined with exoglycosidase array digestions, *Methods in molecular biology (Clifton, N.J)* **347**, 125–143.
25. Dytham, C. (2003) *Choosing and using Statistics: A biologists guide.* Chapter 7, Wiley-Blackwell.
26. Huang, Y. P., Mechref, Y., and Novotny, M. V. (2001) Microscale nonreductive release of O-linked glycans for subsequent analysis through MALDI mass spectrometry and capillary electrophoresis. *Anal. Chem.* **73**, 6063–6069.
27. Royle, L., Mattu, T. S., Hart, E., Langridge, J. I., Merry, A. H., Murphy, N., Harvey, D. J., Dwek, R. A., and Rudd, P. M. (2002) An analytical and structural database provides a strategy for sequencing O-glycans from microgram quantities of glycoproteins. *Anal. Biochem.* **304**, 70–90.
28. White, C. C., and Kennedy, J. F. (1988) *An Introduction to the Chemistry of Carbohydrates.* 3 rd ed., Clarendon Press, Oxford, U.K.

Chapter 12

Assessment of Mucus Thickness and Production In Situ

Lena Holm and Mia Phillipson

Abstract

The nature of the mucus gel layer covering the gastrointestinal tract makes it difficult to study outside its natural site attached to the mucosa. Here, we describe a technique for intravital microscopy studies of the mucus gel layer from the stomach down to the colon in anesthetized rats and mice. Mucus thickness and accumulation rate in each segment of the gastrointestinal tract is measured with a micropipette technique under observation through a stereomicroscope. In this way, the nature of the mucus gel in vivo is readily studied, and effects of interventions or disease on the mucus can be determined in longitudinal studies or by comparing animals. Using this technique, we have been able to demonstrate that there are two forms of mucus gel adherent to the stomach and colon mucosa: one layer which is removable by suction and an underlying firm adherent gel layer, while in the small intestine, all mucus adhering to the mucosa can easily be removed.

Key words: In vivo, Microscopy, Micropipette, Rat, Mouse, Mucus, Stomach, Duodenum, Jejunum, Ileum, Colon

1. Introduction

The gastrointestinal tract is covered by a thin layer of highly hydrated mucus gel. The role of the mucus layer differs between intestinal segments due to diverse luminal challenges, but an overall task is defending the underlying single-layer epithelial cells against luminal threats. The mucus layer in the small intestine can easily be removed, but will rapidly be renewed (1, 7). The mucus layer covering the gastric and colonic mucosa can be divided into two different layers: the most luminal of the two layers, the loosely adherent mucus, can easily be removed by suction, whereas, the inner layer, the firmly adherent mucus, cannot (1) (Fig. 1). Therefore, the loosely adherent mucus layer is most likely rubbed off and mixed with the gastric and colonic contents and thereby functions as a lubricator or for housing commensal colonic bacteria.

Michael A. McGuckin and David J. Thornton (eds.), *Mucins: Methods and Protocols*, Methods in Molecular Biology, vol. 842, DOI 10.1007/978-1-61779-513-8_12, © Springer Science+Business Media, LLC 2012

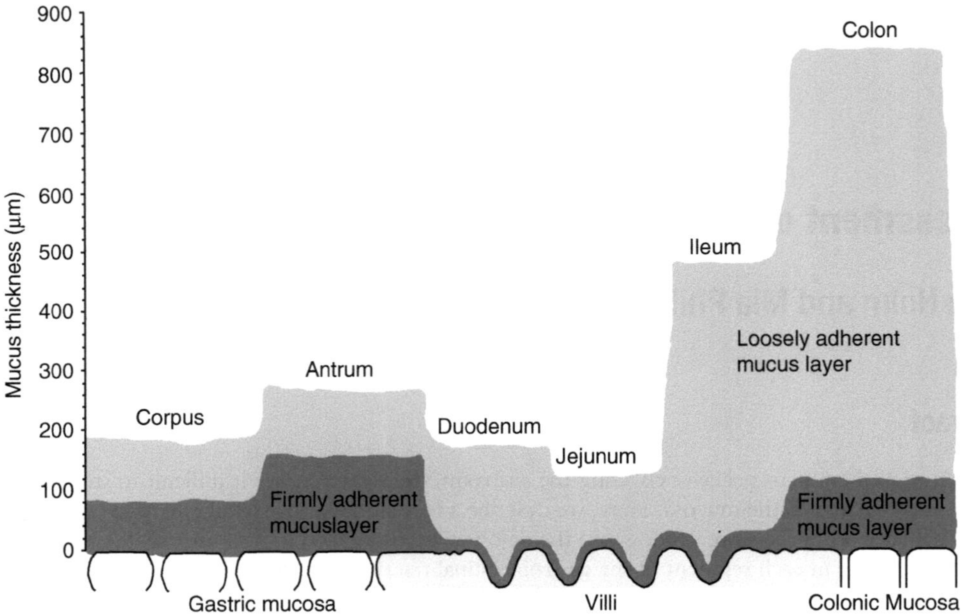

Fig. 1. Mucus thickness of the loosely and firmly adherent mucus throughout the gastrointestinal tract. Adapted from (1) and used with the permission of the American Physiological Society.

Meanwhile, the firmly adherent mucus layer provides the physical barrier protecting the epithelium. In the stomach, the firmly adherent mucus layer creates an unstirred environment where neutralization of hydrogen ions by secreted bicarbonate from the epithelium occurs, maintaining a pH gradient that protects the gastric mucosa from luminal acid (2). In the colon, the mucus has dual functions serving as a barrier to the enormous bacterial population and at the same time enabling bacteria colonization within the mucus. Thinned and disrupted colonic mucus is believed to correlate with bacterial translocation and inflammation occurring during colitis (3). Therefore, studies on mucus thickness and accumulation of the mucus layers are important for the understanding, development, and prevention of disease.

The mucus barrier is a highly hydrated extracellular compartment, with physical properties that limit the use of conventional in vitro and histological methods. The lack of mucus secreting gastrointestinal epithelial cell culture systems together with methodological difficulties in studying mucus from biopsies in vitro, as the mucus becomes dehydrated and eroded (4), restricts the usage of these approaches. Therefore, in vivo mucus measurements are more reliable for measuring thickness, accumulation and degradation of the mucus layer in the gastrointestinal tract (GI) tract. In this chapter, we describe a model system for intravital microscopy studies of the mucus gel thickness and accumulation rate over time in the GI tract of the rat and mouse (1, 2, 5–7).

2. Materials

2.1. Animal Preparation

1. Rat anesthetic Inactin®: Na-5-ethyl-1-(1′-methyl-propyl)-2-thiobarbituric acid (Sigma-Aldrich), administered intraperitoneally (ip) 120 mg/kg body weight (bw). One gram of Inactin is dissolved in 10 mL deionized water and can be stored for up to 1 week at 4°C.
2. Mice are anesthetized through spontaneous inhalation through a breathing mask (Simtec Engineering, Askim, Sweden) of isoflurane (Forene®, Abbott Scandinavia AB, Solna, Sweden) administered by means of an isoflurane pump (Univentor 400 Anesthesia Unit, AgnTho's AB, Lidingö, Sweden). The isoflurane pump delivers a mixture of 40% oxygen, 60% nitrogen and ≈2.2% isoflurane.
3. Rectal thermistor probe connected to a heating pad and temperature regulating unit (Elmedex Elektronik HB, Alunda, Sweden).
4. Rat tracheal tube, sized PE 200 (ID 1.4 mm, OD 1.9 mm, Agn Tho's AB, Lidingö, Sweden). Arterial polyethylene cannulas: PE-50 (rat) or pulled PE-200 (mouse) containing heparin (12.5 IU/mL) dissolved in isotonic 0.9% saline (155 mM NaCl).
5. Blood pressure monitor: pressure transducer connected to PowerLab data acquisition system (ADInstruments, UK).
6. Modified Ringer's solution: 25 mM $NaHCO_3$, 120 mM NaCl, 2.5 mM KCl, 0.75 mM $CaCl_2$ by means of an infusion pump (Harvard apparatus 55-2222).

2.2. Tissue Preparation

1. The animals are placed on an in-house designed and constructed Lucite microscope stage adapted to the microscope used (Fig. 2). In the center of the stage is a truncated cone, on which the tissue is draped, with the following top diameters: 10.2 mm for the rat stomach (corpus and antrum) preparations, 6.4 mm for the rat small and large intestine preparations, and 6.5 mm and 2.9 mm for the mouse stomach and intestinal preparations, respectively.
2. A double bottom Lucite mucosal chamber, with a hole in the bottom corresponding to the position of the cone, is placed on a Lucite water jacket support, which partly encircles the cone. The diameters of the holes in the mucosal chamber are 12.6 and 9.5 mm, respectively for the rat stomach and intestinal preparations and 8 and 4.8 mm, respectively, for the mouse stomach and intestinal preparations. The junction between the Lucite chamber and the exposed mucosa is sealed using silicon grease (Dow Corning High Vacuum Grease, Dow Corning

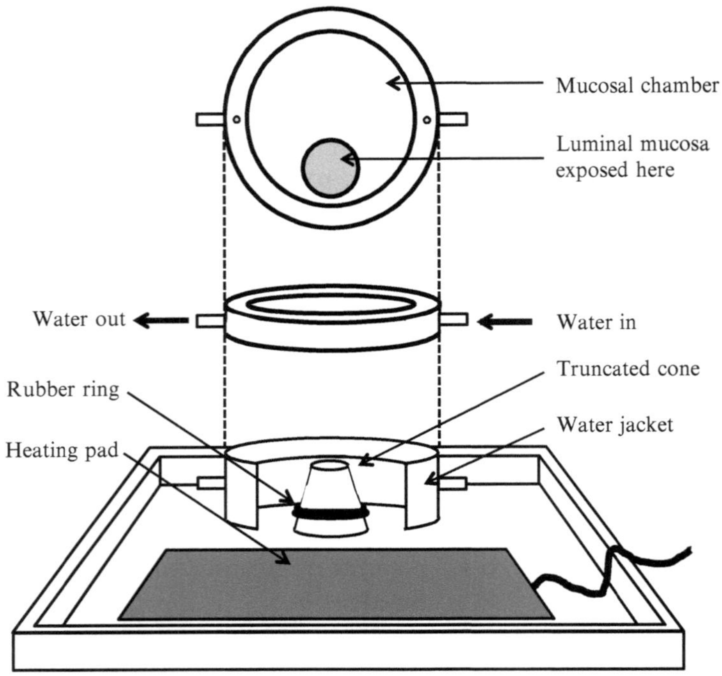

Fig. 2. Schematic drawing of the Lucite microscope stage used to mount the stomach for intravital microscopy. A modified stage with three small pins at the back top edge of the truncated cone is used for the intestine.

GmbH, Weisbaden, Germany). The mucosal chamber is filled with 5 mL (rat) or 3 mL (mouse) of prewarmed (37°C) 0.9% NaCl or the solution of interest.

2.3. Measurement of Mucus with Micropipettes

1. Micropipettes are pulled from custom glass tubing (OD, 1.2 mm; ID, 0.6 mm; Rederick Haer, Brunswick, ME) using a pipette puller (pp-83, Narishige Scientific Instrument Laboratories, Tokyo, Japan) to a tip diameter of 1–2 μm and a tip length of about 150 μm. The tips of the micropipettes are siliconized by dipping them into a silicon solution made up in 25% acetone; subsequently, they are dried at 100°C for 30 min.
2. Graphite particle suspension for mucus visualization: activated charcoal (extra pure, Merck, Darmstadt, Germany) suspended in 155 mM NaCl (0.5–0.75 g/100 mL).
3. The mucosal preparation is studied under a stereomicroscope (Leica MZ12, Leica AG, Heerbrugg, Switzerland) with a maximal magnification of 200× (ocular 10×, zooming objective 2–20×). One of the ocular objectives is equipped with a grid. The mucosal preparation is transilluminated with light from a 150 W light source guided by fiberoptics.

4. The micropipette is attached to a holder on the micromanipulator (Leitz Wetzlar, Germany) and a "digimatic indicator" (IDC Series 543, Mitutoyo Corp., Tokyo, Japan) mounted on the micromanipulator for measurements of distance traveled by the micropipette (see Note 1).

3. Methods

In vivo measurements of gastrointestinal mucus using a micropipette technique will be described in this section. The measurements are described for each segment of the GI tract in the rat and for the stomach and colon in the mouse.

3.1. Animal Preparation (Rats: Steps 1–4, Mice: Steps 5–7)

All animal experiments have to be approved by a properly constituted Animal Research Ethics Committee. Animals should be housed in an appropriate animal facility and kept under standardized conditions of temperature (21–22°C) and illumination (12 h light/12 h dark) unless otherwise dictated by the experiment. Animals should have free access to water and pelleted food and are allowed at least 1 week to adjust to the new environment, after arrival to the animal facility. It is important to minimize the preexperimental stress as much as possible to avoid stress-related changes in the GI mucosa (see Note 2).

1. For gastric studies rats of various strains weighing 180–250 g are fasted over night (18 h) but allowed free access to water before induction of anesthesia. Rats for measurements in the colon are not fasted (see Note 3). If younger or older rats should be used, the size of the microscope stage and the mucosal chamber has to be adjusted (see also Note 11).
2. Anesthesia is induced by intraperitoneal administration of 120 mg/kg body weight Inactin® [Na-5-ethyl-1-(1′-methyl-propyl)-2-thiobarbituric acid] (see Note 4).
3. A rectal thermistor probe connected to a heating pad and temperature regulating unit is used to maintain body temperature at 37–38°C.
4. The rats are tracheotomized to facilitate spontaneous breathing (see Note 5). Systemic arterial blood pressure is continuously monitored through a heparinized catheter placed in the right femoral artery, connected to a pressure transducer. A femoral vein is cannulated for continuous infusion of a modified Ringer's solution at a rate of 1 mL/h.
5. Mice of various strains weighing 20–45 g are anesthetized with spontaneous inhalation of isoflurane (Forene®) without being fasted (see Note 6). Anesthesia is induced with the mouse in a

beaker with lid after which the head of the mouse is placed in a breathing mask. The inhalation gas is continuously administered through the breathing mask by an isoflurane pump (see Note 7). The depth of the anesthesia is continuously checked (see Note 4) and the percent isoflurane changed if necessary.

6. Body temperature is maintained as in step 3 above.
7. A heparinized catheter is placed in the left carotid artery to monitor blood pressure, and the jugular vein is cannulated for continuous infusion of a modified Ringer's solution at a rate of 0.35 mL/h.

3.2. Tissue Preparation

1. The abdomen of the rat is opened through a 2-cm midline incision made with tissue scissors, for all the different gastrointestinal segments. Each segment studied is gently exteriorized, kept moistened, warm and rinsed with saline at 37°C during the subsequent preparative procedure. The organs not studied are kept inside the peritoneal cavity to minimize drying and cooling (see Note 8).
2. Stomach. The gastrohepatic ligaments are cut and the short gastric artery and vein ligated and cut. Using microcautery, an incision is made in the forestomach along the greater curvature for the corpus preparation. For the antrum studies, the stomach is opened with a 1–2 cm incision through its ventral side, close to and along the greater curvature in the corpus, starting just proximal to the junction with the antrum (see Note 9).
3. Duodenum. The common bile duct is catheterized using PE-10 polyethylene tubing close to (2–3 mm) its exit into the duodenum (papilla of Vater) and drained outside the body to avoid leakage of pancreaticobiliary secretions into the preparation. A short incision through the duodenal wall is made approximately 5-mm distal to the papilla of Vater. 2–3 cm of the loop is then opened along the antimesenteric border using microcautery and tissue scissors (see Note 10).
4. Jejunum, Ileum and Colon. The intestine is opened 2–3 cm along the antimesenteric border using microcautery and tissue scissors (see Note 10).
5. When the preparation of the segment of interest to study is complete, the rat is placed on its left side on a heating pad on a Lucite microscope stage designed and built in-house (for details see Fig. 2). At the center of the stage is a truncated Lucite cone on which the segment to study is loosely draped with the mucosal surface facing upward (see Note 11). With the aid of rubber rings mounted around the cone, the forestomach can be held down by two pins inserted through the forestomach and fastened in the rubber rings, thus keeping the corpus in position on the cone. The antrum is held in position

by a pin through the corpus and fastened in the rubber ring. All the segments of the small and large intestine are kept in position by fastening the incision edge on three small pins attached to the top edge of the cone, on the opposite side to the position of the rat.

6. A double bottom Lucite mucosal chamber, with a hole in the bottom corresponding to the position of the cone, is placed over the mucosa. The chamber is resting on a Lucite water jacket support, which partly encircles the cone and together with the body of the animal when in place creates a closed compartment. The diameter of the hole in the mucosal chamber must be at least 2 mm wider than that of the top diameter of the cone including the tissue, to not touch the mucosa and impair blood flow. The gap between the mucosa in position on the cone and the hole in the bottom of the chamber is sealed with silicon grease to allow a solution in the chamber without leakage (see Note 12). The mucosal chamber is filled with 5 mL of 0.9% NaCl maintained at 37°C by means of circulating warm water in the jacket in the double bottom of the chamber as well as in the jacket of the support. By virtue of the water jacket support, the bottom of the mucosal chamber and the animal, the tissue not exposed for studies is confined in an almost closed compartment, which serves to maintain a high humidity and a constant temperature of about 37°C (see Fig. 2) (see Note 13).
7. The mouse corpus and intestine are prepared similarly to the rat (steps 1–6), but the chamber is adapted to the smaller size of the mouse (about 1/10th the size of a rat).
8. The gastrointestinal mucosa of both rat and mouse are observed through a stereomicroscope with a maximal magnification of 200× and transilluminated with light from a 150-W light source guided by fiberoptics.
9. A resting period of at least 30 min (mice) or 60 min (rats) must be allowed after the animals are operated on before any experiment starts. During this period, mean arterial blood pressure must reach steady state (see Note 14).

3.3. Measurement of Mucus Thickness with Micropipettes

Mucus gel thickness is measured using siliconized glass micropipettes held by and maneuvered with a micromanipulator (Fig. 3) (see Note 15). The tips of the pipettes are siliconized to ensure a nonsticky surface. In this way, repeated measurements can be made without the mucus gel adhering to the pipette and subsequently destroying the mucus gel layer. To visualize the otherwise translucent mucus gel surface, graphite particles diluted in saline are instilled onto the gel (see Note 16). Using the micromanipulator, the tip of the micropipette is placed on the surface of the gel and pushed through the gel layer at a constant angle of 30–35° to the

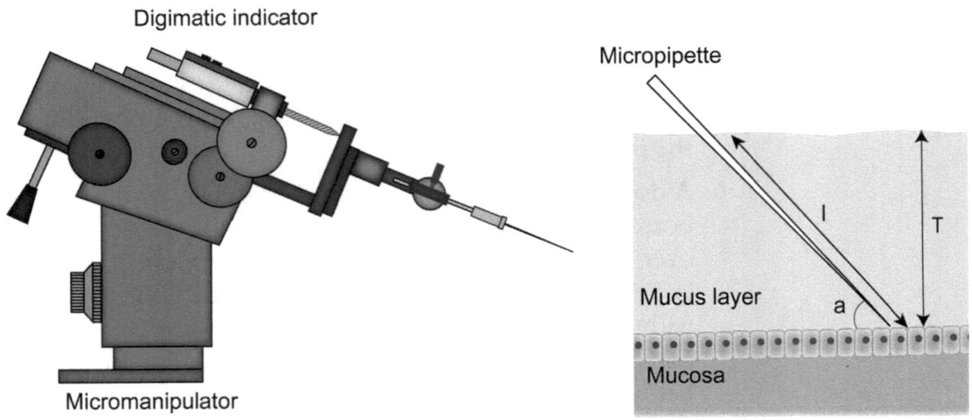

Fig. 3. Mucus thickness measurement setup. Mucus gel thickness is measured using a micropipette held by and maneuvered with a micromanipulator. The actual mucus gel thickness (*T*) is calculated using the formula: $T = l \times \sin a$, where *l* is the distance from the mucus gel surface to the mucosal surface, which is measured with a "digimatic indicator" attached to the micromanipulator and the *a* angle with which the pipette is pushed through the gel layer.

mucosal surface (see Note 17, see also Note 1). This angle is measured with a protractor. The distance (l) from the mucus gel surface to the mucosal surface is measured with a "digimatic indicator" attached to the micromanipulator. Measurements are made at 3–5 different sites or villus tips, and each position is registered and used again if remeasurement is required in an experiment (see Note 18). The mean value of all measurements on every measurement occasion is taken as one thickness value. The distance in the vertical (90° to the surface) direction from the luminal surface of the mucus gel to the cell surface (=mucus gel thickness; *T*) was calculated from the angle ($a = 30$–$35°$) of the insertion and the distance (l) described above. For this calculation we used the formula: $T = l \times \sin a$. The accuracy of this technique is based on the assumption that the mucosal or villus tip surface is resting in a horizontal plane at the measurement position.

3.4. Dynamic Measurement of Mucus Thickness

1. Mucus accumulation is determined by measuring total mucus thickness at regular intervals after the initial stabilization period (see Note 19).
2. As much as possible of the mucus gel is then removed by suction. A vacuum suction pump, or a syringe, connected to a PE-10 cannula is used to remove the mucus, with care taken to avoid contact with the underlying cell layer. This procedure is carried out under observation through the stereomicroscope.
3. Measurements are made immediately after mucus removal, and at regular intervals at the same sites. The protocol may finish with a second removal of the mucus and immediate measurement of the thickness. This procedure can be repeated once in an individual preparation with similar results (see Note 20).

4. Notes

1. Do not adjust angle of micromanipulator; make sure that the electrode is aligned with the manipulator so that extension of the manipulator results in straight movement of electrode.
2. To minimize stress never house rats alone, but always in groups of at least two. Mice on the other hand can be housed alone. Make sure to check the mice so that they do not fight in which case you have to separate them.
3. The main reason for fasting the animals is to have as clean gastrointestinal lumen as possible. The rats for colon studies are not fasted, since the fasting period has to be too long to have a clean colon.
4. It is important that the administration is performed quickly and in a calm atmosphere by personnel familiar to the animals. If the rat becomes excited the sympathetic nervous system is activated and induction of anesthesia might fail. Inactin usually induces good anesthesia for several hours, but if necessary (established by return of corneal reflex, i.e., the rat blinks when the cornea is gently touched, and/or arterial pressure is affected by pinching between the claws) a bolus dose (10% of the original dose) can be given intravenously (i.v.) and repeated if necessary.
5. Make sure that the tracheal tube is not inserted all the way down to the bronchi, which will disturb respiration. If the tube has been inserted too far down retract immediately. Once in place, the tracheal tube must be cut so that it does not extend over the animals mouth.
6. Fasting mice may impair the gastric acid-base balance and one should therefore avoid fasting mice.
7. Make sure that isoflurane spillover is collected and connected to an appropriate exhaust scavenge or extraction system to avoid inhalation by persons in the surroundings.
8. During tissue preparation, care must be taken to avoid twisting or stretching the organs, as this will markedly impair blood flow and the general condition of the tissue.
9. Care must be taken to avoid damaging the right gastroepiploic arteries along the greater curvature. The use of an electric microcautery will minimize bleeding from smaller blood vessels in the corporal mucosa.
10. By holding up the edge of the incision with a pair of forceps, at the antimesenteric border, and introducing a cotton tip, soaked in 37°C saline, into the intestine, the tissue will be protected while using microcautery.

11. The height of the cone should not exceed the level of heart which otherwise could result in increased resistance to blood flow and a reduced blood flow to the tissue segment.
12. If leakage occurs it must immediately be sealed with new silicon grease since the mucus might otherwise collapse by dehydration. If the preparation has great peristaltic movements, leakage occurs more often and might result in a preparation that is not feasible to use for mucus measurements.
13. In order to keep the solution in the mucosal chamber as well as the tissue outside the studied mucosa at around 37°C the temperature of the circulating warm water through the jacket in the double bottomed chamber and in the support must be kept at much higher temperature. Make sure to regularly check that the temperature of the saline solution covering the mucosa is maintained at 37°C since this is important for normal function.
14. To further check that the preparation is healthy, the solution in the mucosal chamber of the rat stomach and antrum can be changed at regular intervals during the stabilization period for measurements of pH: spontaneous acid secretion from the corpus usually starts within 2 h and is a sign of a healthy tissue. The antrum chamber solution should not be acidic showing that no part of the corpus is included in the preparation.
15. The micropipettes should be pulled to have a long slim tip and should not be ground so as not to injure the mucosa.
16. The carbon solution should be shaken to separate the particles and avoid formation of agglomerates. The solution is then dropped onto the chamber solution under observation through the microscope to make sure that not too many particles are applied to the surface of the mucus which will affect mucosal visibility.
17. The cell surface is usually visible through the microscope. In some rats, mostly those with a thicker mucus layer, the surface of the cells is not clearly seen. In such cases, the micropipette has to be gently pushed toward the mucosal surface until the electrode touches the cell surface, which can be seen as sideways displacement of the superficial mucosa. The electrode should then be withdrawn until the mucosa no longer appears displaced.
18. The easiest way to measure mucus thickness is with the aid of an ocular grid in which the mucus under study is divided into fields, which makes it easy to repeatedly measure at the same spots.
19. By using this method we have found that total mucus thickness does not change over a 90-min measuring time in the stomach (corpus and antrum), while it continuously increases in all parts

of the small and large intestine. After removal of as much as possible of the mucus by aspiration, we found a firmly adherent layer, not possible to remove by suction or wiping with a cotton tops, in the corpus, antrum and colon. By contrast, in the small intestine, almost all of the mucus can be removed (see Fig. 1). Immediately after the small intestine is mounted, the mucus gel is translucent and continuous. It has an even surface and does not follow the contours of the villi. However, the mucus gel becomes opaque over time, starting from the surface of the villi and moving upward with the accumulation of mucus, which makes it difficult to measure over longer periods. By contrast, in the stomach and the colon the mucus is always translucent.

20. Small peristaltic movements are usually not a problem for the measurements but greater movements must be avoided. One reason for increased movements of the preparation might be that the tissue is overstretched. Always try to put the animal as close as possible to the chamber and chamber support. Certain underlying conditions causing increased peristalsis may not be possible to eliminate, and in such cases the method for measurements will be less reliable.

Acknowledgments

The authors would like to thank Dr. Joel Petersson for help with the figures. The authors are funded by grants from The Swedish Research Council (55X-08646, 57P-20680, 57X-20675), Nanna Svartz foundation, and Ruth and Richard Julins foundation.

References

1. Atuma, C., Strugala, V., Allen, A., and Holm, L. (2001) The adherent gastrointestinal mucus gel layer: Thickness and physical state *in vivo*. *Am J Physiol Gastrointestinal Liver Physiol* **280,** G922–G929.
2. Phillipson, M., Atuma, C., Henriksnäs, J., and Holm, L. (2002) The importance of mucus layers and bicarbonate transport in preservation of gastric juxtamucosal pH. *Am J Physiol Gastrointestinal Liver Physiol* **282,** G211–G219.
3. Johansson, M. E. V., Gustafsson, J. K., Sjöberg, K. E., Petersson, J., Holm, L., Sjövall, H., and Hansson, G. C. (2010) Bacteria penetrate the inner mucus layer before inflammation in the dextran sulphate colitis model. *PLoS One* **18,** 5(8):e12238.
4. Allen, A., and Flemström, G. (2005) Gastroduodenal mucus bicarbonate barrier: protection against acid and pepsin. *Am J Physiol Cell Physiol* **288,** C1–C19.
5. Henriksnäs, J., Phillipson, M., Petersson, J., Engstrand, L., and Holm, L. (2005) An *in vivo* model for gastric physiological and pathophysiological studies in the mouse. *Acta Physiologica Scandinavica* **184,** 151–159.
6. Holm-Rutili, L., and Öbrink, K. J. (1985) Rat gastric mucosal microcirculation *in vivo*. *Am J Physiol Gastrointestinal Liver Physiol* **248,** G741–G746.
7. Sababi, M., Nilsson, E., and Holm, L. (1995) Mucus and alkali secretion in the rat duodenum: effects of indomethacin, N-nitro-L-arginin, and luminal acid. *Gastroenterology* **109,** 1526–1534.

Chapter 13

Preservation of Mucus in Histological Sections, Immunostaining of Mucins in Fixed Tissue, and Localization of Bacteria with FISH

Malin E.V. Johansson and Gunnar C. Hansson

Abstract

As mucus is highly hydrated, special care has to be taken to preserve this in histological preparations during immunostaining. Here, we describe how to fix tissues in such a way that the mucus is preserved in paraffin-embedded tissue. We also describe how the major macromolecular components in the mucus, the mucins, are immunostained and how bacteria can be localized in preparations with preserved mucus by fluorescent in situ hybridization.

Key words: Mucus, Mucin, Bacteria, Fluorescence, In situ hybridization, Carnoy's solution

1. Introduction

Mucus is highly hydrated and organized around the mucin polymer network (1). This means that the mucus is highly transparent and not visible under normal conditions. To observe the mucus surface, we have added charcoal or beads on top of the mucus in live intestinal tissue preparations under fluid, and observed that the mucus is thicker than expected (2, 3). The high water content also means that it is very sensitive to any method that dries the mucus after which it totally collapses. This has meant that, for example, the two mucus layers of colon, where the inner layer is devoid of bacteria, have been ignored until quite recently when we used appropriate techniques to reveal this (4). The background to this discovery was that we used fixation in water-free Carnoy's solution that does not dehydrate the mucus in the same way as other fixation methods. Using this approach, we could observe a thick inner mucus layer firmly attached to the epithelia in distal colon. Fixing

Michael A. McGuckin and David J. Thornton (eds.), *Mucins: Methods and Protocols*, Methods in Molecular Biology, vol. 842, DOI 10.1007/978-1-61779-513-8_13, © Springer Science+Business Media, LLC 2012

the same tissue with formaldehyde gave only a thin streak of collapsed mucus of less than 1 μm. Here, we describe how the mucus of intestinal tissue is preserved and how to immunolocalize the mucins. We also describe how the bacteria are localized by fluorescent in situ hybridization (FISH). Although these methods have been adapted for intestinal mucus, they should also work for other organs forming mucus.

2. Materials

2.1. Tissue Fixation to Preserve Intestinal Mucus

1. Mouse tissue dissection instruments.
2. Tubes with 5 mL methanol-Carnoy's fixative [60% (v/v) dry methanol, 30% (v/v) chloroform, 10% (v/v) glacial acetic acid] (see Note 1).
3. Methanol, absolute ethanol, absolute ethanol/xylene, xylene.
4. Paraffin embedding and tissue sectioning service.

2.2. Fluorescent In Situ Hybridization

1. Oven, 50 and 60°C.
2. Xylene substitute (Sigma–Aldrich).
3. Ethanol solutions 99.5%.
4. Hybridization solution (20 mM Tris–HCl, pH 7.4, 0.9 M NaCl, 0.1% (w/v) SDS, optional addition of 5–50% (v/v) formamide) heated to 50°C (see Note 2).
5. Fluorescent tag (for example Alexa-555 or Cy-3)-conjugated bacterial 16S rRNA-targeted probe dissolved to 1 μg/μL. For example, use the universal bacterial probe EUB338 (5′-GCTGCCTCCCGTAGGAGT-3′) (5) and a nonsense probe as control (5′-CGACGGAGGGCATCCTCA-3′).
6. Coverslips and a humid incubation chamber for glass slides.
7. FISH washing buffer (20 mM Tris–HCl, pH 7.4, 0.9 M NaCl).

2.3. Immunostaining

1. Oven, 60 and 50°C.
2. Xylene substitute (Sigma–Aldrich).
3. Ethanol solutions (100, 95, 70, 50, and 30%).
4. Microwave oven.
5. Antigen retrieval solution (10 mM citric buffer, pH 6.0).
6. PAP pen used to make circles around tissue sections. It produces a thin hydrophobic film creating a water-repellent barrier.
7. Block solution (5% fetal bovine serum in PBS).
8. PBS.

9. Specific antibody to the mucin (as anti-MUC2 C3 (4)) and fluorescence-conjugated secondary antibody (e.g., Alexa 488 or FITC) (see Note 3). Mounting medium limiting dehydration and bleaching (Prolong® antifade, Invitrogen).
10. Fluorescence or confocal microscope.

3. Methods

3.1. Tissue Fixation to Preserve Intestinal Mucus

1. Wild-type mice are sacrificed according to local regulations. Minimize handling and avoid stressing the animals as intestinal emptying limits the sampling of fecal-filled distal colon (see Note 4).
2. Dissect the intestine, cut pieces of the intestine containing fecal material (cut around the fecal pellets in the colon), and place in a tube with methanol-Carnoy.
3. Fix the tissue for a minimum of 3 h and a maximum of 2 weeks at room temperature.
4. Paraffin embedding is usually automated and tissue sectioning is a technique that demands trained personnel to produce good-quality materials. This is preferably done at a core facility or a company with this expertise. Normal histological processing protocols can be used with a few alterations: wash the fixed tissue two times in dry methanol for 30 min each, followed by two times in absolute ethanol for 20 min each. Finally, incubate in two baths of xylene for 15 min each before paraffin embedding and sectioning in 4-μm-thin sections. The sections are placed on glass after floating on a water bath as per standard procedures.
5. The embedded and sectioned material can be stored at room temperature.

3.2. Fluorescent In Situ Hybridization

1. Dewax the sections by an initial incubation of the slides at 60°C for 10 min. Incubate the slides in xylene substitute solution twice for 10 min with the first solution prewarmed to 60°C.
2. Incubate the sections in 99.5% ethanol for 5 min and let the slides air dry.
3. Mix 50 μL of hybridization solution (prewarmed to 50°C) with 0.5 μL EUB probe (0.5 μg) and add this as drops onto the sections on the slide.
4. Overlay the liquid with a coverslip and place the slide in a humid chamber. Incubate the sections at 50°C overnight.

5. Remove the cover glass and incubate slides in FISH washing buffer at 50°C for 10–20 min. The washing time can vary for each experiment and a quick look in a fluorescence microscope at low magnification allows determination of when the best result is achieved. Minimum background from the tissue and specific staining of the bacteria is the goal (see Note 5).
6. Wash the slide in PBS for 3× for 10 min.
7. Let the slide almost dry completely and mount the sections using Prolong antifade mounting media and let it set at room temperature. Analyze the sections using a fluorescent microscope (see Note 6).

3.3. Immunostaining

The FISH staining as described under Subheading 3.2 should precede the immunostaining if staining with both methods is required. In this case, start at Subheading 3.3, step 4. If only immunostaining is required, start at Subheading 3.3, step 1.

1. Dewax the sections by an initial incubation at 60°C for 10 min and two additional incubations in xylene substitute solution as in Subheading 3.2, step 1.
2. Hydrate the sample in solutions with decreasing concentration of ethanol (100, 95, 70, 50, and 30%). Incubate the slides for 5 min in each bath.
3. Place the slides in antigen retrieval solution, boil in the microwave oven (300 W) for 5 min, allow the solution to stop boiling, and make sure that there is enough solution to cover the slides; repeat the boiling step two more times. Leave the slides in the warm solution for 20 min (see Note 7).
4. Wash in PBS and mark around the sections with a PAP pen.
5. Add block solution and incubate in darkness in a humid chamber for 30 min at room temperature or 4°C if the staining is in combination with FISH.
6. Dilute the specific antibody in block solution and add it to the slide. Incubate in darkness for 4–24 h at 4°C (see Note 8).
7. Wash the slide in PBS for 3× for 10 min.
8. Dilute the secondary antibody in block solution (Anti-rabbit Alexa 488, Invitrogen, is diluted 1/1,500) and add it to the sample. Incubate in the dark in a humid environment at room temperature or 4°C (in combination with FISH) for 2 h.
9. Wash the slide in PBS for 3× for 10 min. Let the slide almost dry completely and mount the sections using Prolong antifade mounting media and let it set at room temperature and analyze the sections using a fluorescent microscope (see Note 9). An example of costaining of bacteria and the Muc2 mucin in mouse is shown in Fig. 1 (see Note 10).

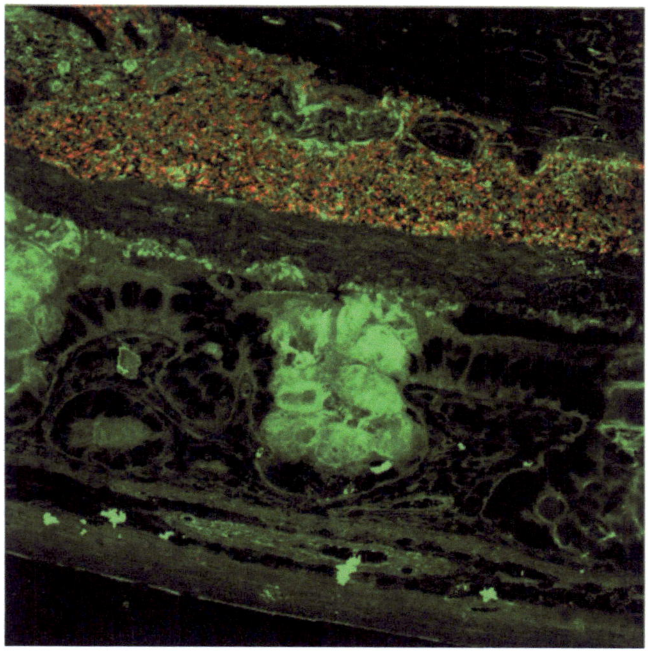

Fig. 1. Costaining of bacteria (*red*) and the Muc2 mucin (*green*) in mouse distal colon. Bacteria are found to be well-separated from the epithelia. Bacteria were stained by FISH using the EUB338 probe and the mucus with an Muc2-specific antiserum (anti-MUC2C3) (4).

4. Notes

1. Water-free methanol should be used as the mucus is preserved in a water-free fixative. Many automated embedding protocols begin with 70% ethanol which disrupts the mucus and should be avoided.
2. Increasing the formamide concentration modulates the hybridization by lowering the melting temperature of the DNA.
3. The fluorescent dye should have a different spectra compared to the dye of the FISH probe and work with the filter settings available on the microscope used to visualize the fluorescent staining. Secondary antibodies, as anti-rabbit-Alexa488 (A-11006, Invitrogen) and anti-mouse-Alexa488 (A-11001, Invitrogen), work well. Lectins could also be used to visualize mucin carbohydrates.
4. The mucus is preserved by this water-free fixation technique, but it is very important not to wash the intestine and to stabilize the mucus by the pressure of luminal contents, for example a colon fecal pellet.
5. If immunostaining is performed after the hybridization, one can accept a higher general background as the slide is further

washed which removes more background during antibody incubations. Less washing of the hybridization is also important to avoid removing the specific staining during these steps. That under some experimental conditions bacteria can penetrate mucus and be found in the underlying epithelium.

6. Care must be taken to remove bubbles from the samples at mounting by applying a pressure on the cover glass slip, without disrupting the tissue or move the loosely fixed luminal material and mucus layer.
7. Different antibodies might react more efficiently with antigens retrieved under various conditions and antigen retrieval methods using both high and low pH can be used to optimize the staining.
8. The specific incubation conditions and also time vary depending on the antibody used and can be optimized by testing a of different conditions.
9. The total amount of bacteria is best visualized in the whole section using a fluorescent microscope or a z-stack of the whole tissue in a confocal microscope. The size of many bacteria only includes them in single sections during confocal microscopy, something that might not give representative views.
10. Muc2 in the goblet cells is densely packed in the granulae and in the inner mucus layer resulting in intense staining. On the other hand, some antibodies cannot reach their epitopes as easily and the staining can be weaker. In the outer mucus layer, the Muc2 is more expanded, but also more diluted. The staining intensity and color, thus, vary depending on the structure and processing of the mucin and the intensity cannot be used to estimate the amount of mucin.

Acknowledgments

This work was supported by the Swedish Research Council (No. 7461, 21027, and 342-2004-4434), The Swedish Cancer Foundation, The Knut and Alice Wallenberg Foundation (KAW2007.0118), IngaBritt and Arne Lundberg Foundation, Sahlgren's University Hospital (LUA-ALF), EU-FP7 IBDase (no. 200931), Wilhelm and Martina Lundgren's Foundation, Torsten och Ragnar Söderbergs Stiftelser, Swedish CF Foundation, and The Swedish Foundation for Strategic Research—The Mucosal Immunobiology and Vaccine Center (MIVAC) and the Mucus-Bacteria-Colitis Center (MBC) of the Innate Immunity Program (2010–2014).

References

1. Johansson, M. E. V., Holmen Larsson, J. M., and Hansson, G. C. (2010) *Proc. Natl. Acad. Sci. USA on-line.*
2. Atuma, C., Strugula, V., Allen, A., and Holm, L. (2001) *Am. J. Physiol. Gastrointest. Liver Physiol. 280*, G922–G929.
3. Johansson, M. E., Gustafsson, J. K., Sjoberg, K. E., Petersson, J., Holm, L., Sjovall, H., and Hansson, G. C. (2010) *PLoS. ONE. 5.*
4. Johansson, M. E. V., Phillipson, M., Petersson, J., Holm, L., Velcich, A., and Hansson, G. C. (2008) *Proc. Natl. Acad. Sci. USA 105*, 15064–15069.
5. Amann, R. I., Binder, B. J., Olson, R. J., Chisholm, S. W., Devereux, R., and Stahl, D. A. (1990) *Appl. Environ. Microbiol. 56*, 1919–1925.

Chapter 14

Ex Vivo Measurements of Mucus Secretion by Colon Explants

Jenny K. Gustafsson, Henrik Sjövall, and Gunnar C. Hansson

Abstract

An explant tissue system for the study and recording of mucus secretion has been developed. Human colon biopsies or tissue from experimental animals are mounted in a horizontal perfusion chamber and the mucus accumulated on the apical side is observed and measured.

Key words: Mucus, Secretion, Biopsy, Ussing chamber, Intestine, Colon

1. Introduction

The organization of the mucus gel of the colon is complex as it is made up of two layers; an inner firmly adherent mucus layer and an outer loose mucus layer (1). These layers were originally observed and their thickness measured in living animals as described by Holm and Phillipson in Chapter 12 of this book. To be able to analyze human mucus secretion in an experimental system that allows manipulation at the same time as it is close to the in vivo situation, we have developed an ex vivo tissue explant culture method. In this, colon biopsies or tissues are mounted in a horizontal perfusion chamber with the apical surface upward (Fig. 1). The mucus is secreted onto the apical surface and the upper surface of the mucus visualized by the addition of charcoal. The mucus secretion thickness can be measured and observed for 1–2 h. The viability of the explants is monitored by a simple potential difference (PD) recording. This allows not only measurements of the thickness of mucus formed, but also manipulation with pharmacological and other reagents. The intestinal explants can be from humans

Michael A. McGuckin and David J. Thornton (eds.), *Mucins: Methods and Protocols*, Methods in Molecular Biology, vol. 842, DOI 10.1007/978-1-61779-513-8_14, © Springer Science+Business Media, LLC 2012

Fig. 1. Photo of the system used for mounting and measuring the mucus formed on *top* of colon explants. The tissue is mounted in the small hole seen in the *middle*.

undergoing normal clinical colonoscopy or from experimental animals, including rats and mice. This method has been used to show that dextran sodium sulfate (DSS) renders the mucus formed on such biopsies permeable to 5 μm beads in contrast to control biopsies not treated with DSS (2).

2. Materials

2.1. Solutions

1. KREB stock solution: 136.2 mM NaCl, 1.5 mM $CaCl_2$, 5.8 mM KCl, 1.6 mM KH_2PO_4, 27.2 mM NaHCO3 and 1.4 mM $MgSO_4$, store at 4°C for up to 3 weeks.
2. KREB transport solution: 115.8 mM NaCl, 1.3 mM $CaCl_2$, 3.6 mM KCl, 1.4 mM KH_2PO_4, 23.1 mM $NaHCO_3$, and 1.2 mM $MgSO_4$, store at 4°C for up to 3 weeks. Before using the transport buffer, set pH to 7.2, oxygenate for 20 min using 95% O_2/5% CO_2 and store on ice.
3. D-Mannitol: 200 mM in H_2O, store at 4°C for up to 1 week.
4. D-Glucose: 200 mM in H_2O, store at 4°C for up to 1 week.
5. Na-L-Glutamate: 102.6 mM in H_2O, store at 4°C for up to 1 week.

6. Na-Pyruvate: 114.5 mM in H_2O, store at 4°C for up to 1 week.
7. KREB-mannitol: Mix KREB stock solution, Na-L-Glutamate, Na-Pyruvate, and D-Mannitol, 17:1:1:1. Prepare a fresh solution for each experiment. Set pH to 7.2, oxygenate for 20 min using 95% O_2/5% CO_2 and store at room temperature (RT).
8. KREB-glucose: Mix KREB stock solution, Na-L-Glutamate, Na-Pyruvate, and D-Glucose, 17:1:1:1. Prepare a fresh solution for each experiment. Set pH to 7.2, oxygenate for 20 min in 95% O_2/5% CO_2 and store at RT.

2.2. Agar Bridges

1. Microbiological agar.
2. NaCl 0.9% in H_2O.
3. Polyethylene tubing (i.d. 2.0 mm), 20-cm long pieces.
4. Magnetic stirrer.
5. Magnet 2 cm.

2.3. Imaging Chamber, Perfusion, and PD Recording

1. Imaging chamber, exposed area 0.017 cm^2, with channels for agar bridges (RC-50, Warner Instruments or similar) (see Fig. 1).
2. Syringe pump (Harvard Apparatus).
3. Temperature controller (Warner Instruments).
4. One pair of reference electrodes, Rf 201 (Radiometer).
5. Polyethylene tubing (i.d. 1.2 mm).
6. 20-mL syringe.
7. 16-gauge needle.
8. Silicon grease.
9. Cover slips (22 mm × 40 mm × 1.5 mm).
10. Voltmeter to record transepithelial potential difference (Warner Instruments or similar equipment used in Ussing chamber systems).

2.4. Measuring Devices

1. Glass capillaries (i.d. 0.6 mm, OD 1.2 mm, length 10 cm) (Sutter instruments).
2. Micropipette puller (Sutter instruments).
3. Activated charcoal.
4. Stereomicroscope, 10× objectives (Leica).
5. Micromanipulator, movable in *x*–*y*–*z* direction (Sutter instruments, Scientifica or similar equipment used in patch clamp experiments).
6. Digimatic indicator 350 serie (Mitutoyo).

3. Methods

3.1. Agar Bridges

1. Prepare 50 mL of 4% agar in 0.9% NaCl. Heat the solution to 150°C during stirring. Once the solution has started to boil, decrease the temperature to 75°C and wait until the solution is completely cleared of air bubbles. Fill the precut PE tubing (i.d. 2.0 mm, 20 cm long) with agar using a syringe and allow to cool to RT. Store in 0.9% NaCl at 4°C. The agar bridges can be reused and are stable at 4°C for up to 3 weeks.

3.2. Buffers and Chamber Assembly

1. The KREB transport buffer is used to store the biopsies in prior to mounting in the image chamber. KREB-mannitol and KREB-glucose are used in the image chamber on the luminal and serosal side, respectively. Prepare fresh solutions for each experiment according to Subheading 2.1, item 2, 7, and 8.
2. Mount the basolateral chamber using silicone grease and a cover slip. Fill the basolateral chamber with KREB-transport buffer and mount an agar bridge in the side channel (Fig. 2). Make sure that there are no air bubbles in the channel since that will block the connection. Fill the apical chamber with KREB-transport buffer and mount an agar bridge in the side channel. Assemble the chamber, connect it to the PD electrodes and start the PD recording. Make sure that the background PD is stable and less than ±0.5 mV.
3. Fill a 20-mL syringe with KREB-glucose, connect it to the PE tubing (i.d. 1.2 mm) and mount it in the syringe pump.

3.3. Tissue Handling and Mounting

1. Take the colonic biopsies one at the time using single-use biopsy forceps (see Note 1). Put biopsies directly into ice-cold oxygenated KREB transport solution and incubate on ice for up to 1 h. Our experience has shown that tissue viability decreases after 1 h of incubation at 37°C.
2. Remove the apical part of the image chamber and flip it around. Put the biopsy on the apical chamber, add a droplet of KREB transport solution on top of the biopsy and spread it with the mucosal side toward the apical chamber. When spreading the biopsy, start with identifying the mucosal side of the biopsy and orientate the biopsy so that the mucosal side is facing the backside of the apical chamber and hold it in that position using forceps. It is important to only touch the edge of the biopsy with the forcep and never pinch the parts that will be in the exposed area. While holding the biopsy in place, remove excessive liquid using a paper tissue. This will make the biopsy stick to the apical chamber. When the biopsy is attached to the apical chamber, gently stretch the biopsy by holding it in the edge with one forcep and spreading it with another. Make sure

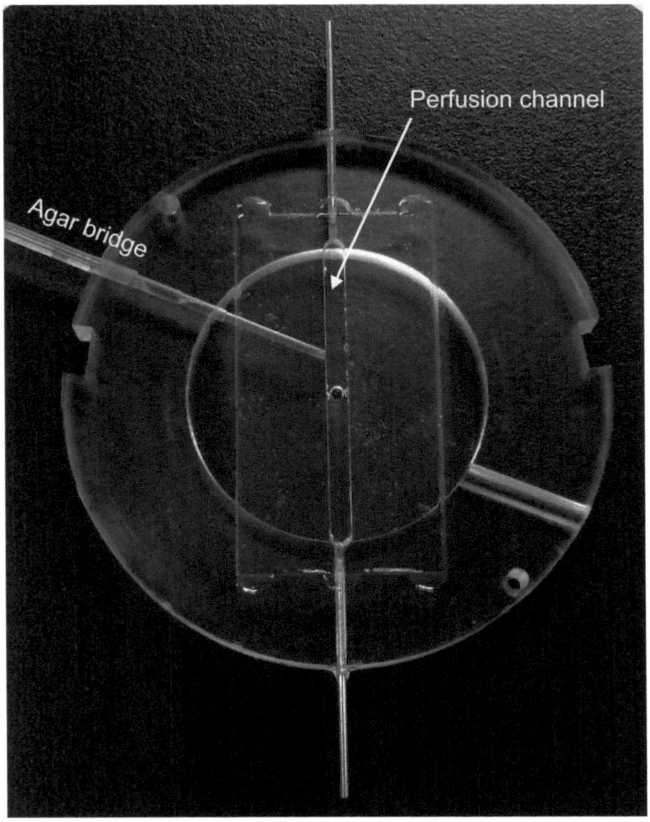

Fig. 2. Photo of the basolateral chamber with an agar bridge to the *left* and an objective glass mounted to close the perfusion channel.

that the biopsy completely covers the opening in the apical chamber, extending 0.5–1 mm outside the hole. Assemble the chamber and add 150 μL of oxygenated KREB-mannitol to the apical chamber.

3. Start the basolateral perfusion (5 mL/h) and turn on the temperature controller. Gradually increase the temperature to 37°C. Check the PD recording to make sure that the basal PD is negative, which tells you that the tissue is viable.

3.4. Mucus Measurement

1. Visualize the surface of the mucus layer by adding 10 μL of a charcoal suspension (activated charcoal mixed with KREB-mannitol) to the mucosal surface. The charcoal will sediment onto the mucus layer and thereby visualize the surface of the mucus layer.
2. Measure the distance to the epithelial surface and the surface of the mucus layer, respectively, using a micropipette (tip diameter 10 μm) mounted into the micropuller. Since the thickness of the mucus layer can vary over the exposed area, make

repeated measurements of the epithelial surface and the surface of the mucus layer to cover the variation. Calculate the average distance and use that as a single measurement (see Note 2).

3. Repeat the measurements with a suitable interval depending on the research question. It is important to measure both the epithelial surface and the mucus layer at each time point since the biopsy is not fixed to the underlying surface and can alter its position during the time of the experiment. In our hands, the biopsies are viable in the chamber for approximately 90 min.
4. In addition to studies on the spontaneous growth of the existing mucus layer, the mucus can be removed by gentle scraping and the thickness of the newly formed mucus can be measured as well as secretagogue-induced mucus release. In case of removing the mucus make sure that you do not harm the epithelium since an intact epithelium is essential for the formation of a normal mucus layer.

4. Notes

1. The approach can be used on normal biopsies taken at routine colonoscopy or from intestinal tissue taken from mice or rats. In the case of working with full thickness tissue, remove the longitudinal muscle layer using a pair of tweezers and a scalpel. This will increase tissue viability due to improved oxygenation of the tissue and reduce problems associated with muscle activity that disturb the experiment.
2. As seen in Fig. 1, the glass needle used for the measurements is at fixed angle (in our case, 40°) and when measuring the distance between the epithelial surface and the surface of the mucus layer the obtained value will not be the vertical distance between the two points due to the angle. To calculate the actual vertical distance between the surface of the epithelium and the mucus surface, the obtained value is multiplied by sinus the incoming angle, which in our case will be sin40°.

Acknowledgments

This work was supported by the Swedish Research Council (No. 7461, 5898, 21027, and 342-2004-4434), The Swedish Cancer Foundation, The Knut and Alice Wallenberg Foundation (KAW2007.0118), IngaBritt and Arne Lundberg Foundation, Sahlgren's University Hospital (LUA-ALF), EU-FP7 IBDase (No. 200931), Wilhelm and Martina Lundgren's Foundation, The

Swedish CF Foundation, Torsten och Ragnar Söderbergs Stiftelser, and The Swedish Foundation for Strategic Research—The Mucosal Immunobiology and Vaccine Center (MIVAC) and the Mucus-Bacteria-Colitis Center (MBC) of the Innate Immunity Program (2010–2014).

References

1. Johansson,M.E.V., Phillipson,M., Petersson,J., Holm, L., Velcich,A., and Hansson,G.C. 2008. The inner of the two Muc2 mucin dependent mucus layers in colon is devoid of bacteria. *Proc. Natl. Acad. Sci. USA* **105**, 15064–15069.

2. Johansson,M.E.V.,Gustafsson,J.K.,Sjoberg,K.E., Pettersson,J., Holm,L., Sjovall,H., and Hansson,G.C. 2010. Bacteria penetrate the inner mucus layer before inflammation in the Dextran sulfate colitis model. *PLoS ONE* **5**, e12238.

Chapter 15

Establishment of Respiratory Air–Liquid Interface Cultures and Their Use in Studying Mucin Production, Secretion, and Function

David B. Hill and Brian Button

Abstract

Primary cultures of human airway bronchial airways represent a valuable tool in understanding the roles of the epithelium, cilia, and the mucus layer in coordinating the clearance of mucus from the airways. The ability to obtain cells from both normal and diseased populations (such as cystic fibrosis and Chronic obstructive pulmonary disease (COPD)) allows researchers to investigate the disease phenotype on these processes. Furthermore, such cultures have provided investigators with a vast source of native airway mucus, devoid of external biological processes that occur in vivo, for biochemical and rheological studies. The primary goal of this chapter is to describe the culturing and use of human airway cultures grown under an in vivo-like air–liquid interface for use in a variety of mucus and mucociliary studies.

Key words: Airway epithelia, Mucus, Cilia, Mucociliary clearance, Mucus rheology

1. Introduction

The thin mucus layer lining the surfaces of the airways is vital for ensuring the sterility of the lungs from the constant bombardment of potentially infectious and toxic substances that are inhaled during normal tidal breathing. Mucus samples obtained from humans, as well as animals, have been invaluable for understanding many aspects of the normal biophysical properties of mucus and how these properties are altered in patients with chronic airway diseases, such as Chronic obstructive pulmonary disease (COPD) and cystic fibrosis. However, our growing understanding of mucus and the role of defects in mucociliary clearance on the pathogenesis of such airways diseases has most recently benefited immensely from the development of a highly differentiated primary human bronchial epithelial (HBE) cell culture system.

Michael A. McGuckin and David J. Thornton (eds.), *Mucins: Methods and Protocols*, Methods in Molecular Biology, vol. 842, DOI 10.1007/978-1-61779-513-8_15, © Springer Science+Business Media, LLC 2012

Based on many years work with these primary cultures, it is clear that the presence of an air–liquid interface is critical to successfully recapitulate the normal airway epithelial biology in vivo. Specifically, when cultured under an air–liquid condition, these cultures differentiate into a state similar to that reached in vivo; i.e. a pseudostratified epithelia with basal cells, ciliated cells, and mucin-secreting cells. As a result of similarities between the well-differentiated primary human airway epithelia cultures grown at the air–liquid interface and the airways in vivo, many studies have focused on their use for investigating various biochemical and biophysical properties of mucus clearance. This chapter describes the techniques used to produce well-differentiated culture human airway epithelia cells for in situ and in vivo studies of mucus and mucociliary clearance.

2. Materials

2.1. Phosphate-Buffered Saline

80 g NaCl, 2.0 g KCl, 20.2 g $Na_2HPO_4 \cdot 7H_2O$, 2.0 g KH_2PO_4 in 10 L of ddH_2O.

2.2. Laboratory of Human Carcinogenesis Basal Medium

Laboratory of human carcinogenesis (LHC) basal medium can be purchased for small-scale production (Invitrogen, cat. no. 12680-013). For large-scale production, LHC basal medium powder can be specially ordered from Sigma-Aldrich (see Note 1). In a 5 L volumetric flask, dissolve the 5 L prepackaged mixture in 4 L of dH_2O. Add 5 g $NaHCO_3$, 150 mL of 200 mM L-glutamine (Sigma-Aldrich, cat. no. G7513), stir, and adjust pH to 7.2–7.4. Bring total volume up to 5 L. Filter into sterile 500 mL bottles using 0.2-μm Vacucap (VWR, West Chester, PA, cat. no. 28143-315). Store at 4°C.

2.3. Bronchial Epithelial Growth Medium

Bronchial epithelial growth medium (BEGM) is prepared using 100% LHC basal medium. For small-scale production, thawed additives are dispensed into media in the top of a bottle top filter unit. Note that some additives are not 1,000× stock solutions. For media made with homemade bovine pituitary extract (BPE) that is difficult to filter, use a 0.4-μm filter unit. For commercial BPE, a 0.2-μm filter is acceptable. To add homemade BPE to media, thawed BPE aliquots are first centrifuged at 1,500 × *g* for 10 min to remove debris and cryoprecipitate, prefiltered through a 0.8-μm syringe filter, and added to the media just as the last few milliliters of media are being filter-sterilized.

2.4. Airway Liquid Interface Medium (see Note 2)

Airway liquid interface (ALI) medium uses a 50–50 mixture of Dulbecco's modified Eagle medium (DMEM) (Gibco, Carlsbad, CA, cat. no. 11995-065) and LHC basal medium as its base. Additives are thawed and dispensed into base media at the proper concentrations. ALI medium is then filtered according to small- or large-scale production. Note that some additives are not 1,000× stock solutions and that base ALI medium omits gentamicin and amphotericin.

2.5. Stock Additives for ALI and BEGM

Additives for media are filtered using 0.2 μM filters (unless product is sterile) and aliquots are stored at −20°C for up to 6 months.

1. Bovine serum albumin (BSA) (300 × 150 mg/mL): Add PBS directly to the BSA (Sigma-Aldrich, St. Louis, MO, cat. no. A7638) container to yield a concentration >150 mg/mL. Gently rock bottle at 4°C for 2–3 h until BSA is dissolved. Transfer to graduated cylinder and set volume to yield a final concentration of 150 mg/mL.
2. BPE (100× stock): Commercially prepared BPE is available from Sigma-Aldrich (cat. no. P1427) and is handled per manufacturer's instructions. It is used at a final concentration of 10 μg/mL. BPE can also be prepared from mature bovine whole pituitaries (Pel Freeze, Rogers, AR, cat. no. 57133-2). Thaw bovine pituitaries, drain, and rinse with chilled 4°C PBS. Add 2 mL of chilled PBS per gram of tissue. In a cold room, mince tissue in a Waring 2-speed commercial blender (Fisher Scientific, Pittsburgh, PA, cat. no. 14-509-17) at low speed for 1 min and then at high speed for 10 min. Aliquot suspension and centrifuge at 2,500 × *g* for 10 min at 4°C. Collect supernatant and centrifuge again at 10,000 × *g* for 10 min. Harvest the final BPE supernatant. Homemade BPE is difficult to filter and needs to be filtered during media preparation as described.
3. Insulin (5 mg/mL; 1,000× stock): Dissolve insulin (Sigma-Aldrich, cat. no. I6634) in 0.9N HCl.
4. Transferrin (10 mg/mL; 1,000× stock): Reconstitute transferrin, human-holo, natural (Sigma-Aldrich, cat. no. T0665) in PBS.
5. Hydrocortisone (0.072 mg/mL; 1,000× stock): Reconstitute hydrocortisone (Sigma-Aldrich, cat. no. H0396) in distilled water (dH_2O).
6. Triiodothyronine (0.0067 mg/mL; 1,000× stock): Dissolve triiodothyronine (Sigma-Aldrich, cat. no. T6397) in 0.001 M NaOH.
7. Epinephrine (0.6 mg/mL; 1,000× stock): Dissolve epinephrine (Sigma-Aldrich, cat. no. E4642) in 0.01N HCl.
8. Epidermal growth factor for BEGM, 50,000× for ALI (25 μg/mL; 1,000× stock): Dissolve human recombinant, culture-grade EGF (Atlanta Biological, Norcross, GA, cat. no. C100) in PBS.
9. Retinoic acid (concentrated stock = 1×10^{-3} M in absolute ethanol, 1,000× stock = 5×10^{-5} M in PBS with 1% BSA): Retinoic

acid (RA) is soluble in ethanol and is light sensitive. First, make a concentrated ethanol stock by dissolving 12.0 mg of RA (Sigma-Aldrich, cat. no. R2625) in 40 mL of 100% ethanol. Store in foil wrapped tubes at −20°C. To prepare the 1,000× stock, first confirm the RA concentration of the ethanol stock by diluting it 1/100 in absolute ethanol. Read the absorbance at 350 nm using a spectrophotometer and a 1 cm light path quartz cuvette, zeroed with a 100% ethanol solution. The molar extinction coefficient of RA in ethanol equals 45,000 at 350 nm. Thus, the absorbance of the diluted stock should equal 0.45. RA with absorbance readings below 0.18 should be discarded. If the absorbance equals 0.45, add 3 mL of 1×10^{-3} M ethanol stock solution to 53 mL PBS and add 4.0 mL of BSA 150 mg/mL stock (s). For absorbance values less than 0.45, calculate the needed volume of ethanol stock as 1.35/absorbance and adjust the PBS volume appropriately.

10. Phosphorylethanolamine (70 mg/mL; 1,000× stock): Dissolve phosphorylethanolamine (Sigma-Aldrich, cat. no. P0503) in PBS.
11. Ethanolamine (30 μL/mL; 1,000× stock): Dilute ethanolamine (Sigma-Aldrich, cat. no. E0135) in PBS.
12. Zinc solution (0.863 mg/mL; 1,000× stock): Dissolve zinc sulfate (Sigma-Aldrich, cat. no. Z0251) in dH_2O. Store at room temperature.
13. Penicillin–streptomycin (100,000 U/mL Pen and 100 mg/mL Strep; 1,000× stock): Dissolve penicillin-G sodium (Sigma-Aldrich, cat. no. P3032) and streptomycin sulfate (Sigma-Aldrich, cat. no. S9137) in dH_2O for a final concentration of (100,000 U/mL and 100 mg/mL, respectively).
14. Gentamicin (50 mg/mL; 1,000× stock): Sigma-Aldrich, cat. no. G1397. Store at 4°C. For BEGM only.
15. Amphotericin B (250 μg/mL; 1,000× stock): Sigma-Aldrich, cat. no. A2942. Used for BEGM only.
16. Salt Stock solution (1,000× stock): Combine 0.42 g ferrous sulfate (Sigma-Aldrich, cat. no. F8048), 122.0 g magnesium chloride (J.T. Baker, Phillipsburg, NJ, cat. no. 2444), 16.17 g calcium chloride-dihydrate (Sigma-Aldrich, cat. no. C3881), and 5.0 mL hydrochloric acid (HCl) to 800 mL of dH_2O in a volumetric flask. Stir and bring total volume up to 1 L. Store at room temperature.
17. Trace elements solution (1,000× stock): Prepare seven separate 100 mL stock solutions. Using a volumetric 1-L flask, fill to the 1-L mark with dH_2O. Remove 8 mL of dH_2O. Add 1.0 mL of each stock solution and 1.0 mL of HCl (conc.). Store at room temperature.

3. Methods

3.1. Establishment of Respiratory Air–Liquid Interface Cultures

HBE cell culture systems serve as a good in-vitro model system of the airway epithelium. As cultures can be developed from both normal and diseased (CF/COPD) lungs, they allow for a direct comparison of differences engendered by lung disease. The HBE cell culture systems have been shown to produce mucus, maintain a 7 μm periciliary layer (PCL) (1), and facilitate coordinated transport of the mucus layer (2, 3). Biophysically, these cell culture systems have been used to assay the effect of shear stress on volume regulation and ATP release (4, 5) as well measuring the force of both individual (6) and patches of cilia (7).The methodology for isolating airway cells from donor lungs and culturing HBE cultures, well described by Fulcher et al. (8), is briefly summarized below. In addition to procuring HBE cells from human donor lungs, there are commercially available sources of human airway epithelial cells.

1. HBE cells are extracted from lungs from potential organ donors that are frequently unsuitable for transplantation but useful for research. HBE cells are also obtained from excess surgical pathology specimens procured through cooperating surgeons and pathologists using protocols in accordance with relevant regulations. In both cases, they are transported to the laboratory in a container of physiologic solution on wet ice, and used within 24 h after removal.
2. Airways from these lungs are manually dissected and all connective tissue removed. Cleaned segments are washed in phosphate-buffered saline (PBS) and placed in a DMEM solution containing antibiotics (100 U/mL penicillin and 100 mg/mL streptomycin), proteases (0.1% (w/v), type XIV), and DNases (1 mg/mL) and gently rocked at 4°C overnight. The epithelial layer is gently scraped and the cell supernatant is harvested. To end the dissociation process, a final concentration of 10% (v/v) fetal bovine serum is added. Cells are centrifuged at 500 × g for 5 min at 4°C. The pellet is resuspended in F-12 medium and counted using a hemocytometer.
3. Cells are plated on Type I/III collagen-coated plastic dishes ($2–6 \times 10^6$ cells per 10 mm dish) in an antibiotic supplemented (100 U/mL penicillin and 100 mg/mL streptomycin) BEGM (8) for 3–5 days.
4. Primary cells are passaged at 70–90% confluence. Rinse cells twice with PBS and add 3 mL of Trypsin (0.1% (w/v))/EDTA (1 mM). Incubate for 5–10 min at 37°C to free cells from the culture dish. Gently tap dish after the incubation and harvest the cells and transfer to a tube containing 3 mL of STI media

(1 mg/mL soybean trypsin inhibitor in F12 culture medium) on ice. Repeat if necessary.

5. Pool harvested cells and centrifuged at 500 × *g* for 5 min at 4°C. The supernatant is then aspirated and the pellet is resuspended in ALI media. The number of viable cells obtained are then determined using trypan blue exclusion or similar technique. At this point, cells can be plated on permeable supports or cryopreserved using an F-12-based freezing media (containing 30 mM HEPES, 10% (v/v) FBS, and 10% (v/v) DMSO).
6. The isolated HBE cells are plated onto collagen Type IV (human placental, Sigma) coated porous supports (Transwell Clear, Corning) at a density of $1.0–2.5 \times 10^5$ cells/cm^2.
7. To facilitate the formation of air–liquid interface cultures, remove the apical medium and rinse the surface of the porous support with gentle washings with PBS (0.5 mL/cm^2) after 24 h following initial plating and daily over the subsequent 7–10 days until confluent.
8. The media in the basolateral compartment is replaced with the appropriate volume of medium as specified for the support. Change the medium every other day until the cells become confluent and generate an air–liquid interface.
9. During the normal maintenance, HBE cultures need to be feed with fresh ALI every 2–3 days. Cultures should receive an apical PBS wash once a week to remove accumulating mucus.
10. After 4–6 weeks, the cultures will differentiate into ciliated cells and mucus-producing goblet cells. While the time to full differentiation is often variable between preparations, this stage can be monitored by observing the presence of beating cilia and the production of a mucus layer (Fig. 1).
11. With proper care, air–liquid interface HBE cultures can be maintained for a period of several months.

3.2. Collection of HBE Mucus for In Vitro Studies

By collection and reconcentrating these washings, mucus from HBE cell culture can be "made to order," thereby serving as a physiologically relevant model system for in-vitro mucus experiments. These systems have been shown to produce mucus, maintain a 7 μm PCL (1), and facilitate coordinate transport of the mucus layer (2, 3). Further, mucus collected from these model systems has been shown to be chemically similar to normal sputum (10), with roughly 75% overlap in the proteins detected in both specimens. Finally, the physical properties of HBE mucus versus % solids (a marker of mucus concentration) fall well within the range of those demonstrated by sputum, while being much more predictable (Table 1, Fig. 2). Further, mucus harvested from the HBE system and concentrated to normal (2.5% solids) and CF-like (8% solids) has been shown to be physiologically relevant model system

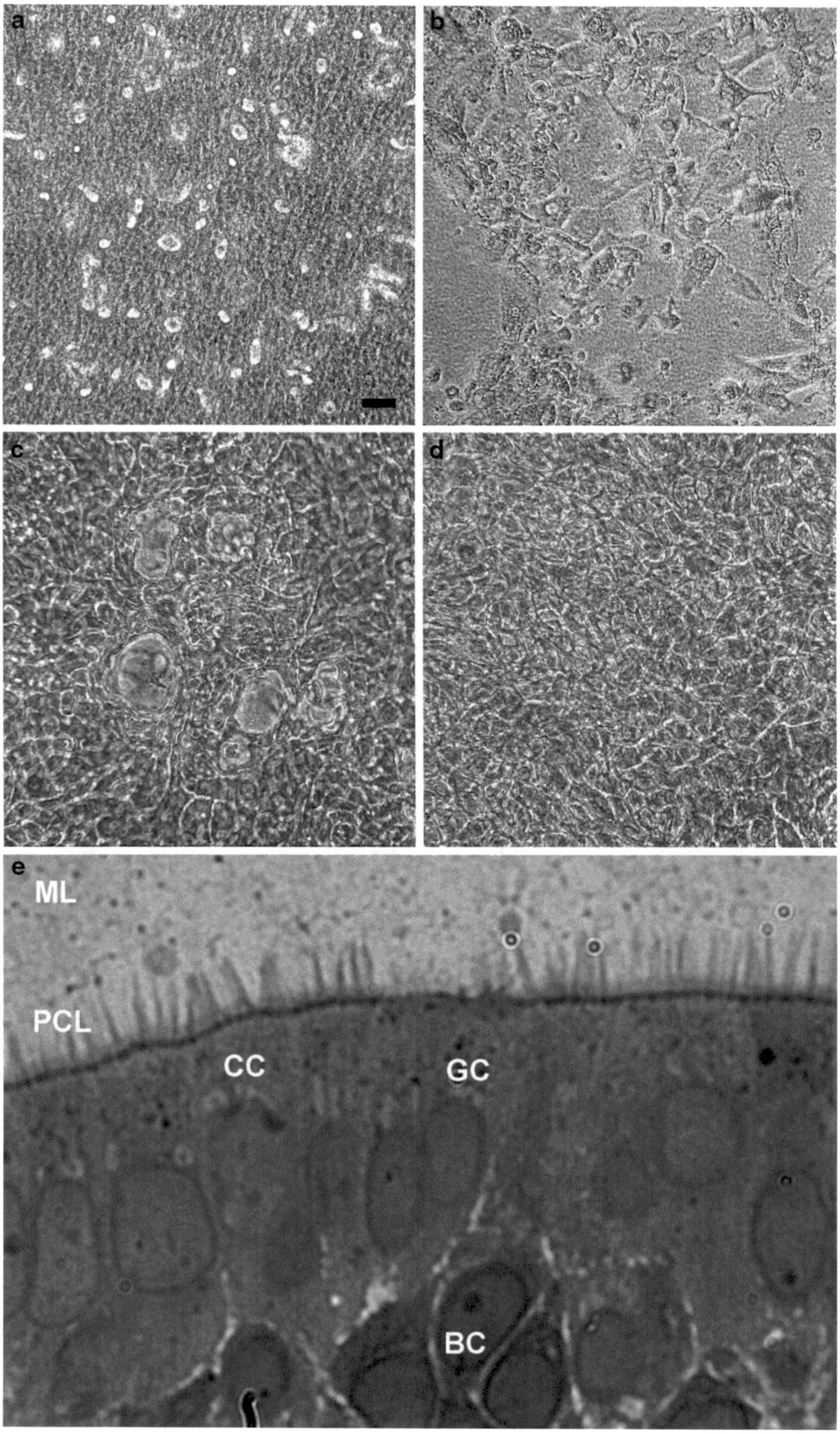

Fig. 1. Growth of Primary HBE cell cultures. Representative XY images of (**a**) freshly plated cells, (**b**) cells at 3–5 days, spreading across the culture support, (**c**) culture at day 10–12, nearly confluent with only small holes left to be filled in, and (**d**) culture at day 14, fully confluent. Bar = 20 μm. (**e**) XZ cross-section of a fully differentiated HBE culture (at 28 days) fixed with Os-PFC (to preserve the overlying mucus layer) and stained with Richardson's (1, 9). *ML* mucus layer; *PCL* periciliary layer; *CC* ciliated cell; *GC* goblet (mucus secreting) cell; and *BC* basal cell.

Table 1
Published rheological properties of human mucus and sputum

References	η_0 (Pa•s)	G_{inf} (Pa)
Puchelle (12), human (recurrent bronchitis)	24.8	6.2
Puchelle (12), human (mild chronic bronchitis)	11.4	0.76
Puchelle (12), human (severe chronic bronchitis)	12.5	0.87
Baconnais (13), human	0.71	N/A
Baconnais (13), human (CF)	0.16	N/A
Puchelle (14), human	24.7	1.7
Dawson (15), human (CF)	60	15.5
Hill (3), 2.5% (w/w) human cell culture mucus	0.12	0.87
Hill (3), 8% (w/w) human cell culture mucus	3.3	4.7

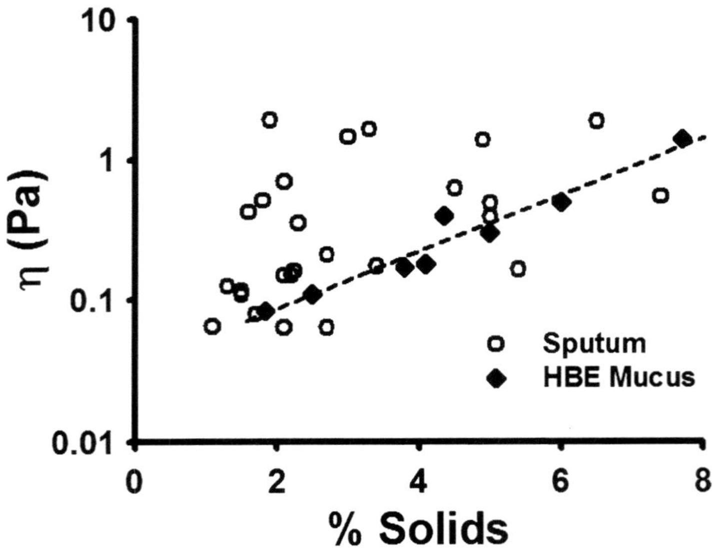

Fig. 2. Comparison of the complex viscosity (η) of HBE mucus (*filled diamonds*) versus sputum samples obtained from various individuals (*open circles*).

for predicting the affect of the biophysical properties of the mucus layer on bacterial biofilm formation (3). Below we describe the methodology used in harvesting and reconstituting mucus from HBE cultures.

1. As a part of routine maintenance, cultures are washed with ~1.5 μL PBS/mm^2 of culture area (i.e., ~1,000 μL per 30 mm diameter dish, ~150–200 μL per 12 mm diameter dish).

2. Once added, PBS is left on the culture for 30 min at 37°C before removal. For cultures that demonstrate heavy mucus production, a second 30 min wash can be beneficial for mucus harvesting. Care must be taken to ensure that cells are not disturbed during collection of the lavage solution.
3. Because these cell cultures can produce mucus for up to 6 months, this procedure can be performed repeatedly on a weekly or biweekly basis. To maintain sterility, it is recommended to perform the lavage procedure in a laminar flow hood using sterile solutions.
4. Once collected, cells and larger debris are collected by a low-speed centrifugation (300 × *g* for 5 min). The pellet is discarded and the washings are pooled and stored at 4°C until sufficient volume is obtained (see Note 3).
5. When one has harvest ~1 L of pooled washing the concentration step can begin. Here, washings are loaded into 3,800 Da molecular weight cut off dialysis bags and loaded into a container filled with polymer absorbent (Spectra/Gel) for 1–5 days at 4°C to concentrate the mucus. Absorbent is replaced daily.
6. Midway through concentration, mucus is next dialyzed against PBS containing 500 μM $MgCl_2$, and 800 μM $CaCl_2$ at 4°C to establish the proper salt balance (3).
7. Once mucus has reached its final concentration (typically when 200- to 1,000-fold reduction in volume is achieved), it is once again dialyzed against PBS.
8. The concentration of the final product is determined by placing a sample aliquot (~50–100 μL) on a preweighed piece of foil which is then placed in an 80°C oven overnight to dry the sample. The final concentration of solid material (% solids) is then determined.

Once the dry weight of the sample is determined, the concentration of mucins in mucus is assessed by differential refractometry (11). A 500 μL sample is chromatographed on a Sephacryl S-1000 column (Amersham Pharmacia) and eluted with 200 mM sodium chloride/10 mM EDTA at a flow rate of 0.5 mL/min. The column effluent is passed through an in-line Dawn EOS laser photometer coupled to a Wyatt/Optilab DSP inferometric refractometer to measure light scattering and sample concentration, respectively. The concentration of the mucin is calculated by integrating the refractive index peak associated with the material eluted in the void volume of the column and employing a value for the refractive-index increment (dn/dc) of 0.165 mL/g, which has been measured previously at 650 nm. The reproducibility of this procedure is typically within 5%. The total protein content of a given mucus sample is determined by similar methods, but with a Sephadex

G-25 column and a $dn/dc = 0.170$. The non-mucin content of the mucus sample is then determined by subtraction of the mucin content from the total protein (mucin and non-mucin) content.

3.3. In Vitro Measures of Mucociliary Transport

The lung is continually under assault by pathological and noxious materials inhaled during every breath that we take. Lung health is maintained because these inhaled particles and pathogens are trapped within a blanket of mucus and swept from the lung by a continuous flow of mucus generated by the beating of cilia. The failure of mucus clearance in the environmentally damaged lungs, in COPD or asthma, leads to severe health problems as the lung tissue is destroyed by the inflammation response to microbial infections that cannot be cleared.

While insight into airway surface liquid and cilia beat parameters (height regulation, mucus secretion, beat frequency, etc.) have been obtained, the understanding of the full phenomena of mucus clearance has remained elusive. Key to understanding potential links between genetic disorders and failure of mucus transport may lie in the two essential components of the mucociliary clearance system, cilia and mucus. However, animal models have been the only source of true clearance models. This is due to the lack of an established methodology for promoting coordination of beating cilia across the length of a cell culture system in a suitable geometry.

One of the key advantages of this well-differentiated human airway culture system is the ability to generate cultures with coordinated cilia beating that produces vectoral mucus transport. We have observed that anywhere from 10 to 25% of well-differentiated HBE cell cultures develop millimeter sized patches in which the cilia spontaneously coordinate their beating in a circular pattern, yielding rotational (the so-called "mucus hurricanes"), reflecting the circular boundary of the culture support for flowing mucus (see Figure 3, below). The protocol below is designed to facilitate the production of cell cultures with transporting mucus and the approach used by our laboratory to measure the rate of transport.

1. These studies rely on the use of well-differentiated HBE cultures prepared using the approach listed in the section above. To facilitate differentiation into well-ciliated cultures necessary for mucus transport studies, initial cell seeding densities as high as $250–500 \times 10^3$ cells/cm^2 can be used.
2. A primary requirement for producing cultures which are capable of generating transporting mucus is the degree of ciliation. Typically, cultures with >70% of the surface area with beating cilia are required (see Note 4).
3. The mucus layer that is produced in HBE cultures is not "cleared" as it is in vivo, due to the constraints of the fixed Transwell culture support. The result is a continual accumulation of mucus. In order to maintain a "flowable" mucus layer during ciliogenesis, it is necessary to remove accumulated

mucus on a frequent basis. The next few steps are designed to maximize the chances of generating cultures with rotational transport.

(a) First, the accumulated mucus is regularly removed from the luminal surface of the HBE cell cultures. This is accomplished by three incubations with an isotonic salt solution (e.g., PBS) for 10 min at 37°C. At the completion of the washing procedure, any remaining fluid is suctioned from the apical surface and the cultures returned to the incubator. Typically, this procedure is performed three times per week following confluence.

(b) Second, because mucus layer on the surface of an HBE culture is very thin (typically between 10 and 50 μm, depending on washing interval), and therefore very small volumes (microliters), it is very susceptible to dehydration, even in incubators at 90% humidity (see Note 5).

4. Following culture confluence, it can take anywhere from 3 to 5 weeks for cultures to reach >70% surface area ciliation.

5. As noted above, rotational coordination of cilia typically occur in less than 25% of all well-ciliated cultures. Therefore, prior to transport studies it is necessary to prescreen for cultures exhibiting rotational transport. This can be done by observing the movement of cellular detritus following the addition of a small volume of fluid (~20 μL/cm^2) to the apical surface in the absence of any exogenous label. Only cultures which exhibit rotational mucus transport are used for the transport studies.

6. Depending on the type of study being performed, endogenous mucus is allowed to accumulate for 1–3 weeks prior to the transport assay. During this time, only the basolateral media is changed.

7. To better visualize transport of mucus during the experiments, 0.5–1.0 μm fluorescent microspheres (FluoSpheres, Invitrogen) are added to the luminal surface 24 h prior to the transport studies (see Note 6). To resolve single microspheres, the microspheres are diluted 1:10,000 in PBS and added to the cultures (at 20 mL/cm^2). The excess fluid is absorbed over the 24 h. The rate of mucus transport during the experiment is determined from time-lapse images of the microspheres using a low power (10×) fluorescent microscope. The microspheres appear as streaks when taken as 2–5 s exposure images (Fig. 3a).

8. During the experiment, images are taken at desired time points and/or after the addition of mucus altering agents, depending on the experiment being performed.

9. Because the mucus layer is quite thin (10s of micrometer), the mucus layer is susceptible to evaporation during data analysis which can significantly alter the transport measurements. Such evaporation can be prevented by the addition of perfluorocarbon

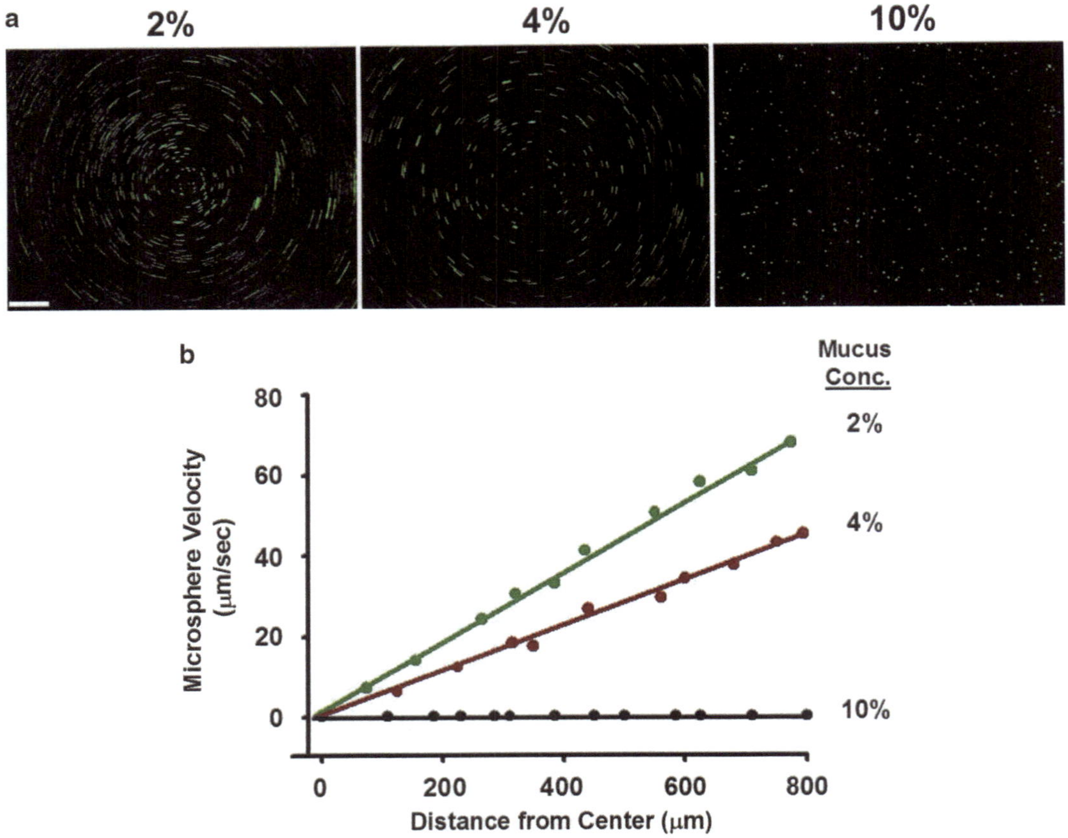

Fig. 3. Determination of mucus transport rates in HBE cultures. (**a**) Example 2-s time-lapse images of 1.0 μm fluorescent microspheres (1.0 μm, *green*) in cultures with 2, 4, and 10% mucus (% solids). (**b**) Plot of microsphere velocity (μm/s).

(such as Fluorinert FC-3283, 3M) to the luminal surface (100 μL volume/cm^2). This can be done with no noticeable alteration in transport rate and can be easily removed to add additional test agents.

10. Because the coordination of mucus transport is rotational, the microsphere-streaks in the center of the culture will have a different length than at the periphery. Image analysis software (such as Image J, NIH) can be used to determine the length of the streaks.

11. Linear regression analysis of the length of the streak plotted versus the distance from the center of the transporting mucus (Fig. 3b) results in an angular velocity of the beads. For general comparison, transport rates can be normalized at a given distance from the center of rotation (e.g., 0.5 mm).

12. In the example in Fig. 3a, b, the transport rate versus distance is shown for three different mucus concentration ranging from normal airway mucus (~2% solids) to cystic fibrosis-like (10% solids), where mucus transport ceases (17). Here, the time-lapse image of the microspheres appears as spots rather than streaks.

4. Notes

1. If anticipated usage of LHC basal medium exceeds 550 L/year, powdered stock can be custom ordered from Sigma-Aldrich.
2. For more details regarding LCH, ALI, and BGEM Media, see Fulcher et al. (8).
3. Degradation of mucins during storage is a potential shortcoming of this mucus preparation methodology. In general, we observe a roughly 25% decrease in the molecular weight and radius of gyration of mucins, as determined by dynamic light scattering, over a 1 month period. To prevent this effect from becoming larger, it is recommended that mucus be prepared from pooled washings that have been stored less than 1 month.
4. During differentiation into ciliated cells and associated ciliogenesis, coordination of cilia beating pattern is necessary in the formation into mucus transporting cultures. While we do not know how coordination of cilia is formed, recent work in fish and amphibian development indicates that ciliary orientation at the apical membrane of vertebrate ciliated cells is a two-step process beginning with tissue patterning, followed by a flow-based refinement phase in which the effective stroke of the beat aligns tightly with the direction of fluid flow (16).
5. To maintain proper mucus flow during ciliogenesis, it is necessary to use well-humidified incubator, which can be accomplished by having multiple water pans in the incubator, as the humidity is proportional to the surface are of the water exposed to the air. Using this approach, we observe near saturating conditions. It is useful to regularly measure the humidity in an incubator using a hygrometer.
6. Alternatively, small volumes (nanoliter) containing microspheres can be aerosolized onto the culture using a ultrasonic nebulizer (Aeroneb, Aerogen) to avoid the addition of a relatively large volume (microliter) which might alter mucus properties.

Acknowledgments

The authors would like to thank Leslie Fulcher for advice and technical information related to culturing human airway epithelial cells. This work was supported by the Cystic Fibrosis Foundation and the National Institutes of Health.

References

1. Matsui H, Grubb BR, Tarran R, Randell SH, Gatzy JT, Davis CW, Boucher RC (1998) Evidence for periciliary liquid layer depletion, not abnormal ion composition, in the pathogenesis of cystic fibrosis airways disease. Cell **95**, 1005–1015
2. Matsui H, Randell SH, Peretti SW, Davis CW, Boucher RC (1998) Coordinated clearance of periciliary liquid and mucus from airway surfaces. J Clin Invest **102**, 1125–1131
3. Matsui H, Wagner VE, Hill DB, Schwab UE, Rogers TD, Button B, Taylor RM, 2nd, Superfine R, Rubinstein M, Iglewski BH, Boucher RC (2006) A physical linkage between cystic fibrosis airway surface dehydration and Pseudomonas aeruginosa biofilms. Proc Natl Acad Sci USA **103**, 18131–18136
4. Button B, Boucher RC (2008) Role of mechanical stress in regulating airway surface hydration and mucus clearance rates. Respir Physiol Neurobiol. **163,** 189–201
5. Button B, Picher M, Boucher RC (2007) Differential effects of cyclic and constant stress on ATP release and mucociliary transport by human airway epithelia. J Physiol-London **580**, 577–592
6. Hill DB, Swaminathan V, Estes A, Cribb J, O'Brien ET, Davis CW, Superfine R (2010) Force Generation and Dynamics of Individual Cilia under External Loading. Biophys J **98**, 57–66
7. Teff Z, Priel Z, Gheber LA (2007) Forces applied by cilia measured on explants from mucociliary tissue. Biophys J **92**, 1813–1823
8. Fulcher ML, Gabriel S, Burns KA, Yankaskas JR, Randell SH (2005) Well-differentiated human airway epithelial cell cultures. Methods Mol Med **107**, 183–206
9. Sims DE, Westfall JA, Kiorpes AL, Horne MM (1991) Preservation of tracheal mucus by nonaqueous fixative. Biotech Histochem **66**, 173–180
10. Kesimer M, Kirkham S, Pickles RJ, Henderson AG, Alexis NE, Demaria G, Knight D, Thornton DJ, Sheehan JK (2009) Tracheobronchial air-liquid interface cell culture: a model for innate mucosal defense of the upper airways? Am J Physiol Lung Cell Mol Physiol **296**, L92–L100
11. Kirkham S, Sheehan JK, Knight D, Richardson PS, Thornton DJ (2002) Heterogeneity of airways mucus: variations in the amounts and glycoforms of the major oligomeric mucins MUC5AC and MUC5B. Biochem J **361,** 537–546
12. Puchelle E, Zahm JM, Aug F (1981) Viscoelasticity, protein content and ciliary transport rate of sputum in patients with recurrent and chronic bronchitis. Biorheology 18 (3–6): 659–666
13. Baconnais S, Tirouvanziam R, Zahm JM, de Bentzmann S, Peault B, Balossier G, Puchelle E (1999) Ion composition and rheology of airway liquid from cystic fibrosis fetal tracheal xenografts. Am J Respir Cell Mol Biol **20**, 605–611
14. Puchelle E, Zahm JM, Duvivier C (1983) Spinability of bronchial mucus. Relationship with viscoelasticity and mucous transport properties. Biorheology **20**, 239–249
15. Dawson M, Wirtz D, Hanes J (2003) Real-time multiple nanoparticle tracking of gene carriers in cystic fibrotic sputum. Mol Ther 7, S373–S374
16. Mitchell B, Jacobs R, Li J, Chien S, Kintner C (2007) A positive feedback mechanism governs the polarity and motion of motile cilia. Nature **447**, 97–101
17. Boucher RC (2002) An overview of the pathogenesis of cystic fibrosis lung disease. Adv Drug Deliv Rev **54**, 1359–1371

Chapter 16

Studying Mucin Secretion from Human Bronchial Epithelial Cell Primary Cultures

Lubna H. Abdullah, Cédric Wolber, Mehmet Kesimer, John K. Sheehan, and C. William Davis

Abstract

Mucin secretion is regulated by extracellular signaling molecules emanating from local, neuronal, or endocrine sources. Quantifying the rate of this secretion is important to understanding how the exocytic process is regulated, and also how goblet/mucous cells synthesize and release mucins under control and pathological conditions. Consequently, measuring mucins in a quantitatively accurate manner is the key to many experiments addressing these issues. This paper describes procedures used to determine agonist-induced mucin secretion from goblet cells in human bronchial epithelial (HBE) cell cultures. It begins with primary epithelial cell culture, offers methods for purifying MUC5AC and MUC5B mucins for standards, and describes five different microtiter plate binding assays which use various probes for mucins. A polymeric mucin-specific antibody is used in standard and sandwich ELISA formats for two assays while the others target the extensive glycosylated domains of mucins with lectin, periodate oxidation, and antibody-based probes. Comparing the data derived from the different assays applied to the same set of samples of HBE cell cultures indicates a qualitative agreement between baseline and agonist stimulated mucin release; however, the polymeric mucin-specific assays yield substantially lower values than the assays using nonspecific molecular reporters. These results indicate that the more nonspecific assays are suitable to assess overall secretory responses by goblet cells, but are likely unsuited for specific measurements of polymeric mucins, per se.

Key words: Mucin, Secretion, Exocytosis, Human bronchial epithelial cell culture, Goblet cell

1. Introduction

Polymeric mucins are a crucial component of the mucus that lines the normal airway epithelium and provides a first line of defense for the lungs against inspired aerosols, particulates, and pathogens (1). Their exocytic release from goblet cells and submucosal glands in the airways marks the beginning of the maturation of mucins

Michael A. McGuckin and David J. Thornton (eds.), *Mucins: Methods and Protocols*, Methods in Molecular Biology, vol. 842, DOI 10.1007/978-1-61779-513-8_16, © Springer Science+Business Media, LLC 2012

into luminal mucus. The mucins are heavily glycosylated proteins (~80–90% carbohydrate), with monomeric molecular weights of ~2 MDa, and polymeric molecular weights measured in the 10–100 s of MDa (2). Mucins are synthesized and dimerized in the ER and glycosylated and oligomerized in the Golgi complex. They are packaged in, and released from, the trans-Golgi network as large, 1-μm-sized secretory granules, which are stored in the apical pole of goblet or mucous cells until they are released by exocytosis following activation by agonist (3, 4). Essential to lung defense in normal physiology, mucins and mucus are overproduced in all of the airway inflammatory diseases, including asthma, chronic obstructive pulmonary disease, and cystic fibrosis, to such an extent that mucus plugging, gas trapping, and infection are major problems for patients and their pulmonologists (5). As a consequence, understanding the regulation of mucin biosynthesis and secretion is a long-sought goal, toward which substantial progress has been made in the past two decades. Our laboratory has been involved in the effort to delineate the regulation of mucin secretion from goblet cells (4) for most of this period, and in this chapter we share our current methods for studying mucin secretion from human bronchial epithelial (HBE) cell primary cultures.

2. Materials

2.1. Human Bronchial Epithelial Cell Culture

1. *Bronchial epithelial growth medium (BEGM)*. LHC Basal Medium (Biosource, Camarillo, Cat. # P118-500), supplemented with the growth factors, antibiotics, and other ingredients detailed in the procedures and Tables 1 and 2 of ref. 6 (see Note 1).
2. *Air–liquid interface (ALI) medium.* 50:50 Mixture of LHC Basal Medium and DMEM-H (Gibco, Carlsbad, CA, Cat. # 11995-065), supplemented with the growth factors, antibiotics, and other ingredients detailed in the procedures and Tables 1 and 2 of ref. 6 (see Note 1).
3. *Human airway epithelial cells.* Cells may be harvested by proteolytic digestion of human turbinates, polyps, trachea, or bronchi obtained in conjunction with surgeons and/or pathologists, following the procedures detailed in ref. 6. Alternatively, they can be purchased from commercial sources, e.g., Cell Applications, Inc. (http://www.cellapplications.com), Lonza Group (formerly, Clonetics, Cambrex; http://www.lonza.com), or Lifeline Cell Technology (http://www.lifelinecelltech.com/docs/SPC_AirwayEpi.pdf; see Note 2).
4. Corning Costar Transwell® Clear (polyester) cell culture inserts, 12 or 24 mm diameter (see Note 3).

2.2. HBE Cell Culture Mucin Secretion Experiments

1. Foam pads with cutouts for culture plates (see Note 4).
2. Dermabond®, a long-chain, nontoxic cyanoacrylate adhesive, available from hospital pharmacies (see Note 3).

2.3. Preparing Mucin Standards

1. 8 M GuHCl Buffer is prepared by adding guanidine hydrochloride (spectrophotometric grade, e.g., Acros Organics, Cat. # 120230010) in 10 mM phosphate buffer, containing 5 mM EDTA, pH 6.5. Add one mini protease inhibitor cocktail tablet (PICtab-mini) for each 10-mL volume used, just before use (see step 2).
2. Complete® protease inhibitor cocktail tablets (PICtab), Roche Applied Sciences, use 1 tablet/50 mL of solution (Cat. # 11873580001) or one mini tablet/10 mL (Cat. # 11836153001).
3. Spectra/Gel® Absorbent (Spectrum Laboratory Products, Cat. # 888–16582).
4. Optilab Interferometric Refractometer with a filter at 680 nm (Wyatt, CA, USA).
5. Astra 4.9 (or later) data acquisition and processing software (Wyatt, CA, USA).
6. Sepharose CL2B (Sigma, MO, USA) or Sephacryl 1000 (GE Bio-sciences Uppsala Sweden) (12 × 2.5 cm, ~12 mL) gel-permeation chromatography column.
7. 7 Port V7 manual injection valve (GE Bio-sciences Uppsala Sweden).
8. Pump, Rheos 2000 (Flux instruments, Thermo Scientific, USA) or equivalent solvent delivery pump.
9. Running buffer: 200 mM NaCl, 10 mM Tris, 10 mM EDTA, pH 7.0. Filter and degas by sonication (for longer storage, add 0.1% w/v Na azide).

2.4. Microtiter Plate Mucin Assays

1. Flat Bottom, High Binding, 96-well microtiter plates (Costar, Cat. # 3590).
2. Phosphate-buffered saline (PBS).
3. Wash buffer: PBS containing 0.05% v/v Tween-20 (PBST).
4. Blocking solution: 5% w/v milk, 1% w/v BSA, or 0.1% w/v gelatin, as specified, in PBST.
5. Phosphate citrate buffer: Citric acid 3.36 g, sodium phosphate monobasic 9.88 g, adjust pH to 5.0, bring to 1 L. Store at 4°C; alternatively, use phosphate-citrate buffer tablet, Sigma, Cat. # P4809.
6. Substrate solution: O-phenylene diamine dihydrochloride (OPD; Pierce, Cat. # 34062) in citrate phosphate buffer (*make fresh for each use*), 40 mg OPD in 100 mL citrate phosphate buffer, add 40 μL 30% (v/v) hydrogen peroxide (Sigma, Cat. # 1009).

2.4.1. Conventional ELISA

1. Primary antibody, monoclonal antibody (H6C5, hybridoma supernatant) made against intact mucins purified from CF sputum ((7); see Note 5).
2. Secondary antibody, goat anti-mouse-HRP conjugated (Jackson ImmunoResearch, Cat. # 115-035-020), from which a stock, stored at –20°C, is made by diluting 1:1 in glycerol.

2.4.2. Mucin Subunit Ab ELISAs

1. The mucin subunit antibody is a rabbit, polyclonal antibody made against purified, reduced, and alkylated salivary MUC5B; however, as detailed in Note 5, the antibody appears to detect all vertebrate polymeric mucins.
2. Goat anti-rabbit-HRP (Jackson ImmunoResearch, Cat. # 111-035-003), from which a stock, stored at –20°C, is made by diluting 1:1 in glycerol.
3. Non-binding 96-well plates (Corning, Cat. # CLS3641) or 1-mL cluster tubes (Costar, Cat. # 4408).
4. 0.1 M Tris buffer, pH 8.0.
5. Dithiothreitol (DTT; Sigma, Cat. # D9163), 10× stock, 100 mM, 15.42 mg/mL, in Tris buffer (*make fresh for each use*).
6. Iodoacetamide (Sigma, Cat. # I1149), 10× stock, 250 mM, 46.25 mg/mL, in Tris buffer (*make fresh for each use*).

2.4.3. WGA ELLA

1. Lectin from Triticum vulgaris (WGA), Peroxidase labeled, Sigma, Cat. # 3590, lyophilized powder reconstituted in sterilized PBS, 1 mg/mL.

2.4.4. Periodic Acid-Biotin-Hydrazide Assay

1. Sodium acetate buffer, 100 mM, EDTA, 5 mM, pH 5.5. Dissolve 13.61 g sodium acetate·$3H_2O$ and 1.46 g EDTA in 500 mL water. Adjust pH to 5.5 with acetic acid.
2. Periodic acid, 1 mM. Dissolve 2.28 mg periodic acid in 10 mL of acetate buffer (*make fresh for each use*).
3. Biotin hydrazide: Sigma, Cat. # B-7639, stock 50 mM in DMSO. Dissolve 12.9 mg of biotin hydrazide in 1 mL DMSO (need to warm for complete solubility). Store in 50-µL aliquots at –20°C; *make a fresh* working solution by adding a 50-µL aliquot to 15 mL acetate buffer.
4. Streptavidin–HRP, Sigma, Cat. # 3892. Supplied as a 1 mg/mL solution.

3. Methods

3.1. HBE Cell Culture

HBE cell culture techniques that yield good-quality primary cultures with a degree of differentiation that mimics adult airway epithelia have been available for several years. These cultures have proven quite useful for the study of regulated mucin secretion. However, for laboratories just engaging in the practice, it must be pointed out that good primary culture techniques are essential to ensure high-quality results. The techniques are considerably more laborious than those used for the growth and maintenance of cell lines; shortcuts are few. That said, the techniques required are outside the scope of this review with its focus on techniques for using the cultures in secretion experiments. Consequently, we offer a brief overview, but otherwise refer the reader to ref. 6 which offers both a historical perspective and detailed procedures of the culture system and to Chapter 15.

HBE cell cultures are primary cultures that begin, ideally, with epithelial cells digested proteolytically from airway tissues (see Note 6). When these cells are seeded on plastic or the type of permeable insert used for epithelial cell culture, only the basal cells adhere—the occasional ciliated and/or goblet cell that might be observed in such cultures the first few days after seeding has been introduced into the culture as a clump that includes basal cells (see Note 7). The seeded cells proliferate for 2–3 days in BEGM medium to form a confluent monolayer, at which point the luminal liquid is removed and the culture medium switched from BEGM to ALI, the classic conditions of "ALI" cultures (see Note 8). Under ALI conditions in which the cells are fed with ALI medium, selectively, at their basolateral surfaces three times per week, the cells differentiate over a 4- to 6-week period to assume a mucociliary phenotype.

3.2. HBE Cell Culture Mucin Secretion Experiments

The methods described here for determining the mucins secreted from HBE cell cultures evolved over several years of trial and error in our laboratory. Despite extensive previous experience with mucin secretion assays using SPOC1 and other cell lines (8), and with HBE cell cultures in airway biology experiments (9–11), our initial experience in HBE cell culture mucin secretion experiments was so disappointing that our first published results used HBE cells grown instead in tracheal xenografts, in the backs of nude mice (7). The problems we experienced using HBE cell cultures were reflected in poor secretory responses, or a complete lack of response, to purinergic agonists (ATP, UTP, ATPγS). In subsequent troubleshooting, we found from periodic acid-Schiff (PAS)-stained cultures that goblet cell mucin stores were discharged during the wash procedure that preceded agonist exposure, likely resulting from mechanical stimulation associated with *mis*handling the cultures. A serious complicating requirement in these experiments is that *several* successive washes are required to remove accumulated mucins from culture surfaces in order to be able to detect those freshly secreted—one or two washes are unsatisfactory, as shown below. As a consequence, we developed the following protocol to preserve HBE mucins stores during preparation for experiments.

It is based on an experimental paradigm of a wash procedure to remove extracellular mucins from HBE cell cultures, followed by shorter baseline and agonist secretion periods. One way to emphasize the importance of a gentle wash procedure is to note that it *consumes ~4.5 h of time, relative to a typical experimental period of ~1 h.*

3.2.1. Prewash, 48–72 h Prior to the Beginning of an Experiment

6- or 12-well cluster plates bearing cell cultures are removed from the incubator and placed individually in foam pads (see Note 4). The lumen (top) of each culture is washed, gently, with PBS, and the ALI medium in the basal (bottom) compartment is replaced. At this time, Dermabond® is used to glue the inserts in place by applying a small drop between the rim of each cell culture insert and the top of the respective well (see Note 3). The cultures, in their foam pads, are placed, unstacked, in the incubator (see Note 9).

3.2.2. Careful Wash Procedure, Four Washes, Total Time ~4.5 h

In all subsequent handling steps, plates are prepared one at a time. A foam pad bearing a culture is carefully picked up and removed from the incubator, carried to the hood without any sudden, jarring motions, and gently placed on the deck—placement on the deck is best done, before sitting, by placing the rear edge (toward the back of the hood) of the foam pad down first at a slight angle, and then lowering the pad slowly to rest on the deck surface.

1. Plates are subjected to a gentle wash, defined as follows: raising the rear edge of the foam pad to an ~45° angle and holding it there with a foam block or other device, the lid is removed, gently, the pipette tip, held vertically, is inserted into the front corner of the cell culture to slowly remove and add luminal liquid, following which the pad is relowered gently to the deck surface. These actions allow surface liquids to flow slowly over culture surfaces to, and from, an isolated corner during removal and addition, respectively. "Gently" cannot be overemphasized in all movements involving the HBE cell culture wash procedure.
2. Gentle wash 1: Add warm (37°C) DMEM (400 μL for 12-mm inserts, 1 mL for 24-mm inserts) to culture lumens, gently reposition the dish and pad to the horizontal, and replace in the incubator for 10 min.
3. Gentle wash 2–4: Collect and save the surface liquid, add a new volume of *warm* DMEM, and return to incubator for 1 h (see Note 10).

3.2.3. Baseline Secretion Period

Procedure is the same as in gentle wash 2–4, but during this incubation the mucins secreted and collected at the end of the 1-h period are taken to represent those released at baseline (see sample results at the end of Subheading 3).

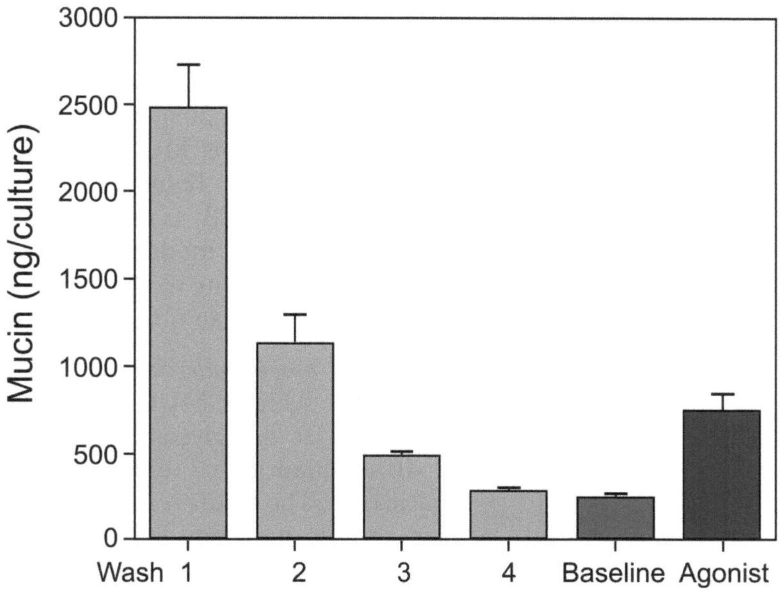

Fig. 1. Mucins recovered from HBE cell cultures (12-mm inserts) during a careful wash procedure, and the subsequent baseline and agonist (ATPγS, 100 μM) secretion periods (mean ± SE, *n* = 6). The conventional mucin subunit ELISA was used for these assays.

3.2.4. Experimental Period

Most experiments have an experimental procedure during the next and generally final incubation. The mucins released, collected at the end of the period, are frequently normalized to those released during the baseline secretion period. In the sample results illustrated below, ATPγS was used to stimulate regulated mucin secretion via $P2Y_2$ purinoceptors (4), and we suggest this condition as a good control for most experiments.

3.2.5. Illustrative Data

Figure 1 shows the results of a typical agonist secretion experiment with HBE cell cultures grown on 12-mm Transwell Clears using ATPγS. During the wash procedure, ~4,300 ng of mucin was removed from the cultures. While the majority of this mucin was removed during the initial 10-min wash period, significant quantities remained that were removed during the next 3 h-long washes. Note that the mucins secreted during the baseline period were quantitatively similar to those from the final wash period. In other experiments (not shown), in which a second baseline period follows the first instead of an incubation with agonist, the mucins from both baseline periods were similar to those from the final wash period; i.e., with this procedure, a steady-state release of mucins at baseline is achieved. The approximately three-fold stimulation of mucin secretion by a maximal concentration of ATPγS is typical; the multiplicity (agonist/baseline) in our hands ranges from ~2 to 4.

3.3. Preparing Mucin Standards

Ready sources of mucus from which mucins may be purified for use as standards are HT29 cell cultures (3), saliva (12), and HBE cell cultures (13), which secrete, respectively, MUC5AC, MUC5B, and both MUC5AC (~20%) and MUC5B (~80% (13); see Note 11). Saliva also contains MUC7, but this smaller, nonpolymeric "contaminant" is easily removed by size-exclusion chromatography on Sepharose CL2B, as described below. Mucins are purified according to their buoyant density using density-gradient centrifugation in CsCl, according to the technique first described in 1983 by Carlstedt and Sheehan ((14); see Note 12).

1. *HT29 cell mucus collection.* HT-29-A1 cells, or another subclone that secretes MUC5AC as the only polymeric mucin detectable at the glycoprotein level, are cultured in RPMI 1640 medium with L-glutamate containing 10% (v/v) fetal calf serum. The media removed from the cultures during feeding and/or passaging is saved by adding PICtab (1 tablet/50 mL), and pooled (4°C) until a sufficient quantity is achieved (1 L is a reasonable minimum). The pooled material is concentrated in dialysis bags using Specta/Gel® to ~1/3 its original volume, and then GuHCl is added to a final concentration of 4 M.
2. *Saliva collection.* Collect (whole) saliva by chewing pieces of Parafilm® and expectorating the saliva into ice-cold 8 M GuHCl buffer, to which PICtab (1 tablet/50 mL) is freshly added, until the GuHCl is diluted to 4 M (150 mL is a reasonable minimum). This material is subjected to a crude separation by chromatography on a Sepharose CL2B column (see Note 13) eluted with 4 M GuHCl; the void volume containing MUC5B is collected while the included volume containing other proteins and glycoproteins, such as MUC7 and DMBT1, is discarded.
3. *HBE cell culture mucus collection.* Beginning ~7–10 days following confluence, HBE cell cultures secrete significant quantities of MUC5AC and MUC5B in a ratio of ~1:5 (13). These mucins are collected at the times of culture feeding by placing 0.5 or 1.0 mL of PBS in the lumens of 12- or 24-mm Transwell cultures, gently pipetting it up and down 2–3 times, then removing the material, adding PICtab (1 tablet/50 mL), and pooling it (4°C) until a sufficient quantity is achieved (250 mL is a reasonable minimum). GuHCl is added to a final concentration of 4 M (see Note 14).
4. *Purification of mucins by cesium chloride density gradient* (see Note 15). To any of the above mucin-rich materials in 4 M GuHCl, add CsCl to a density of 1.4 g/mL using the equation: $x = v(1.347\ \rho - 0.0318M - 1.347)$, where x is CsCl (g), v is the volume of material, ρ is the final density (1.4 g/mL), and M is the molarity of GuHCl (14). The material is centrifuged to equilibrium at 38,000 rpm at 15°C in a Beckman 50.2

Ti rotor (131,242 × *g*) for 3 days, the tubes fractionated into 2.5-mL volume, and each fraction is analyzed for density (by accurately weighing 1 mL), absorbance at 260 nm (DNA) and 280 nm (protein), and PAS staining (mucins; by slot blotting, (15)). The mucin-rich, PAS-positive fractions, which should have densities of ~1.4–1.5 g/mL, are pooled and dialyzed exhaustively against 0.2 M GuHCl, 5 mM EDTA in 10 mM phosphate buffer, pH 6.5. Next, the material in 0.2 M GuHCl is prepared for the second centrifugation in CsCl, centrifuged, the tubes fractionated, and fractions analyzed exactly as the first time. The mucin-rich, PAS-positive fractions are pooled and dialyzed against PBS for 3–4 days with two buffer changes each day.

5. *Determination of mucin concentrations in the standards* (see Note 16). Refractive index is used to measure the absolute concentration of mucins that are being used as standards. The refractive index of a solution receives contributions from the solvent and the solute. The contribution from the solute is generally a linear function of the concentration and is described as the refractive index increment, d*n*/d*c*, which for mucins is 0.165 mL/g. Using a gel filtration column with large exclusion media (Sepharose CL2B or Sephacryl S1000), mucins are separated from other proteins and can be quantified in a single step as the eluant from a size-exclusion chromatography column is simply passed through a refractive index detector.
6. Precondition the column with at least two-column volumes of running buffer or until getting a stable baseline.
7. Dilute a sample of the purified mucin stock solution 10–20 times with the running buffer, and inject 500 μL of the diluted standard to the column.
8. Elute the sample at a 0.5 mL/min flow rate and collect data from the in-line Optilab refractometer for 40 min with 5-s collection intervals.
9. Using ASTRA software, define the mucin peak at the void (V_0) volume, and use the d*n*/d*c* ratio of 0.165 mL/g to calculate the absolute quantitation of mucin in the injected volume. Multiply by the dilution factor to calculate mucin concentration in the stock standard solution.
10. Store purified mucin in aliquots at −20°C.

3.4. Microtiter Plate Mucin Assays

Below, we offer five different microtiter plate assays for detecting mucins in secretions from HBE cell cultures obtained during experiments in which the luminal surfaces are carefully prewashed to remove mucus that may have accumulated prior to the experiment. Illustrative data are presented below from an agonist challenge experiment for each of the assays using each assay on the same set of samples.

3.4.1. Enzyme-Linked Immunosorbent Assay

These are conventional microtiter plate binding assays that use a primary mucin-specific antibody for detection and a secondary antibody conjugated to horseradish peroxidase (HRP) as a reporter. We offer three enzyme-linked immunosorbent assay (ELISA) variants: (1) *conventional ELISA*, which uses a conventional antibody, monoclonal, in this case; (2) *mucin Subunit ELISA*, which uses the "mucin subunit" antibody which requires the mucins to be reduced with DTT for detection (see Note 5); and (3) *mucin subunit sandwich ELISA*, which uses the same mucin subunit antibody in a sandwich format.

3.4.2. WGA Enzyme-Linked Lectin Assay (see Note 17)

Lectins are carbohydrate-binding proteins, some of which are useful for detecting mucins. Because lectins are glycan specific, they are not mucin specific and their rigorous use therefore requires validation (e.g., 16–18). Our laboratory uses WGA, specific for GlcNAcβ1-4GlcNAcβ1-4GlcNAc and Neu5Ac (sialic acid), to detect mucins from HBE cell cultures and mouse tracheas.

3.4.3. Periodic Acid-Biotin-Hydrazide Assay

This assay (7) is based on the same principle as the PAS stain for mucins; however, it uses a biotin conjugate of hydrazide, instead of Schiff's reagent, to react with the ketones formed from the periodate oxidation of the vicinal diols of susceptible sugars—which in mucins are typically terminal sialic acid residues. The advantage of periodic acid-biotin-hydrazide (PABH) over PAS is that the staining is amplified by streptavidin-conjugated HRP, increasing the sensitivity of the assay many fold.

3.4.4. Mucin Binding Assays

Because these assays have many steps in common, the procedure that follows, for brevity, is offered as a linear list of steps, each declared as being common to all the assays or specific to a particular one.

Techniques Common to All Assays (see Note 18)

1. The dilution factor used for preparation of samples and standards depends on the relative concentrations of mucins; consequently, it is therefore determined empirically in preliminary assays (see Note 19).
2. All incubations are done in a humidified chamber(s).
3. All washes are with PBST, 200 μL/well; multiple washes are indicated by an "*x*" factor, e.g., wash = single wash; wash 2*x* = 2 washes.
4. All blocks are incubations with 0.1% (w/v) or 1% (w/v) gelatin, 5% (w/v) milk, or 1% (w/v) BSA, in PBST, as specified below, 200 μL/well, 1 h at 37°C.

Mucin Subunit Antibody, Sandwich ELISA

For the Sandwich ELISA, this is the first step; skip for all other assays.

1. Coat plates with WGA lectin (10 μg/mL) in PBS (see Note 17), 100 μL/well. Incubate overnight at 4°C or 2 h at 37°C; wash 4×.

2. Block in 1% gelatin and wash 2×, or block, store in blocking buffer overnight at 4°C; wash 4×.

   Since the mucin subunit antibody detects mucins whose disulfide bonds have been reduced, the samples must be reduced ***and*** alkylated prior to detection (see Note 20); however, because reduction harms WGA binding, this step needs to occur before sample plating for the sandwich ELISA. Reduction and alkylation can be done on samples distributed on separate nonbinding microtiter plates or in cluster tubes.
3. To each sample and standard (1–100 μL; see Note 18), add Tris buffer such that the total volume = 180 μL (e.g., for a 20 μL sample, add 160 μL Tris buffer).
4. Add 20 μL DTT stock (final concentration, 10 mM); mix and incubate for 15–20 min at 37°C.
5. Add 25 μL iodoacetamide stock (final concentration ~25 mM; iodoacetamide should be ~2.5-fold molar excess over DTT); mix and incubate, *in the dark*, for 30 min at RT.

Common Procedures: Plate and Wash

For all assays:

1. Plate samples and standards diluted in PBS, as necessary, 100 μL/well (see Note 18). Incubate for 2 h at 37°C or overnight at 4°C.
2. Wash 4×.

Blocking and Detection: Assay Specific

1. Conventional ELISA (H6C5 or other antibody).
   (a) Block with 5% w/v milk in PBST; wash 2×.
   (b) Incubate with H6C5 or other primary antibody (generally, 1:1,000 in 1% w/v milk in PBST), 100 μL/well, overnight at 4°C or 2 h at 37°C; wash 4×.
   (c) Incubate with an appropriate secondary, HRP-conjugated antibody (generally, 1:1,000 in PBSt), for 1 h at 37°C.
2. *WGA ELLA* (see Note 21):
   (a) Block with 0.1% gelatin in PBST; wash 2×.
   (b) Incubate with WGA: HRP (2.5 μg/well) in PBST, 100 μL/well, 1 h at 37°C.
3. *Mucin subunit Ab-specific ELISAs, conventional and sandwich formats*: Prior to blocking for the conventional mucin subunit ELISA, the mucins need to first be reduced and alkylated (see Note 20); for the sandwich ELISA, skip steps (a)–(c), as the samples were reduced and alkylated prior to plating.
   (a) Add DTT stock diluted 1:10, 100 μL/well (final concentration 10 mM), and incubate for 10–15 min at RT; pour off the DTT solution (invert and shake plate over sink).

(b) Add iodoacetamide stock diluted 1:10, 100 μL/well (final concentration 25 mM), incubate *in the dark* for 30 min at RT.

(c) Wash 4×.

(d) Block with 1% BSA in PBST; wash.

(e) Incubate with mucin subunit antibody (1:5,000 in 1% BSA in PBST), 100 μL/well, 2 h at 37°C or overnight at 4°C; wash 4×.

(f) Incubate with an appropriate secondary, HRP-conjugated antibody (1:2,000 in PBST), for 1 h at 37°C.

4. *PABH assay*:

(a) Incubate with 1 mM periodic acid (100 μL/well) at RT for 10 min *in the dark*.

(b) Add biotin hydrazide solution (50 μL/well), mix well, and incubate for 45 min at RT.

(c) Wash 4× with 5-min incubations between each wash (to more completely remove the DMSO from the biotin hydrazide solution), and incubate with streptavidin–HRP in PBST, 100 μL/well, for 20 min.

Common Procedures: Wash and Develop

For all assays:

1. Wash 4×.
2. Develop with OPD substrate, 150 μL/well, 10–15 min at room temperature; stop OPD reaction with 4 M H_2SO_4, 50 μL/well.
3. Determine ODs at 490 nm in plate reader (e.g., Spectra Max Plus, Molecular Devices) and use an appropriate software program to construct standard curve and calculate sample values.

Sample Data

Figure 2 shows data for baseline and agonist exposure periods from an experiment with HBE cell cultures, with the samples being analyzed by each of the five assays described above. Note that the two mucin subunit antibody-based assays (conventional and sandwich ELISAs) yielded similar data, with baseline- and agonist-stimulated mucins of ~200 and ~700 ng/culture, as well as multiplicities (fold increases) >3.5. The other assays, in contrast, had values several fold higher, with baseline- and agonist-stimulated values >1,000 and >3,000 ng/culture, and proportionately lower multiplicities. These differences most likely reflect probe specificities: WGA, PABH, and the H6C5 antibody bind indiscriminately to mucins of all types, as well as to other glycan-bearing compounds, whereas the mucin subunit antibody is specific to polymeric mucins, MUC5AC and MUC5B in this case. Nonetheless, the less specific assays do yield results qualitatively similar to the more specific, so their use may be justified for situations, where the intent lies more in studying whether or not a goblet cell secretory response is initiated

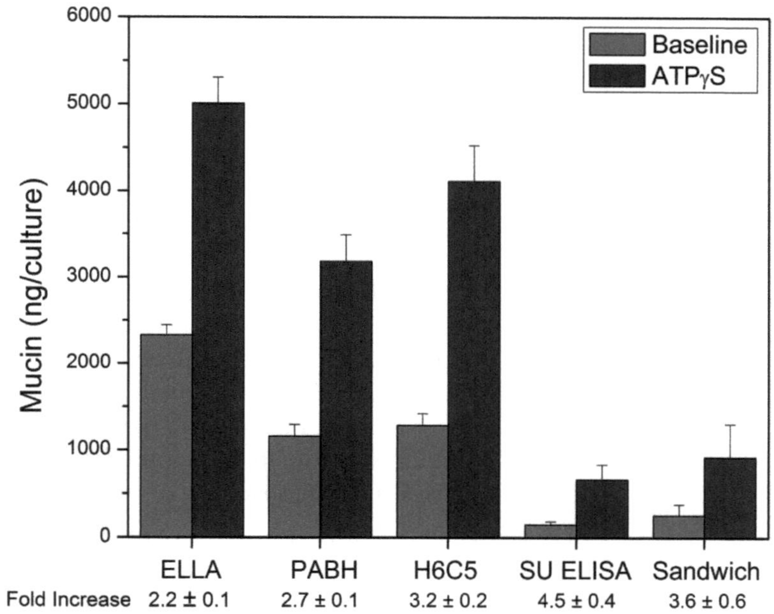

Fig. 2. Mucins sampled following baseline and agonist (ATPγS, 100 μM) periods, as measured, on the same set of samples, by the binding assays described above. SU = mucin subunit antibody, for both the ELISA and sandwich (ELISA) assays indicated. Results expressed as the mean ± SE, $n = 3$; fold increase = agonist stimulated/baseline.

or modulated. For questions relating to the release of polymeric mucins, per se, mucin-specific probes, such as the mucin subunit antibody, are required.

4. Notes

1. The culture media used for HBE cell culture, BEGM, and ALI have a substantial history, which is outlined in ref. 6 along with detailed instructions on their compositions. This reference is the result of many years of experience in HBE cell culture by our UNC colleague, Dr. Scott Randell, who directs the operation of our Cell Culture Facility, and it should be read in detail by anyone contemplating HBE cell culture, at any scale.
2. Commercial sources for human epithelial cells appear to change dynamically, as companies are bought, sold, or traded by large corporations. Hence, if the Web links indicated do not work, we suggest searching the Web using the suppliers' names (e.g., Cambrex).
3. We use Transwell® inserts because they hang from the rims of the wells in a cluster plate. This configuration allows them to be mechanically stabilized by gluing them in place with Dermabond® just before an experiment. Other brands of hanging,

polyester or PTFE inserts, should work as well. If Millicel® standing inserts are desired, we have found that the 10-mm PTFE inserts can be placed snugly into (recycled) 12-mm Transwell units after removing the membrane. Dermabond® is packaged in "single-use" plastic capsules with an applicator at one end—the adhesive is contained in an internal, thin-walled glass capsule, from which it is generally released for use by crushing the sides of the nested capsules. We instead remove the adhesive by cutting through the capsule walls with scissors, just below the applicator, and pouring it off into a microfuge tube in which it is stored until consumed. Dermabond is applied in 3 μL droplets to the Transwell® in three spots, placing the tip of the pipette to the grove between the underside of the insert rim and the top of the well to allow the adhesive to wick into the space.

4. The foam pads are essential for successful mucin secretion experiments. During an experiment, they provide a convenient means of handling plates, they reduce significantly the mechanical stimuli that can elicit secretion by cushioning the stresses caused by handling, and they provide insulation against temperature changes while plates are removed from an incubator. Unfortunately, those used in our laboratory (Fig. 3, inset) are no longer available and we have been unable to find commercial replacements of equal quality. It is quite reasonable, however, to make them from EPDM foam rubber, adhesive backed, 0.5″ thick, 6″ wide, 25′ long (~1.2 × 15.2 cm × 8 m long) cut into 7″ (17.5 cm) lengths. Half of the foam pieces are used, as is, as bottom pieces. The others, which form the top pieces, first have 2.1″ (2.54 cm) holes cut through with a cylindrical cutter (e.g., a cork borer), and then a 5 × 3 5/16″ (12.7 cm × 8.4 cm) center rectangle is cut out, as indicated in Fig. 3. The adhesive faces of the two pieces are placed/glued together, with careful positioning, after removing the adhesive backings. The adhesive material on the bottom piece, exposed in the center of the top cutout, can be coated with an innocuous powder.

5. The conventional ELISA is suitable for most mucin antibodies that are likely to be used. For the mucin subunit ELISA, we use a polyclonal antibody made originally against purified cervical mucins that had been reduced and alkylated (19), a procedure that not only opens up all the D-domains and Cys-domains, but also depolymerizes the mucins. The antibody binds to epitopes exposed by disulfide reduction, and it has proved to detect all the polymeric mucins from every vertebrate species against which we have tested it, hagfish included (Abdullah, LH, CW Davis, and JK Sheehan, unpublished observations). To improve specificity for the Cys-rich and/or

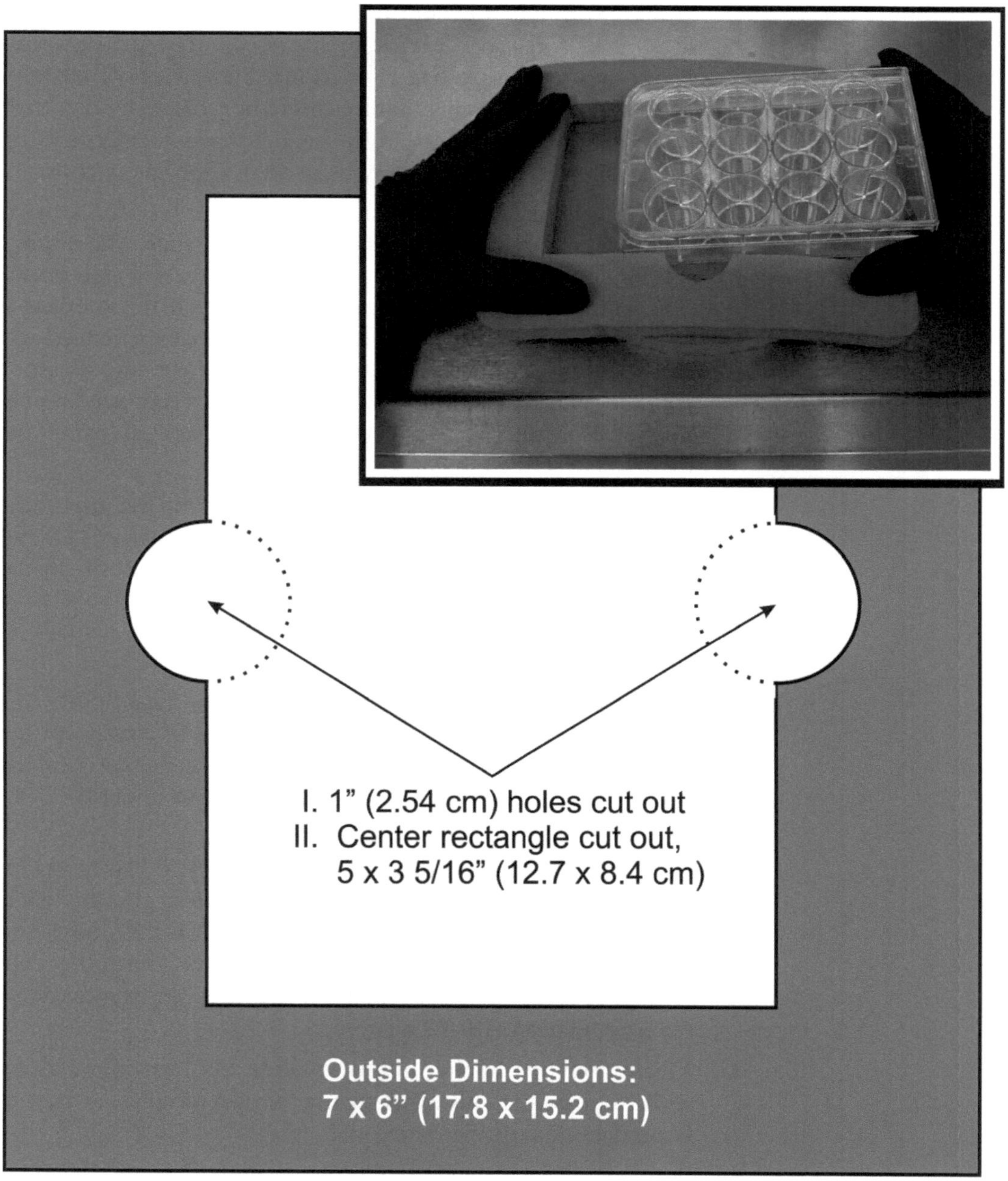

Fig. 3. Foam pad, top piece, to hold 6- or 12-well cluster plates during HBE cell culture wash procedures. The center cutout should hold the plate quite firmly, i.e., it should be necessary to work the plate into the cutout by applying pressure to all sides of the plate as it is inserted. *Inset*: Image of a cluster plate being inserted into an assembled foam pad.

D domain peptides exposed upon reduction, the antibody is preabsorbed against intact, nonreduced salivary MUC5B, 10 μg mucin/100 μL antibody in nondiluted serum, and incubated overnight at 4°C (the proportions may need to be adjusted for other lots of antibody). The solution is centrifuged at 80,000 rpm in a Beckman TL-1000 tabletop ultracentrifuge for 2.5 h, and the supernatant harvested. The pre-absorbed antibody preparation has a functional lifetime of ~2 months. Note that the requirement for mucins to be reduced for detection with this antibody offers an unusual element of specificity, in that nonreduced samples may be used as a specificity control.

6. Airway cells received from commercial sources may, or may not, be derived by proteolysis of bronchial tissues. The original system described by Lechner and LaVeck harvested cells grown out from explanted airways tissue (20), and if this method is used commercially, the cells received will have experienced several series of cell division prior to seeding in the investigator's laboratory. If this is the case, the cultures derived may be phenotypically quite different from those described using the methods described in ref. 6.
7. Basal cells are not only the primordial cell of the superficial airway epithelium in the large airways of humans (21), but they also serve as the anchor for columnar cells. Both ciliated and goblet cells possess basal protrusions which contact the basement membrane; however, their anchorage is actually to basal cells (22).
8. HBE cells grown in culture media other than the BEGM/ALI system have been studied for various aspects of airway epithelial biology (see ref. 6); however, regulated mucin secretion has been studied nearly exclusively from cells grown under BEGM/ALI culture conditions.
9. HBE cultures exposed to maximal doses of ATP (100 μM) for 30–40 min typically secrete all of their mucin stores; the stores take a minimum of 48 h to recharge fully (L.H. Abdullah and C.W. Davis, unpublished observation). Hence, the goblet cells should be fully charged when the cultures are removed for experimentation, 48–72 h later.
10. Mucin contents in the luminal washes are determined along with the baseline and experimental period samples, as part of good data quality assurance.
11. Purified mucins from HBE cell cultures contain both MUC5AC and MUC5B, but are useful nonetheless as a mucin standard when using the mucin subunit antibody since it recognizes both mucins or when using nonspecific assays, such as an ELLA or the PABH binding assay.
12. Other methods for purifying mucins, e.g., size-exclusion chromatography on Sepharose CL2B, are unsuitable for the preparation of standards because they result in dilution, rather than concentration, of the mucins during purification.
13. "Crude separation" is used to mean a low-resolution Sepharose CL2B chromatography procedure, in which up to 10% of the

total column volume is loaded onto the column. It is used merely to separate the polymeric MUC5B from the monomeric MUC7.

14. HBE cell cultures may be established solely for the purpose of mucus collection (e.g., see ref. 23), as they maintain a good mucin production for 2–3 months after confluence. In such cases, the larger, 24-mm culture inserts are a better choice than the smaller ones.

15. The reader is referred to refs. 14, 24 for a full description of mucin purification by double isopycnic density-gradient centrifugation in CsCl. In the first centrifugation, in 4 M GuHCl, mucins and DNA (if present) comigrate to densities of ~1.4–1.5 g/mL. When the GuHCl is reduced to 0.2 M for the second centrifugation, DNA reassociates into a double-stranded form, which makes it migrate to lighter, final densities of ~1.2–1.3, whereas the mucins still migrate to a final density of ~1.4–1.5 g/mL.

16. An alternative to the use of refractive index in quantifying mucins is to use gravimetric analysis, in which a small volume of sample is dialyzed extensively against water and 100 μL is dried completely in an 80°C oven and weighed to the nearest μg. Though simpler to perform, both conceptually and mechanistically, this method is not preferred for it is subject to a variety of artifacts, including incomplete drying, and absorption of moisture as the sample is cooled and weighed.

17. Two styles of multichannel pipette are essential for good mucin binding assays on microtiter plates. First, a "single-stepper," multichannel pipette enhances the transfer of test samples and standards to the microtiter plate, accurately measuring and transferring the solutions from a set of nonbinding cluster tubes (Costar 8-strip tubes, Cat. # 4408) arranged in the same pattern as the microtiter plate (we use either a 2.5–25 μL Brand Transferpette-8® or a 20–200 μL Costar 8-Pette®). Second, a "multistepper," multichannel pipette allows wash solutions and other reagents to be applied rapidly to an entire plate (we use an Ovation BioNatural Pipette).

18. Sample and standard volumes applied to microtiter plates need to be adjusted according to the quantity of mucins they contain. As a guide, samples from HBE cell cultures grown in 12-mm Transwell inserts, in our hands, contain 50–150 ng/mL mucin at baseline, so for the subunit ELISA, our primary assay system, we use standards with the following dilution series: 500, 250, 125, 62.5, 31.3, 15.6, and 7.8 ng/mL HBE mucin.

19. Wheat germ agglutinin, WGA, is commonly used to bind sialic acid residues. In this sandwich ELISA, it is used to coat the

microtiter plate surfaces and it functions by binding the mucins to the surface via the mucin-glycosylated repeat domains (as well as any other sialic acid glycoproteins or glycolipids present). Hence, in the end, the mucins are "sandwiched" between WGA and the detecting reagent, the mucin subunit antibody in this case.

20. Alkylation blocks reduced Cys residues, preventing disulfide reassociation when DTT is removed. Because the iodoacetamide consumes the excess DTT, the resulting mixture of sample and reagents/reactants can be applied safely to WGA-coated plates in the sandwich ELISA. Important: Do not wash before alkylation, as Tween-20 interferes in the reaction.
21. The blocking step is optional for the WGA ELLA. In our experience, skipping the step has no effect on the outcome of the assay; however, some laboratories may feel more comfortable including the step.

Acknowledgments

Studies in the authors' laboratories which lead to the development of the techniques described were supported by National Institutes of Health grants, HL-063756 and HL084934, and grants from the Cystic Fibrosis Foundation.

References

1. Knowles, M. R. and Boucher, R. C. (2002) Mucus clearance as a primary innate defense mechanism for mammalian airways. *J. Clin. Invest* **109**, 571–577.
2. Thornton, D. J., Rousseau, K., and McGuckin, M. A. (2007) Structure and function of the polymeric mucins in airways mucus. *Annu. Rev. Physiol.* **70**, 459–86.
3. Sheehan, J. K., Kirkham, S., Howard, M., Woodman, P., Kutay, S., Brazeau, C., Buckley, J., and Thornton, D. J. (2004) Identification of molecular intermediates in the assembly pathway of the MUC5AC mucin. *J. Biol. Chem.* **279**, 15698–15705.
4. Davis, C. W. and Dickey, B. F. (2008) Regulated airway goblet cell mucin secretion. *Annu. Rev. Physiol* **70**, 487–512.
5. Danahay, H. and Jackson, A. D. (2005) Epithelial mucus-hypersecretion and respiratory disease. *Curr. Drug Targets. Inflamm. Allergy* **4**, 651–664.
6. Fulcher, M. L., Gabriel, S., Burns, K. A., Yankaskas, J. R., and Randell, S. H. (2005) Well-differentiated human airway epithelial cell cultures. *Methods Mol. Med.* **107**, 183–206.
7. Conway, J. D., Bartolotta, T., Abdullah, L. H., and Davis, C. W. (2003) Regulation of mucin secretion from human bronchial epithelial cells grown in murine hosted xenografts . *Am. J. Physiol Lung Cell Mol. Physiol* **284**, L945–L954.
8. Davis, C. W. (2002) Regulation of mucin secretion from in vitro cellular models. *Novartis. Found. Symp.* **248**, 113–125.
9. Matsui, H., Davis, C. W., Tarran, R., and Boucher, R. C. (2000) Osmotic water permeabilities of cultured, well-differentiated normal and cystic fibrosis airway epithelia *J. Clin. Invest* **105**, 1419–1427.
10. Matsui, H., Grubb, B. R., Tarran, R., Randell, S. H., Gatzy, J. T., Davis, C. W., and Boucher, R. C. (1998) Evidence for periciliary liquid layer depletion, not abnormal ion composition,

in the pathogenesis of cystic fibrosis airways disease. *Cell* **95**, 1005–1015.

11. Matsui, H., Randell, S. H., Peretti, S. W., Davis, C. W., and Boucher, R. C. (1998) Coordinated clearance of periciliary liquid and mucus from airway surfaces. *J. Clin. Invest 102*, 1125–1131.
12. Thornton, D. J., Khan, N., Mehrotra, R., Howard, M., Veerman, E., Packer, N. H., and Sheehan, J. K. (1999) Salivary mucin MGl is comprised almost entirely of different glycosylated forms of the MUC5B gene product. *Glycobiology* **9**, 293–302.
13. Holmen, J. M., Karlsson, N. G., Abdullah, L. H., Randell, S. H., Sheehan, J. K., Hansson, G. C., and Davis, C. W. (2004) Mucins and their O-Glycans from human bronchial epithelial cell cultures. *Am. J. Physiol Lung Cell Mol. Physiol* **287**, L824–L834.
14. Carlstedt, I., Lindgren, H., Sheehan, J. K., Ulmsten, U., and Wingerup, L. (1983) Isolation and characterization of human cervical-mucus glycoproteins. *Biochem. J.* **211**, 13–22.
15. Thornton, D. J., Carlstedt, I., and Sheehan, J. K. (1994) Identification of glycoproteins on nitrocellulose membranes and gels. *Methods Mol. Biol.* **32**, 119–128.
16. Abdullah, L. H., Davis, S. W., Burch, L., Yamauchi, M., Randell, S. H., Nettesheim, P., and Davis, C. W. (1996) P2u purinoceptor regulation of mucin secretion in SPOC1 cells, a goblet cell line from the airways. *Biochem. J.* **316**, 943–951.
17. Jackson, A., Kemp, P., Giddings, J., and Sugar, R. (2002) Development and validation of a lectin-based assay for the quantitation of rat respiratory mucin. *Novartis. Found. Symp.* **248**, 94–105.
18. Clancy, S. M., Yeadon, M., Parry, J., Yeoman, M. S., Adam, E. C., Schumacher, U., and Lethem, M. I. (2004) Endothelin-1 inhibits mucin secretion from ovine airway epithelial goblet cells. *Am. J. Respir. Cell Mol. Biol.* **31**, 663–671.
19. Sheehan, J. K., Boot-Handford, R. P., Chantler, E., Carlstedt, I., and Thornton, D. J. (1991) Evidence for shared epitopes within the 'naked' protein domains of human mucus glycoproteins. A study performed by using polyclonal antibodies and electron microscopy. *Biochem. J.* **274**, 293–296.
20. Lechner, J. F. and LaVeck, M. A. (1985) A serum-free method for culturing normal human bronchial epithelial cells at clonal density. *J. Tiss. Cult. Meth.* **9**, 43–48.
21. Roomans, G. M. (2010) Tissue engineering and the use of stem/progenitor cells for airway epithelium repair. *Eur. Cell Mater.* **19**, 284–299.
22. Evans, M. J., Van Winkle, L. S., Fanucchi, M. V., and Plopper, C. G. (2001) Cellular and molecular characteristics of basal cells in airway epithelium. *Exp. Lung Res.* **27**, 401–415.
23. Matsui, H., Verghese, M. W., Kesimer, M., Schwab, U. E., Randell, S. H., Sheehan, J. K., Grubb, B. R., and Boucher, R. C. (2005) Reduced three-dimensional motility in dehydrated airway mucus prevents neutrophil capture and killing bacteria on airway epithelial surfaces. *J. Immunol.* **175**, 1090–1099.
24. Thornton, D. J., Sheehan, J. K., Lindgren, H., and Carlstedt, I. (1991) Mucus glycoproteins from cystic fibrotic sputum. Macromolecular properties and structural 'architecture'. *Biochem. J.* **276**, 667–675.

Chapter 17

Assessment of Intracellular Mucin Content In Vivo

Lucia Piccotti, Burton F. Dickey, and Christopher M. Evans

Abstract

Airway mucus presents a first line of defense against inhaled materials. It also, however, is a significant pathological contributor to chronic lung diseases, such as asthma, cystic fibrosis, and chronic obstructive pulmonary disease. Thus, gaining a better understanding of the mechanisms of mucus production and secretion is an important goal for improving respiratory health. Mucins, the chief glycoprotein components of airway mucus, are very large polymeric glycoproteins, and measuring their production and secretion in experimental animals presents significant technical challenges. Over the past several years, we have developed assays for accurately quantifying mucin production and secretion using histological and biochemical assays. These methods are described here.

Key words: Airways, Asthma, Cystic fibrosis, Chronic obstructive pulmonary disease, Goblet cell, Lungs, Mouse, Mucin, Mucous, Mucus

1. Introduction

The amount of mucin present within secretory cells of the airway epithelium reflects a balance between two tightly regulated processes—mucin production and mucin secretion (1–3). The predominant secreted mucins produced in mouse airways are Muc5ac and Muc5b (4). Muc5b is produced constitutively, but its rate of production may be further increased during lung inflammation (4–7). Muc5ac is scarcely produced at baseline, but its production is greatly increased during inflammation (4, 7, 8).

In the healthy respiratory tract, mucins are continuously secreted at a low basal rate, but the rate of secretion can be greatly stimulated by an increase in the concentration of extracellular secretagogues (9). Basal secretion reflects tonic activity of a regulated

Michael A. McGuckin and David J. Thornton (eds.), *Mucins: Methods and Protocols*, Methods in Molecular Biology, vol. 842, DOI 10.1007/978-1-61779-513-8_17, © Springer Science+Business Media, LLC 2012

secretory machinery, as indicated by the spontaneous accumulation of intracellular mucin when a key regulatory protein, Munc13-2, is genetically deleted (6). The rate of basal secretion is sufficiently high that very little mucin accumulates intracellularly when mucins are produced at basal rates. Indeed, intracellular mucin content under these conditions is so low that mucins are not detected by histochemical stains, such as alcian blue-periodic acid Schiff's (AB-PAS) and periodic acid fluorescent Schiff's (PAFS) (9), even though significant amounts of Muc5b are detected by more sensitive enzyme-linked immunolabeling probes (6, 7).

Increased mucin production results in histochemically visible intracellular mucin accumulation traditionally termed "mucous metaplasia." The most efficacious stimuli for mucous metaplasia are agents that induce IL-13 predominant allergic inflammation (3). The principal endogenous ligand regulating mucin secretion, both basal and stimulated, appears to be ATP (10).

Despite the fact that intracellular mucin content levels depend upon two variables, the rate of production and the rate of secretion, the measurement of intracellular mucin content can be useful to interrogate the rate of just one of these if the rate of the other is held at steady state. For example, increased intracellular mucin content indicates increased production if the rate of secretion is not altered (9). Conversely, increased intracellular mucin content indicates decreased secretion if the rate of production is not altered (6, 9, 11, 12).

Assessment of intracellular mucin content has the following three advantages.

1. *For measuring mucin production*, it is simple to measure and correlate with other phenotypic parameters in pathophysiological models (9).
2. *For measuring basal secretory rate*, small differences are magnified by progressive intracellular mucin accumulation, which makes these differences detectable (6).
3. *For measuring stimulated secretion*, fractional release can be readily determined (see Subheading 3.7, below), thereby relating stimulated secretory function to initial intracellular content to avoid artifacts from changes in intracellular pool size (9, 12).

Disadvantages of the measurement of intracellular mucin content to assess mucin production or secretion are the lack of strict linearity of image-based techniques, and the lack of sensitivity of lectin-based detection on blots following electrophoresis when measuring basal rates of mucin production and secretion.

2. Materials

Because mice are very frequently used for experimental respiratory research, many details below (e.g., identifying anatomical structures during dissection) focus on them. While mice differ from humans and other large animals anatomically (e.g., airway sizes and gland presence) and morphologically (e.g., epithelial shape and stratification), the applications for detection of mucin production and secretion described below can be applied across species. We do not specify a particular disease model or exposure system because these vary widely. Indeed, mucous metaplasia is a phenotype common to allergen exposed, viral infected, and cytokine-treated mice (2). Details for establishing these models are contained in the following refs.: 9, 13–19. The materials and methods suggested below are suitable for use in numerous animal species, and they can also be applied to human pathological, segmental challenge, or primary cell culture studies.

2.1. Histology

1. Neutral buffered formalin: Histological-grade formalin (37% formaldehyde content) diluted 1:10 (v/v) in 0.1 M phosphate buffer or phosphate-buffered saline (PBS) (pH 7.4). Store at room temperature.
2. Paraffin-embedded tissue sections can be prepared by the investigator or by an institutional histopathology core. Tissue sections should be collected on positively charged glass slides to allow for strong adhesion should heated antigen retrieval is necessary in downstream immunohistochemical labeling experiments.
3. Periodic acid solution: Prepare 1% (w/v) fresh for each usage by dissolving electrophoresis-grade periodic acid in ddH_2O.
4. Fluorescent Schiff's reagent: Prepare at least 48 h prior to use by dissolving acriflavine hydrochloride in an appropriate volume of ddH_2O to obtain a final concentration 0.5% (w/v) in the final product. Once acriflavine is dissolved, concentrated HCl (~10 N) is added to obtain a final concentration of 1% (v/v), and sodium metabisulfite is added to obtain a final concentration of 1% (w/v). The solution is mixed well, stoppered tightly, and stored at room temperature in the dark for at least 48 h. Fluorescent Schiff's reagent can then be used or stored at 4°C and is good for at least 1 month (see Note 1).
5. Acid alcohol solution: Dilute concentrated HCl to 1% (v/v) with 70% ethanol.
6. Mounting medium: Make a 1:1 v/v mixture of Canada balsam (Fisher Scientific, Pittsburgh, PA) and methyl salicylate (Fisher). Store at room temperature.

2.2. Mucin Extraction and Specimen Preparation

1. Guanidinium buffer: Dissolve 6 M guanidinium chloride (Sigma), 0.1 M Tris–HCl, pH 8.0, 5 mM ethylenediamine tetraacetic acid (EDTA) in ddH_2O. Store at 4°C.
2. Protease inhibitor buffer: Prepare fresh by dissolving 1 Complete Protease Inhibitor Cocktail Tablet (Roche Applied Science, Indianapolis, IN) in 7 mL of guanidinium buffer.
3. Urea buffer: Dissolve 6 M urea (Sigma), 0.1 M Tris–HCl, pH 8.0, 5 mM EDTA in ddH_2O. Store at 4°C.
4. Loading buffer (10×): Dissolve 77 mg dithiothreitol (DTT) and 25 mg bromophenol blue (Fisher Scientific, Fair Lawn, NJ) in 5 mL of 50% (v/v) glycerol and 1% (w/v) sodium dodecyl sulfate (SDS) in urea buffer. Aliquot and store at −20°C.
5. Alkylation buffer (10×): Dissolve 250 mM iodoacetamide in water. Prepare fresh.

2.3. SDS-Agarose Gel Electrophoresis

1. Tris–acetate–EDTA (TAE) buffer (50×): Dissolve 242 g Tris base in 750 mL deionized water, 57.1 mL glacial acetic acid, and 100 mL of 0.5 M EDTA (pH 8.0) and adjust the solution to a final volume of 1 L (final pH 8.5). Store at room temperature. Dilute to 1× with deionized water.
2. Electrophoresis buffer: 0.1% (w/v) SDS in TAE buffer. Store at room temperature.

2.4. Vacuum Blotting

1. Saline-sodium citrate (SSC) buffer 20×: 3.0 M NaCl and 0.3 M Na citrate in ddH_2O. Adjust to pH 7.0 with HCl. Store at 4°C.
2. Transfer buffer: 0.2% (w/v) SDS in 4× (SSC) buffer. Store at 4°C.
3. Reducing transfer buffer: Dissolve DTT in transfer buffer to a final concentration of 10 mM. Prepare fresh.
4. Immobilon-NC 0.45 μm nitrocellulose membrane (Millipore, Billerica, MA) and blotting paper (Bio-Rad Laboratories, Hercules, CA).
5. PBS (10×): 1.37 M NaCl, 27 mM KCl, 80 mM Na_2HPO_4, and 20 mM KH_2PO_4 in ddH_2O. Sterilize by autoclaving. Adjust the final volume to 1 L. Store at room temperature. Dilute to 1× with ddH_2O (final pH 7.4).
6. 0.3% (v/v) Tween 20 (Bio-Rad Laboratories, Hercules, CA) in 0.01 M Tris–HCl, pH 6.8, and 0.1% (v/v) Tween 20 in 0.01 M Tris–HCl, pH 6.8.
7. Blocking solution and antibody dilution buffer: 5% (w/v) Blotting Grade Non-Fat Dry Milk (Bio-Rad Laboratories, Hercules, CA) and 0.1% (v/v) Tween 20 in PBS (1×). Prepare fresh.
8. Reagents for enhanced chemiluminescence (ECL) detection: SuperSignal® West Pico Chemiluminescent Substrate (Thermo Scientific, Rockford, IL) and Bio-Max Light Film (Kodak, Rochester, NY). Films are developed in a Konica Film Processor.

3. Methods

The two methods for measuring mucin production in vivo that are described below utilize microscopic imaging and biochemical assays.

Histological specimens provide a useful tool to determine the localization and degrees of mucin production and secretion. For measuring intracellular mucin content, a technique, such as transmission electron microscopy, is exquisitely sensitive. However, instrument expense and sample preparation time can be prohibitive when large numbers of samples are assessed. For these reasons, we have relied more heavily on light microscopy and the use of inexpensive technologies to measure mucin production and secretion. Images obtained at relatively low magnification (e.g., using a 40× specimen objective) can be analyzed and compared at different anatomical locations in the same slide, provide adequate sample sizes (10s to 100s of cells per image), and display sufficient detail to determine whether there is heterogeneity among cells. Numerous image analysis software tools make quantitation of staining simple and inexpensive.

Immunoblotting is also an efficacious approach for measuring the airway mucin content. Due to particular biochemical properties of polymeric mucins, treatment with a chaotropic agent, such as guanidinium chloride, is necessary for breaking noncovalent bonds and solubilization (20). Polymeric mucins are also held together by disulfide bonds. Therefore, it is important to reduce these with agents, such as DTT, prior to electrophoresis and transfer (21). These procedures permit resolution of single bands of monomeric mucins. Mucins can then be blotted and detected using selective probes. Their heavy glycosylation has made use of specific antibodies difficult in some instances (22), but this same property makes mucins suitable for detection with lectins, a class of highly specific sugar binding proteins. Results can be compared relative to each other within blots or across different vacuum blots when a standard curve and appropriate internal controls are also applied.

3.1. Animal and Tissue Preparation

1. To preserve airspace morphology, the lungs are preserved via intracheal instillation of fixative. Simple immersion of lungs in fixative results in alveolar and airway collapse, tissue folding, and accumulation airway debris. For surgical preparation, small laboratory animals are deeply anaesthetized and tracheostomized by removing a patch of skin from the neck surface. The trachea is isolated by removing connective, glandular, and infrahyoid muscle tissues. Using rounded forceps, the trachea is then separated from the esophagus, and a 30-cm length of surgical thread is folded in half, passed underneath the trachea, and pulled through such that half of the double-length thread is on

either side of the trachea. The thread is then cut at the center point releasing two 15-cm pieces that are used to secure and stabilize the tracheal cannula. A partial incision is made in ventral half of the trachea using microdissecting spring scissors, and the animal is cannulated. For mice, a blunt syringe catheter (18–20 G) is suggested, and for larger animals (e.g., rats and guinea pigs) 2–3 mm O.D. polyethylene tubing is suggested. Once inserted, the cannula is secured by tying the threads above and below the insertion site.

2. Once cannulated, animals are opened at the abdomen, the descending aorta is cut, and the animals are euthanized by exsanguination.
3. Another incision is made longitudinally along the sternum, and then the diaphragm is separated from the thorax by cutting along its rib attachment edges.
4. To clear blood from the pulmonary vasculature, saline solution is slowly injected through the right heart circulation by placing a needle into the right cardiac ventricle. Upon successful exsanguination, the lungs turn from pink to white. For mice, 3–10 mL are required. This is optional, and is primarily done to eliminate excessive blood from samples, which in many cases can interfere with downsteam immunohistochemical stains. Use of fixative for perfusion is possible, but not required for optimal airway fixation.
5. The lungs are fixed in situ by instilling 4% neutral buffered formalin intratracheally. For consistent results, the fixative should be instilled by siphoning from a large vessel held at a specific height above the mouse to keep the lungs inflated while maintaining constant pressure during fixation. For volume estimation, this is essential, and 5–30-cm H_2O is physiologically relevant. Importantly, once a set pressure is chosen, this should be kept constant across experiments.
6. After 30 min, the lungs should be removed and kept in fixative overnight at 4°C.
7. To prepare tissues for paraffin embedding, the right and left lungs should be separated from the trachea at the main stem bronchi. In mice, the left lung is a single lobe, and it is the largest. Therefore, it is commonly used for histology. Once separated, it is first divided into three pieces by making cuts in cross section with a razor or scalpel blade. As shown in Fig. 1, the first cut is made ~1 mm cranial to the root of the lung (the entry point of the lobar bronchus), and the second cut is made immediately cranial to the diaphragmatic curvature. The central portion

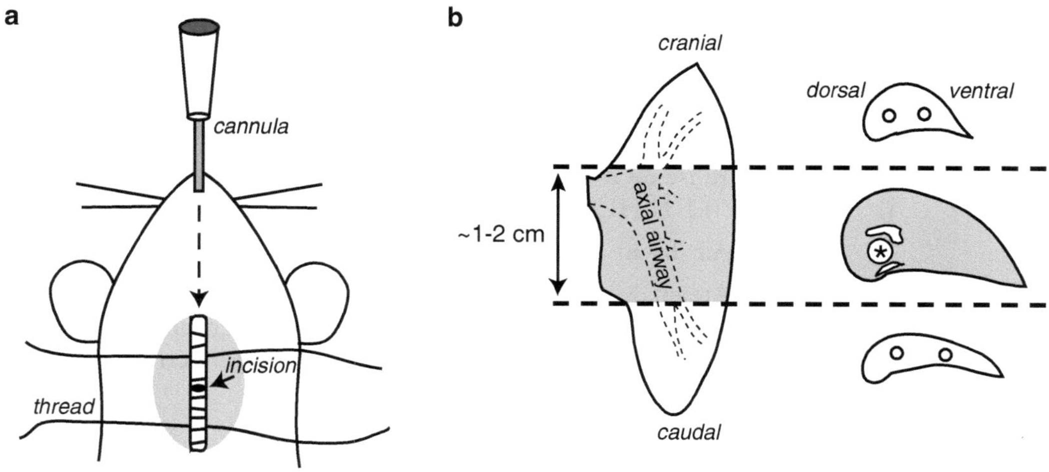

Fig. 1. Surgical preparation, lung dissection, and isolation of mucin-producing airways. Anesthetized mice are tracheostomized by dissecting superficial tissues, isolating the trachea, and inserting cannula through a small partial incision made in the trachea (**a**). For histology, the left lung is isolated and sectioned into three large pieces at the areas depicted by the dashed lines using a razor blade (**b**). The middle piece containing the main axial airway (*gray*) is further sectioned into 2–3-mm-thick "filets." This region contains an easily identifiable airway situated between the pulmonary artery and vein and running along the dorsal aspect of the lung. This is the axial airway. Filets should be embedded in paraffin while keeping this airway positioned for cross-sectioning onto microscope slides. Nonmucin-producing regions (*white*) can be discarded or embedded separately and used for other staining procedures to fit investigators' needs.

contains the axial bronchus, which is the predominant location of mucin production in mouse lungs (6, 9, 23). This can be carefully dissected into ~2-mm thick "filets" that can be embedded together in the same paraffin blocks to provide adequate sampling of mucin-producing cells along the proximal–distal axis of the bronchus in a slide specimen.

8. Dissected tissues should be embedded in paraffin, cut into 5-μm-thick sections, and collected on positively charged glass slides according to standard microtome procedures.

3.2. PAFS Staining

All procedures in this section should be performed at room temperature.

1. Slides are dewaxed in an organic solvent, such as xylene, toluene, or a nontoxic substitute, such as Histo-Clear® by incubating 2× for 10 min.
2. Slides are then rehydrated through graded ethanol solutions—100% (2× for 2 min), 95% (2× for 2 min), 90% 1× for 1 min; 70% 1× for 1 min—and then submerged in ddH_2O until staining. Once rehydrated, do not let the tissues dry until staining is complete (see Note 2).

3. Oxidize tissues for 10 min in freshly prepared 1% w/v periodic acid dissolved in ddH_2O.
4. Rinse 3× for 5 min in ddH_2O.
5. Treat with fluorescent Schiff's reagent for 20 min (see Note 3).
6. Rinse 3× for 5 min in ddH_2O and 2× for 5 min in acid alcohol (0.1 N HCl in 70% ethanol).
7. Air dry in a dark dust-free place.
8. Attach a coverslip with Canada balsam mounting medium. Keep slides on benchtop for 2 h to overnight in the dark prior to imaging. Slides are ready to image after setting overnight (see Note 4).

3.3. Fluorescence Microscopy

1. Place a slide on the microscope stage and obtain a focused image of a bronchial airway section at low magnification. Position the airway and use a computerized random number generator (e.g., http://www.random.org/), or a dodecahedral die, to identify a number between 1 and 12. This number is used to reposition the specimen and center it to the corresponding hour graduations on an analog clock.
2. For observation of intracellular mucin and other glycoconjugates, use Texas red excitation–emission. In our experience, this works best with an FITC/Texas red filter set (No. 51006, Chroma, Bellows Falls, VT). Samples are obtained under dual 500/573-nm peak excitation with image acquisition using dual emission with peaks at 531 nm (green) and 628 nm (red). Acriflavine Schiff's reagent binds covalently to the aldehydes formed during periodate-mediated oxidation, and when excited over a broad range (380–580 nm) this label fluoresces red (600–650-nm emission). In this staining procedure, acriflavine also intercalates within nucleic acids as an acridine agent, and this fluoresces green.
3. Two images are acquired. The first is a green and red two-color image that demonstrates all cells and structures. This is used to make a boundary length measurement—in this case, the length of the basement membrane—as described in the morphometry section below, see Subheading 3.4.
4. The second image is acquired as a red-only image. This can be attained by manually inserting a long-pass (600–∞ nm) or a band-pass (600–680 nm) filter or by digitally altering camera acquisition settings. Since the green emission due to intercalated acriflavine binding to nucleic acids overlies the cytosolic and nuclear compartments almost completely, the remaining signal in the epithelium is predominantly labeled glycoconjugates in cytoplasmic organelles, especially secretory granules. It is critical to obtain images that are suitable for image analysis below (see Subheading 3.4). The acquisition settings should

be determined such that a largely nonmucin-producing airway (e.g., a naïve mouse bronchus or a terminal bronchiole) has few or no visible red pixels.

5. For image analysis, files are best saved in formats, such as a raw (.raw), bitmap (.bmp), or tagged image file format (.tif). "Lossy" compression files, such as JPEG (.jpg), can be used, but care should be taken to ensure that data loss is minimized by not resaving JPEG files.

3.4. Morphometry

1. Choose appropriate software. For morphometric analyses, there are numerous software packages available. A popular choice is NIH *ImageJ*. This freeware is Java based, so it can be used on multiple computing platforms. It is available for download at http://rsbweb.nih.gov/ij/. ImagePro (Media Cybernetics, Bethesda, MD) and other software packages for morphometry are commercially available. Similar functions to those described below are applicable to most software packages, but care should be taken in determining which is most suitable for the individual user.
2. Set image analysis parameters. Because pixel density, image intensity, and software/video signal amplification are highly variable among labs, it is critical that steps are taken to set consistent image acquisition standards. This is done by establishing a working range wherein background signals from nonmucin-producing airways are as low as possible during initial image acquisition. Using a terminal bronchiole, for example, the area of red pixel staining is assessed at all pixel intensity levels, and the lower limit of the pixel intensity histogram is raised above 0 to the highest value in which no pixels are detected on the image. In ImagePro, the lowest pixel value is usually set to 45 on a 0–255 scale. The precision of this is then assessed in separate nonmucin-containing samples. Afterward, a section containing abundant mucin is measured in order to ensure that all observable contents are detected using exactly the same settings (see Note 5).
3. Measure the sample boundary length (Fig. 2a). Open a two-color image, and ensure that the correct spatial calibration setting is selected for the objective used to acquire the image. Select the appropriate length measurement tool in the software being used, and measure and record the length of the basal lamina.
4. Measure the area of mucin staining in the sample (Fig. 2b). Open the red-only one-color image corresponding to that used in Step 2 above. Measure the area of cytoplasmic red (number of red pixels) in the conducting airway epithelium. Stained fibers in the submucosa is also highlighted. These need to be omitted from the final count. Record the remaining area of red staining.

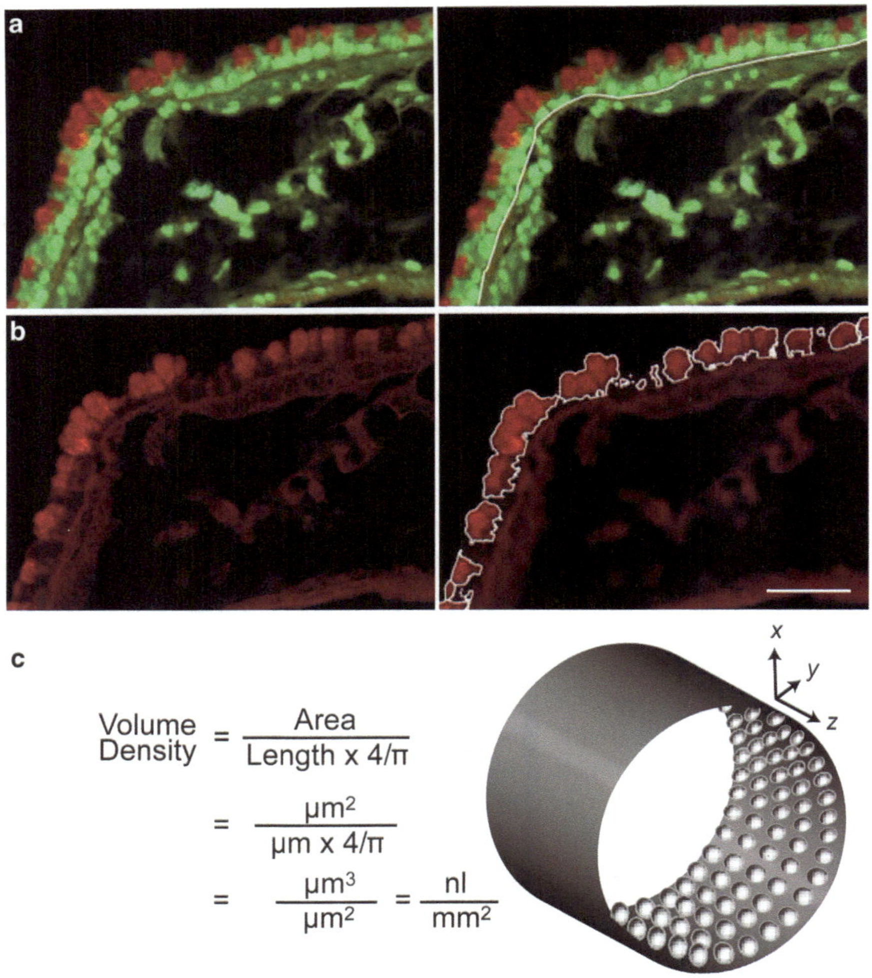

Fig. 2. Image analysis and quantitation of intracellular mucin content in PAFS-stained airways. Tissues from antigen-challenged mice were stained with PAFS and imaged using an Olympus BX-61 upright microscope under FITC/Texas Red dual fluorescence. In two color images (**a**), the full length of the airway region to be analyzed is measured (*white line* along basal lamina). In *red-only images* (**b**), the *area* of *red* in PAFS-stained epithelium is identified and measured digitally. The area and length of staining are applied to the formula for volume density calculation (**c**) for conversion into a three-dimensional estimate of mucin volume per square millimeter of airway epithelium. *Scale bar* is 20 μm in (**a**) and (**b**).

5. Calculate mucin volume density (Fig. 2c). Divide the area of mucin staining by the product of the boundary length and $4/\pi$. The resulting value is a volume density with units of volume of epithelial mucin per area of basement membrane. For a typical image measured on a micrometer scale, this value has units of nL/mm^2 (9) (see Note 6).

3.5. Extracted Mucins Specimen Preparation, Electrophoresis, and Transfer

1. Perform surgical preparations as described above in Subheading 3.1, steps 1–3. Using a syringe, inflate lungs with guanidinium buffer containing protease inhibitors (0.035 mL buffer/g of body weight when inflating both lungs) (see Note 7).

2. Remove thoracic organs, and carefully remove the heart and thymus.
3. Homogenize the lungs in 1 mL of guanidinium buffer and incubate at 4°C overnight mixing gently. Centrifuge at 16,000 × *g* for 30 min at 4°C, and collect supernatants.
4. Dialyze the supernatants against urea buffer using Slide-A-Lyzer Mini Dialysis Units, 2–10 kDa MWCO (Thermo Scientific, Rockford, IL) overnight at 4°C with gentle stirring.
5. Measure protein concentration by bicinchoninic acid (*BCA*) *protein* assay (Thermo) (see Note 8).
6. Prepare loading buffer. To reduce disulfide bonds and denature mucins, add eight volumes of sample to one volume of 10× loading buffer. Incubate at 95°C for 20 min.
7. Alkylate free thiol moieties. Add one volume of alkylation buffer and incubate for 30 min at room temperature in the dark. Keep at 4°C until agarose gel electrophoresis.
8. Prepare 1% agarose (GenePure LE, ISC Bioexpress, Kaysville, UT) in TAE buffer and submerge in electrophoresis buffer (see Note 9). Load specimens and run at 90 V for 90 min or until the blue dye front has run ~90% of the gel length (see Note 10).
9. Wash the gel in 4× SSC for 5 min and then incubate for 20 min at room temperature in 4× SSC containing 10 mM DTT.
10. Transfer proteins to nitrocellulose membrane with a Vacuum Blotter (Model 785, Bio-Rad) at a pressure of 10 mm in Hg for 4 h at room temperature in 4× SSC buffer.

3.6. Detection of Blotted Mucins with Antibodies and Lectins

1. After transfer, wash the membrane two times in PBS and then block with the appropriate protein and/or detergent solution for the chosen detection system. When using antibodies, 5% (w/v) nonfat milk or bovine serum albumin in PBS is usually effective for blocking. However, when using lectins, these can cause nonspecific signal in the background. When using the *Ulex europaeus* agglutinin UEA-1 to detect fucosylated mucins, preincubation with a solution of 0.3% (v/v) Tween 20 in PBS is adequate.
2. Incubate with primary antibody or lectin either for 2 h at room temperature or overnight at 4°C.
3. Wash 3× for 10 min with 0.1% (v/v) Tween 20 in PBS.
4. Incubate with horseradish peroxidase-conjugated secondary antibody for 2 h at room temperature. Alternatively, incubate in the dark with a fluorochrome-conjugated secondary antibody (see Note 11).
5. Wash 3× for 10 min with 0.1% (v/v) Tween 20 in PBS.

6. Detect and analyze signals. Perform ECL detection using commercially available kits per manufacturer's instructions (see Subheading 2). Scan films and perform densitometry using image analysis software (e.g., *ImageJ*). Alternatively, chemiluminescent or fluorescent signals may be directly detected using an advanced imaging system [e.g., Odyssey system (LI-COR Biosciences)] to image and measure the signal intensities.

3.7. Interpretation of Data for Assessment of Mucin Secretion

1. For assessment of a change in basal secretion rate, the most straightforward method is to compare intracellular mucin content (measured either by PAFS image analysis or probing of vacuum blots) between control and experimental groups. This yields fold increases of intracellular mucin content for a loss-of-function phenotype (see Note 12), and fractional decreases for a gain-of-function phenotype (Fig. 3a). An additional comparator should be provided for loss-of-function experiments by including a positive-control specimen with robust mucous metaplasia (e.g., antigen-challenged wild-type mouse samples). Thus, the additional statement can be made that intracellular mucin accumulation due to the loss of secretory function results in a value equal to some fraction of that measured in mucous metaplasia.
2. For assessment of a change in stimulated secretory function, the most straightforward method is to first induce mucous metaplasia in control and experimental groups, and then to compare intracellular mucin content between control and experimental groups after treatment with a maximal (100 mM) aerosolized ATP stimulus (Fig. 3a). In antigen-challenged wild-type mice, 60–75% of intracellular mucin is acutely released on average (see Note 13). A loss-of-function phenotype is indicated by a reduced fractional release compared to control (Fig. 3b, left), and a gain-of-function phenotype is indicated by an increased fractional release (see Note 14). Data may be further analyzed as "release efficiency," calculated as the fractional release of the experimental group divided by the fractional release of the control group. This yields numbers less than 1 for a loss of function and greater than 1 for a gain of function

Fig. 3. (continued) mucin content measured in each of these conditions (naïve, metaplastic, degranulated) by any of these methods is illustrated for wild-type (WT) mice (*black bars*), and for mice with a loss-of-secretory function (LOF, *grey bars*) or gain-of-secretory function (GOF, *white bars*) due to genetic mutation or pharmacologic treatment. (**b**) For further comparison of differences in stimulated secretory function between WT and LOF or GOF animals, "fractional release" may be calculated as mucin content in degranulated animals as a percentage of mucin content in metaplastic animals (*left*) or "release efficiency" may be calculated as fractional release by LOF and GOF animals divided by fractional release by WT animals (*right*).

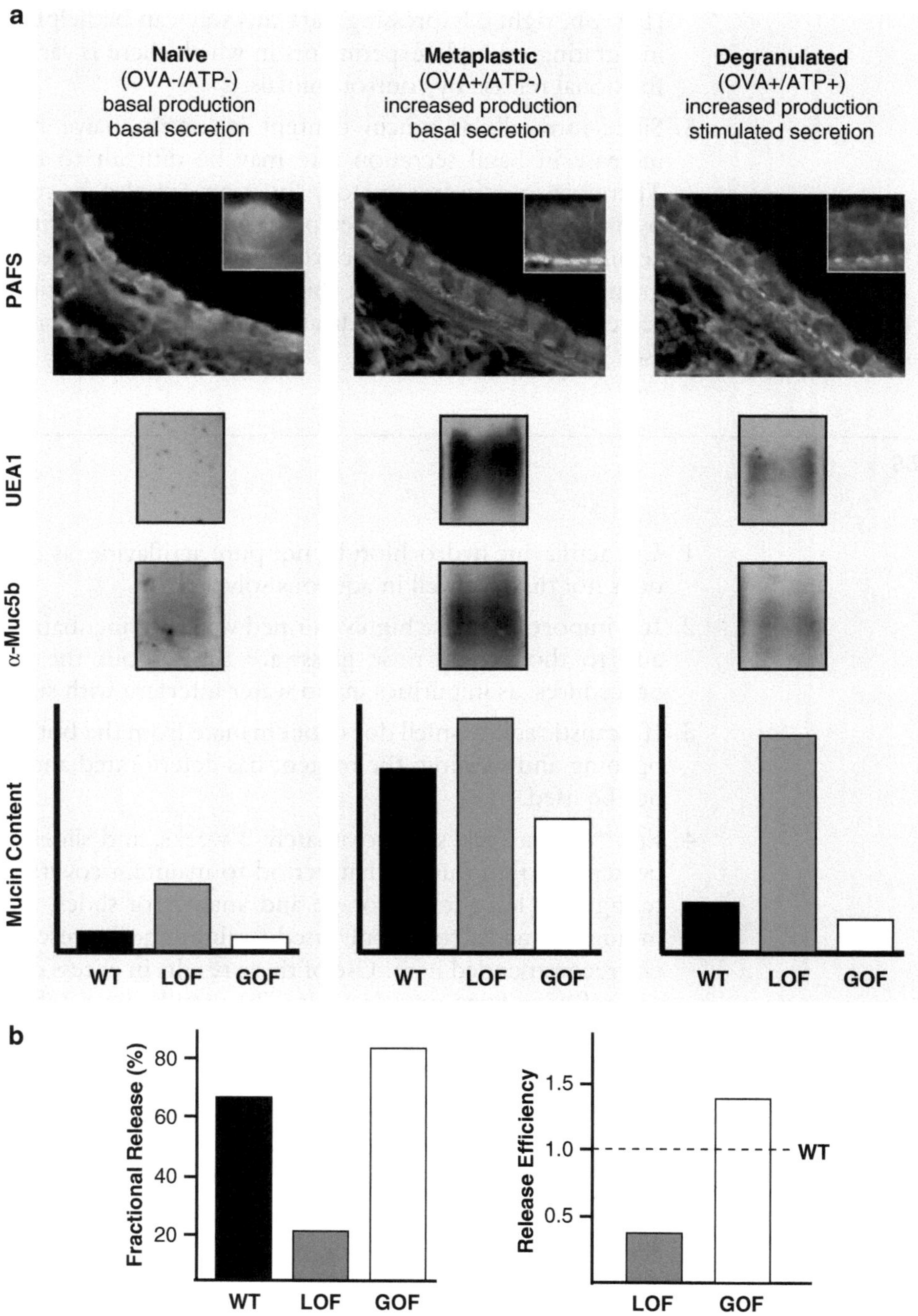

Fig. 3. Assessment of mucin secretion in vivo by measurement of intracellular mucin content. (**a**) Airway epithelial mucin content is measured in three conditions—naïve, metaplastic, and degranulated. These three conditions can be induced by treatment or not with ovalbumin immunization and airway challenge (OVA), followed or not by airway challenge with ATP. Treatment with OVA results in increased mucin production, and treatment with ATP induces acute mucin secretion. Intracellular mucin content after treatment can be measured by image analysis of tissue sections stained with PAFS or lung tissue extracted with guanidinium hydrochloride, electrophoresed through 1% agarose, then transferred to nitrocellulose, and probed with UEA1 lectin to detect Muc5ac or antibodies against Muc5b (all shown for wild-type mice only). Intracellular

(Fig. 3b, right). Expressing data this way can be helpful when integrating multiple experiments in which there is variation in fractional release by control animals.

3. Since intracellular mucin content is low in naïve mice, an increase in basal secretion rate may be difficult to measure. Therefore, a gain in secretory function can also be measured by first inducing mucous metaplasia, and then comparing intracellular mucin content between control and experimental groups after 7–14 days. In this case, a gain-of-function basal secretory phenotype is indicated by less intracellular mucin (see Note 15).

4. Notes

1. Use acriflavine hydrochloride, not pure acriflavine, as the latter does not dissolve well in aqueous solvents.
2. It is important to use highly purified water for incubation steps and to thoroughly rinse glassware throughout the staining procedures, as impurities in tap water interfere with staining.
3. If a caustic sulfite smell does not emanate from the bottle upon opening and swirling, the reagent has deteriorated and should not be used.
4. Full "drying" takes approximately 2 weeks, and slides should be kept upright during that period to maintain coverslip centering. For long-term storage and analysis of slides, aqueous mounting media commonly used for immunofluorescence are not recommended here. Use of these results in excess diffusion of acriflavine from nucleic acids. They can be useful, however, for short-term colocalization experiments using immunofluorescence in combination with PAFS.
5. Video output varies among monitors and is not necessarily a reliable gauge of actual pixel values. If separate workstations are being used for image acquisition and analysis, it is highly recommended that users determine analysis parameters for the appropriate workstation.
6. This formula is a simplified version of that described by Weibel (24) and derived by Harkema et al. (25).
7. It is possible to use one lung for PAFS staining and the other for vacuum blotting, or one lung for measurement of intracellular mucin by either technique and the other for other assays, such as measurement of transcript levels or bronchoalveolar lavage fluid cell counts and cytokines. To do so, apply a clamp to the main bronchi connected to the lung to be preserved before starting to inflate. Alternatively, a thread can be used to

make a knot to block guanidinium buffer flow to one of the lungs.

8. At high concentrations, urea can interfere with the BCA assay. Therefore, it is recommended to dilute protein standards (usually, ovalbumin) with the same buffer in which the specimens to be read are dissolved. For example, if a small portion of the specimen is diluted 1:5 in water before the BCA assay, it is suggested to use a 1:5 diluted urea buffer to prepare the standard. It is also important to keep in mind that excessive dilution in water can trigger mucin aggregation.
9. It is suggested to fill a 20 × 20-cm tray with 100 mL of 1% agarose and to utilize two combs to have two rows of 12 or 20 wells.
10. A working range of 50–100 μg of protein per specimen to be subjected to electrophoresis is suggested.
11. It is recommended to utilize a high-molecular-weight internal loading control to minimize loading errors. One example of a suitable housekeeping protein is adenomatous polyposis coli (APC) which is 312 kDa, and antibodies to APC can be purchased conjugated to horseradish peroxidase (Santa Cruz Biotechnology).
12. Muc5ac production is low in naïve mice, and UEA1 lectin staining of blotted proteins is usually not sufficiently sensitive to detect Muc5ac protein in these mice, complicating the quantitation of mucin accumulation.
13. If less than 50% of intracellular mucin is released in control animals, it may be best to discard the experiment, since this suggests a problem with aerosol delivery of intact secretagogue to the bronchial airways.
14. Since stimulated secretion is so high in wild-type mice, a gain of function may be best uncovered using submaximal stimulation with a 10 mM ATP aerosol.
15. This technique may also be used to measure a subtle loss-of-function phenotype as increased intracellular mucin retention.

Acknowledgments

This work was supported by NIH grants HL080396 (C.M. Evans), HL094848 (B.F. Dickey), and HL097000 (B.F. Dickey), American Heart Association Grant 10GRNT4200070 (C.M. Evans), and Cystic Fibrosis Foundation grant 08GO (B.F. Dickey). The authors thank C. William Davis for instruction in the performance of vacuum blotting of mucins.

References

1. Davis, C. W. and Dickey, B. F. (2008) Regulated airway goblet cell mucin secretion. *Annu. Rev. Physiol.* **70**, 487–512.
2. Evans, C. M. and Koo, J. S. (2009) Airway mucus: the good, the bad, the sticky. *Pharmacol. Ther.* **121**, 332–348.
3. Fahy, J. V. and Dickey, B. F. (2010) Airway mucus function and dysfunction *N. Engl. J. Med.* **363**, 2233–2247.
4. Young, H. W., Williams, O. W., Chandra, D., Bellinghausen, L. K., Perez, G., Suarez, A., Tuvim, M. J., Roy, M. G., Alexander, S. N., Moghaddam, S. J., Adachi, R., Blackburn, M. R., Dickey, B. F., and Evans, C. M. (2007) Central role of Muc5ac expression in mucous metaplasia and its regulation by conserved 5' elements *Am J Respir Cell Mol Biol.* **37**, 273–290.
5. Chen, Y., Zhao, Y. H., and Wu, R. (2001) In Silico Cloning of Mouse Muc5b Gene and Upregulation of Its Expression in Mouse Asthma Model. *Am. J Respir. Crit Care Med* **164**, 1059–1066.
6. Zhu, Y., Ehre, C., Abdullah, L. H., Sheehan, J. K., Roy, M., Evans, C. M., Dickey, B. F., and Davis, C. W. Munc13-2−/− baseline secretion defect reveals source of oligomeric mucins in mouse airways. (2008) *J Physiol.* **586**, 1977–1992.
7. Nguyen, L. P., Omoluabi, O., Parra, S., Frieske, J. M., Clement, C., mmar-Aouchiche, Z., Ho, S. B., Ehre, C., Kesimer, M., Knoll, B. J., Tuvim, M. J., Dickey, B. F., and Bond, R. A. (2008) Chronic exposure to beta-blockers attenuates inflammation and mucin content in a murine asthma model. *Am J Respir Cell Mol. Biol.* **38**, 256–262.
8. Zudhi Alimam, M., Piazza, F. M., Selby, D. M., Letwin, N., Huang, L., and Rose, M. C. (2000) Muc-5/5ac mucin messenger RNA and protein expression is a marker of goblet cell metaplasia in murine airways. *Am. J. Respir. Cell Mol. Biol.* **22**, 253–260.
9. Evans, C. M., Williams, O. W., Tuvim, M. J., Nigam, R., Mixides, G. P., Blackburn, M. R., DeMayo, F. J., Burns, A. R., Smith, C., Reynolds, S. D., Stripp, B. R., and Dickey, B. F. (2004) Mucin is produced by clara cells in the proximal airways of antigen-challenged mice. *Am J. Respir. Cell Mol. Biol.* **31**, 382–394.
10. Davis, C. W. and Lazarowski, E. (2008) Coupling of airway ciliary activity and mucin secretion to mechanical stresses by purinergic signaling. *Respir. Physiol Neurobiol.* 163, 208–213.
11. Singer, M., Martin, L. D., Vargaftig, B. B., Park, J., Gruber, A. D., Li, Y., and Adler, K. B. (2004) A MARCKS-related peptide blocks mucus hypersecretion in a mouse model of asthma. *Nat. Med.* 10, 193–196.
12. Tuvim, M. J., Mospan, A. R., Burns, K. A., Chua, M., Mohler, P. J., Melicoff, E., Adachi, R., mmar-Aouchiche, Z., Davis, C. W., and Dickey, B. F. (2009) Synaptotagmin 2 couples mucin granule exocytosis to Ca2+ signaling from endoplasmic reticulum. *J Biol. Chem.* **284**, 9781–9787.
13. Grunig, G., Warnock, M., Wakil, A. E., Venkayya, R., Brombacher, F., Rennick, D. M., Sheppard, D., Mohrs, M., Donaldson, D. D., Locksley, R. M., and Corry, D. B. (1998) Requirement for IL-13 independently of IL-4 in experimental asthma. *Science* **282**, 2261–2263.
14. Lappalainen, U., Whitsett, J. A., Wert, S. E., Tichelaar, J. W., and Bry, K. (2005) Interleukin-1beta causes pulmonary inflammation, emphysema, and airway remodeling in the adult murine lung. *Am J Respir Cell Mol. Biol.* **32**, 311–318.
15. Tomkinson, A., Cieslewicz, G., Duez, C., Larson, K. A., Lee, J. J., and Gelfand, E. W. (2001) Temporal association between airway hyperresponsiveness and airway eosinophilia in ovalbumin-sensitized mice. *Am J Respir Crit Care Med.* **163**, 721–730.
16. Trifilieff, A., Ahmed, E., and Bertrand, C. (2000). Time course of inflammatory and remodeling events in a murine model of asthma: effect of steroid treatment. *Am. J. Physiol Lung Cell Mol. Physiol* **279**, L1120–L1128.
17. Walter, M. J., Morton, J. D., Kajiwara, N., Agapov, E., and Holtzman, M. J. (2002) Viral induction of a chronic asthma phenotype and genetic segregation from the acute response. *J. Clin. Invest* **110**, 165–175.
18. Wills-Karp, M., Luyimbazi, J., Xu, X., Schofield, B., Neben, T. Y., Karp, C. L., and Donaldson, D. D. (1998) Interleukin-13: central mediator of allergic asthma. *Science* **282**, 2258–2261.
19. Zheng, T., Zhu, Z., Wang, Z., Homer, R. J., Ma, B., Riese, R. J., Chapman, H. A., Shapiro, S. D., and Elias, J. A. (2000) Inducible targeting of IL-13 to the adult lung causes matrix metalloproteinase- and cathepsin-dependent emphysema. *J. Clin. Invest* **106**, 1081–1093.
20. Carlstedt, I., Lindgren, H., Sheehan, J. K., Ulmsten, U., and Wingerup, L. (1983) Isolation and characterization of human cervical-mucus glycoproteins. *Biochem. J.* **211**, 13–22.

21. Aksoy, N., Thornton, D. J., Corfield, A., Paraskeva, C., and Sheehan, J. K. (1999) A study of the intracellular and secreted forms of the MUC2 mucin from the PC/AA intestinal cell line. *Glycobiology* **9**, 739–746.
22. Thornton, D. J., Rousseau, K., and McGuckin, M. A. (2007) Structure and function of the polymeric mucins in airways mucus. *Annu. Rev Physiol.* **70**, 459–86.
23. Reader, J. R., Tepper, J. S., Schelegle, E. S., Aldrich, M. C., Putney, L. F., Pfeiffer, J. W., and Hyde, D. M. (2003) Pathogenesis of mucous cell metaplasia in a murine asthma model. *Am. J. Pathol.* **162**, 2069–2078.
24. Weibel, E. R. (1979) *Sterological Methods* Academic Press Inc. Ltd., London.
25. Harkema, J. R., Plopper, C. G., Hyde, D. M., and St George, J. A. (1987) Regional differences in quantities of histochemically detectable mucosubstances in nasal, paranasal, and nasopharyngeal epithelium of the bonnet monkey. *J. Histochem. Cytochem.* **35**, 279–286.

Chapter 18

Techniques for Assessment of Interactions of Mucins with Microbes and Parasites In Vitro and In Vivo

Yong H. Sheng, Sumaira Z. Hasnain, Chin Wen Png, Michael A. McGuckin, and Sara K. Lindén

Abstract

Most mammalian pathogens and parasites infect their hosts via the mucosal surfaces. The first barrier they encounter in all mucosal tissues is a layer of viscous mucus which can be modulated by immune responses to the pathogen or parasite. The major macromolecular constituents of mucus are secreted mucin glycoproteins which give mucus its viscous properties. Underneath the mucus layer, the mucosal epithelial cells have a cell surface glycocalyx that is rich in transmembrane mucin glycoproteins. Both the cell surface and secreted mucins present a vast array of potential binding sites for pathogens and parasites and both forms of mucins are involved in protecting the host from infection. However, many pathogens and parasites have evolved mechanisms to subvert the mucin barrier. Thus, studying mucin interactions with pathogens and parasites is critical to understanding host–pathogen interactions at the mucosal surfaces. In this chapter, we describe methods for studying the interactions between mucins and pathogens and parasites, methods for studying the degradation of mucins by pathogens and parasites, and in vitro and in vivo methods for exploring the functional significance of the mucins in host defence from infection.

Key words: Mucin, Pathogen, Parasite, Mucin degradation, Mucin binding, Mucin knockout, siRNA

1. Introduction

Most mammalian pathogens and parasites infect their hosts via the mucosal surfaces and infections in the gastrointestinal, respiratory, and urinary tracts represent a large proportion of the infectious disease burden in the Western and developing world. The first barrier encountered by pathogens and parasites in all mucosal tissues is a layer of viscous mucus which can be modulated by immune responses to the pathogen or parasite (1). The major macromolecular constituents of mucus are secreted mucin glycoproteins which give mucus its viscous properties and retain anti-microbial

Michael A. McGuckin and David J. Thornton (eds.), *Mucins: Methods and Protocols*, Methods in Molecular Biology, vol. 842, DOI 10.1007/978-1-61779-513-8_18, © Springer Science+Business Media, LLC 2012

molecules close to the mucosal surface (2). Mucus is difficult to work with in experiments because of its viscous properties and the mucin glycoproteins themselves present challenges because of their large size accentuated by formation of oligomers, heterogeneity, and complex patterns of O-glycosylation.

Underneath the mucus layer, the mucosal epithelial cells have a cell surface glycocalyx that is rich in transmembrane mucin glycoproteins. Cell surface mucins are typically also very large and like the secreted mucins contain large extended densely O-glycosylated domains. These mucins are also usually cleaved into two subunits during biosynthesis and the extracellular subunit can be released/shed from the cell surface, further complicating analysis of these mucins.

Both the cell surface and secreted mucins present a vast array of potential binding sites for pathogens and parasites via their complex O-linked glycan chains. Individual mucins can carry many different glycans (around 200 for human intestinal MUC2 (3)) and mucin glycosylation varies between tissues and with infection and inflammation (4). Binding between mucin glycans and a range of different types of pathogens has been demonstrated and these interactions are important in host defence from infection (4). Both secreted and cell surface mucins are involved in protecting the host from infection. In addition to their barrier function, some mucins have direct antimicrobial properties or can directly bind host antimicrobial molecules demonstrating that these glycoproteins are an integral component of host defence from infection (reviewed in ref. 4). However, many pathogens and parasites have evolved mechanisms to subvert the mucin barrier, including production of mucin-degrading enzymes (4). Studying mucin interactions with pathogens and parasites and exploring the outcome of infections when mucins are deficient are critical to understanding host–pathogen interactions at the mucosal surfaces.

In this chapter, we describe methods for studying the interactions between mucins and pathogens and parasites using purified mucins and ELISA-based techniques. Methods for studying the specific degradation of mucins by bacteria and parasites are also described; more detailed techniques for studying mucin degradation by bacteria are also dealt with in Chapter 12. We also describe generic methods for exploring the functional significance of mucins in host defence from infection using pathogen host-cell in vitro cell co-culture with siRNA knockdown of mucin genes, and in vivo using mucin knockout mice. The protocols are presented as optimized for particular pathogens and parasites, but should be easily adapted to other organisms.

2. Materials

2.1. Preparation of Bacteria

2.1.1. Helicobacter pylori

1. We use strain J99 and mutants derived from this strain, including J99 ΔCagA and J99 ΔBaba/ΔSaba mutants, but the methods are applicable to other strains.
2. HBA plate (blood agar base supplemented with 5% defibrillated horse blood, 2.5 μg/mL amphoteracin B, 10 μg/mL vancomycin, and 5 μg/mL trimethoprim).
3. Brain heart infusion (BHI) broth.
4. Phosphate-buffered saline (PBS), pH 7.4.
5. Spectrophotometer.

2.1.2. Campylobacter jejuni

1. Strain 81116 or 81–126 wild type, 81–126 ΔCDT-β mutant.
2. Columbia agar plate (2% Columbia agar, 1% bacteriological agar, supplemented with 5% defibrinated horse blood, Skirrow Selective Media).
3. Brucella broth (Oxoid).

2.1.3. Citrobacter rodentium

1. Strain ICC169.
2. Lauria–Bertani (LB) agar plate or McConkey's selective agar plate.
3. Lauria–Bertani broth (Oxoid).

2.2. Preparation of Trichuris muris Parasites

1. Eggs initially need to be sourced from a collaborator. Infect nude or any immunodeficient mice with 200 embryonated eggs by oral gavage.
2. After 35 days, cut open the caecum of infected mice and rinse in PBS containing 5% penicillin/streptomycin. For sufficient egg harvest, use at least 6–10 infected mice.
3. Transfer caecum into a Petri dish with 10 mL of RPMI-1640 with 5% penicillin/streptomycin and remove the worms using blunt forceps.
4. Transfer worms into 6-well plates with 4 mL of RPMI-1640 and 5% penicillin/streptomycin.
5. Incubate the 6-well plate placed in a humid box (wet cloth placed at the bottom of a box) overnight at 37°C.
6. Transfer media containing eggs into a 50-mL Falcon tube and centrifuge at 200 × *g* in a microfuge for 15 min at room temperature.
7. All the steps hereafter must be performed in a sterile cell culture hood. Remove supernatant and resuspend eggs in sterile dH_2O.

8. Pass resuspended eggs through a 100-μm filter into a sterile 50-mL tube.
9. Centrifuge at 200 × *g* in a microfuge for 5 min at room temperature, remove supernatant, and resuspend eggs in 50 mL of sterile dH_2O.
10. Repeat step 8 three times.
11. Transfer resuspended eggs into a non-filter top tissue culture flask and top up the flask (50%) with sterile dH_2O.
12. Wrap up the flask in tin foil and store in the dark for 8 weeks for embryonation to occur as evidenced by the microscopic appearance of larvae within the egg. Ensure that the stored eggs are checked regularly visually for contamination by microbes.

2.3. Preparation of Cultured Epithelial Cells

1. The gastric epithelial cell line MKN7 (Riken Cell Bank, Japan) expressing a high level of MUC1.
2. The intestinal epithelial cell line LS513 and LIM2463 expressing high levels of MUC13.
3. All the epithelial cells were cultured in RPMI containing 10% FCS, 2 mM L-glutamine, 100 U/mL penicillin G sodium, and 100 μg/mL streptomycin.
4. For transfection and co-culture, the medium was changed to antibiotic-free medium 24 h prior to the treatment.

2.4. Buffers, Reagents, and General Materials

1. Biotinylation buffer: 0.2 M carbonate buffer, pH 8.3.
2. Biotin-XX-NHS (Roche Diagnostics).
3. Blocking Reagent for ELISA (Boehringer Mangleheim).
4. PBS-Tween: PBS with 0.05% (v/v) Tween 20.
5. Streptavidin–HRP.
6. ABTS: 2,2′-Azinobis(3-ethylbenzothiazoline)-6-sulphonic acid as a substrate (0.550 g/L in citrate phosphate buffer, pH 4.3).
7. ELISA plates for bacterial binding assays—we find that NUNC polysorb plates work well.
8. M75 minimal bacterial culture medium.
9. Anthrone reagent test: Dissolve 1 g of anthrone in 500 mL of 72% sulphuric acid.
10. Mucin-specific and control siRNA, e.g. Dharmacon Smart-POOL.
11. Lipofectamine 2000.
12. Gentamicin.
13. Flow Cytometer (e.g. LSRII, BD Biosciences, San Jose, USA) and analysis software (e.g. Diva, BD Biosciences, San Jose, USA).

14. Fluorescence mounting medium, e.g. Prolong Gold (Invitrogen).
15. Confocal microscope, e.g. Zeiss LSM510.
16. RPMI-1640/FBS: RPMI-1640 supplemented with 10% (v/v) foetal bovine serum.
17. Opaque-walled luminescent plates.
18. Luminescence Detection Reagent: Cell-Titre Glo (Promega).
19. Luminometer.
20. FastPrep® (MP Biomedicals).
21. Stereo microscope, e.g. Leica M165C.

3. Methods

3.1. Adhesion Assays and Identification of Microbial Ligands for Mucins

This protocol uses biotinylation of bacteria (see Note 1) to track binding of the bacteria to purified mucins in vitro using either of two ELISA-based techniques.

3.1.1. Biotinylation of Bacteria

1. Wash freshly harvested bacteria twice in biotinylation buffer, centrifuge at 2,500 × *g* for 3 min, and resuspend in the same buffer (see Note 2).
2. Dissolve biotin-XX-NHS in DMSO and add 125 μg/2 × 10^9 bacteria in 1 mL biotinylation buffer.
3. Incubate for 15 min, in the dark, with continuous tube rotation.
4. Wash twice in PBS-Tween.
5. Prepare at required concentration in 1% blocking reagent for ELISA.

3.1.2. Direct Microbial ELISA on Plate-Bound Mucins

1. Dilute the mucin and control samples to 3 μg/mL in PBS or 4 M GuHCl (see Note 3).
2. Coat micotitre plates (polysorb 96-well plate, NUNC) with 100 μL/well and incubate overnight at 4°C.
3. Wash three times with PBS-Tween.
4. Block wells with 150 μL 1% blocking reagent for ELISA in PBS-Tween 20 for 1 h.
5. Incubate wells with 100 μL biotinylated bacteria (OD 0.15 diluted 1:50, see Note 4) in PBS-Tween.
6. Wash three times with PBS-Tween.
7. Incubate wells with 100 μL streptavidin–HRP (1:1,000), 1 h, room temperature (RT).

8. Wash three times with PBS-Tween.
9. Add 100 μL/well of ABTS substrate and incubate for 30 min. Read OD at 405-nm absorbance using a microtitre plate reader.

3.1.3. Sandwich ELISA: Mucin Capture Detected by Bacteria

1. Coat parallel wells of ELISA plates with mucin-specific (e.g. BC2 for MUC1) or isotype-control antibodies at 3 μg/mL in PBS overnight at 4°C.
2. Wash three times with PBS-Tween.
3. Block wells with 150 μL 1% BSA in 1% blocking reagent for ELISA in PBS-Tween for 1 h at RT.
4. Add crude or purified mucins (see Note 5) in 1% blocking reagent for ELISA in PBS-Tween, and incubate for 1 h, RT.
5. Wash three times with PBS with 0.05% Tween.
6. Empty wells, add 100 μL biotinylated *H. pylori*/well or for positive-control biotinylated BC2/well, and incubate at 37°C on shaking board for 2 h.
7. Wash three times with PBS with 0.05% Tween.
8. Incubate wells with 100 μL streptavidin–HRP (1:1,000), 1 h, RT.
9. Wash three times with PBS with 0.05% Tween.
10. Add 100 μL/well of ABTS substrate. Read OD at 405-nm absorbance using plate reader.

3.2. Degradation of Mucins by Bacteria and Parasites

3.2.1. Degradation of Mucins by Mucolytic Bacteria

This protocol adds purified mucins to bacterial cultures in minimal media and measures utilization of the mucin and expansion of the bacteria when mucin is the only potential carbon source. Any source of mucin can be used, but the purified mucins should be dialyzed against water or a physiological buffer that does not influence bacterial growth.

1. Inoculate 2×10^8 CFU of bacteria into 1 mL of supplemented M75 medium containing 1 mg of mucin (see Note 6).
2. Grow bacteria under appropriate anaerobic, microaerophilic, or aerobic conditions at 37°C for 48 h.
3. Measure the amount of mucin degraded in defined culture with mucin as the sole carbon source by the loss of hexose using the anthrone reagent test with dilutions of galactose used as standards ensuring the standard curve extends above the starting concentration of mucin and below the minimal concentrations reached experimentally. Bacteria that can degrade >25% of mucin/hexose are defined as mucolytic.
4. Centrifuge aliquots of culture at 14,000 × *g* in a microfuge for 10 min to pellet bacteria. Extract total bacterial genomic DNA

using standard techniques [we use FastDNA® Spin kit for soil (Bio 101, USA)].

5. Determine the fold change in abundance of bacteria using quantitative PCR and 16 s ribosomal RNA primers specific for the bacteria under study or universal primers, comparing the Ct values generated from the initial inoculum and the culture at 48-h incubation. This enables determination of whether bacterial growth has occurred when mucin is the only carbon source.

3.2.2. Mucin Degradation by Parasitic Exo-products

This protocol uses exo-products secreted by parasites into in vitro culture and examines their ability to degrade purified mucins in vitro.

1. Culture live parasites in serum-free culture medium in vitro overnight.
2. In a sterile tissue culture hood, remove medium and filter through a 0.2-μm filter, concentrate supernatant using 1-kDa cut-off membrane centrifugal concentrators.
3. Transfer medium into dialysis membrane (high-grade pre-treated, molecular weight cut-off of 1 kDa) and leave overnight at 4°C, stirring in sterile PBS. Change the PBS at least three times over the 24-h time period.
4. Filter the supernatant containing the parasitic exo-products again using a 0.2-μm filter in a sterile hood.
5. Determine protein concentration of parasitic exo-products using a standard BCA assay.
6. Add titrated concentrations of parasitic exo-products to non-reduced purified mucins (see Chapter 2) for varying time periods at 37°C. In parallel, as a negative control, add parasitic exo-products deactivated by heating to 100°C for 10 min.
7. Analyze the mucins using rate-zonal centrifugation, which separates proteins based on mass and shape—for a method, see ref. 5.
8. Fractionate the rate-zonal centrifugation gradients, and analyze fractions after slot blotting by PAS or mucin antibody staining. A clear change in the sedimentation profile determines whether the parasitic exo-products are capable of degrading mucins (see Note 7).

3.3. Co-culture of Bacteria and Epithelial Cells with and Without the Cell Surface Mucins

When co-culturing bacteria with epithelial cells, there are a few important issues to consider as follows: (a) The co-culture system will more closely resemble the in vivo situation if a cell line that forms a confluent layer and polarizes is used. (b) Cell lines are highly variable in what mucins they express and mucin glycosylation also varies between lines (6). Many cancer cell lines do not express secreted mucins and, therefore, do not form a mucus barrier and many cell lines express mucins not normally expressed in their tissue of origin (6). (c) The apical density of the cell surface mucins is dependent on the density of the cells; therefore, it is important to make the comparisons between wells with the same cell density. (d) Oxygen tolerance of the bacteria and the mammalian cell should be considered. Some microaerophiles, such as *H. pylori*, may vary in their tolerance to oxygen. *H. pylori*'s ability to induce mammalian cell responses changes substantially in aerobic vs. microaerobic conditions (7). Therefore, if comparing different strains of *H. pylori*, it may be safer to do the comparison under microaerobic conditions to ensure that what one measures is actually differences between strains/virulence factors, and not the comparison of healthy vs. stressed bacteria. Most cell lines tolerate microaerobic conditions well; in fact, the oxygen tension in the tissue in most places of the body is also microaerobic.

3.3.1. Preparation of Bacteria for Co-culture

H. pylori

1. Immediately before use, thaw the glycerol stock, and spread the stock evenly over the HBA plate with a sterile spreader.
2. Put two pieces of paper towel to cover the base of a gas jar and soak thoroughly with water (see Note 8); place the plates in the jar face up.
3. Insert a microaerobic gas sachet (e.g. Campygen) into the side of the jar ensuring that it is not in contact with the wet paper towel or any other moisture (see Note 8), and immediately place the lid on the jar.
4. Incubate at 37°C for 3 days in 5% O_2 and 15% CO_2 (microaerobic conditions).
5. If the cultures look pure (*H. pylori* should form glassy colonies), subculture using a wire loop onto as many as HBA plates as required, and incubate at 37°C for 2 days in microaerobic conditions.
6. Harvest *H. pylori* for broth culture or co-culture (see Note 9):
   (a) Add 2 mL of BHI broth without supplements or PBS to each plate.
   (b) Use a sterile spreader to detach the cells from the agar surface.
   (c) Carefully remove the suspension using a sterile pipette and transfer it to a sterile tube.
   (d) Wash *H. pylori* three times in BHI or PBS (see Note 5) by centrifugation at 3,500 rpm in a microfuge for 30 min.
7. Add harvested *H. pylori* from two plates into 150 mL of BHI broth culture.

8. Place the flask into a gas jar with a Campygen sachet generating gas, and grow for 1 day at 37°C with 180 rpm shaking.
9. Harvest the cells as in step 6.
10. Prepare the inoculums by resupending the washed bacteria into PBS to a concentration, where $OD_{600nm} = 0.1$ (equivalent to ~1×10^8 CFU/mL).
11. Quantify the starting inoculums used in each experiment by serially diluting the sample and plating on HBI plate.

C. jejuni

12. Grow *C. jejuni* on Columbia agar at 42°C for 2 days in microaerophilic conditions.
13. Harvest *C. jejuni* for co-culture as in steps 6–11, except using Brucella broth.

C. rodentium

14. Grow *C. rodentium* on LB agar or McConkey's selective agar at 37°C for 24 h under aerobic conditions.
15. Harvest *C. rodentium* for co-culture as in steps 6–11, except using LB broth.

3.3.2. Knock-Down of Cell Surface Mucin Gene Expression in Mucosal Epithelial Cells by siRNA

1. Plate epithelial cells into 90-cm culture dishes with a confluence of ~50% in RPMI containing 10% FCS and 2 mM L-glutamine without antibiotics 1 day before transfection.
2. Transect with 100 nM siRNA specific for mucin as well as off-target-control siRNA using Lipofectamine 2000 according to the manufacturer's instructions (see Note 10).
3. Transfer the cells into 96-well plates 24 h after transfection at $1\text{–}5 \times 10^4$/well.
4. Assay the level of knock-down 48 h after transfection (see Note 11) by:
   (a) Real-time PCR using relative expression of the mucin gene from mucin siRNA vs. scrambled siRNA control
   (b) Flow cytometry after anti-mucin antibody labelling the cell surface mucins, and comparing the median fluorescence intensity (MFI) of mucin siRNA-treated cells with the scrambled control after subtracting the MFI of cells stained with an isotype-control antibody
   (c) Western blot using relative protein expression of the mucin corrected for a housekeeping protein, such as β-actin vs. scrambled siRNA control

3.3.3. Co-culture of Bacteria and Mucosal Epithelial Cells

1. Gently wash the epithelial cells once in RPMI containing 10% FCS and 2 mM L-glutamine without antibiotics 48 h after transfection.

2. Co-culture epithelial cells with bacteria at multiple of infection (MOI) 1:1–1:20 (mammalian cell:bacteria) in microaerobic condition (see Note 12).
3. Cells can then be harvested at different time points and assayed for desired functional readouts (e.g. level of apoptosis, necrosis, cell cycle events, proliferation) or, as described below, bacterial adhesion and invasion and mucin-bacterial binding.

3.3.4. Analysis of Bacterial Adhesion and Invasion in Co-culture with Epithelial Cells

1. Gently wash off non-adherent bacteria 3× with PBS—keep the supernatant and first wash from each treatment and pool to ascertain viable bacterial numbers by agar culture.
2. Culture a replicate set of wells of the entire experiment with 400 μg/mL gentamicin for 2 h at 37°C, and then wash three times with PBS to remove the antibiotic.
3. Lyse both non-treated and gentamicin-treated cells with 0.02% (v/v) Triton X-114 in PBS for 10 min.
4. Plate bacteria from non-antibiotic-treated cell lysates at three concentrations (equivalent to 0.05, 0.5, and 45 μL of lysate) onto small *H. pylori* plates and lysates from gentamicin-treated cells onto one plate (without dilution).

3.3.5. Assessment of Mucin-Bacterial Binding by Flow Cytometry

1. Harvest bacteria from the culture medium of epithelial cell co-cultures by first sedimenting non-adherent mammalian cells (300 × *g*, 5 min) and then sedimenting bacteria (5,000 × *g*, 10 min).
2. Wash the bacteria three times with cold PBS (5,000 × *g*, 3 min).
3. Block non-specific binding by the bacteria with 1% BSA in PBS for 60 min at 4°C.
4. Label the bacterial with anti-mucin antibodies (e.g. BC2 for MUC1) and isotype-control antibodies at 5 μg/mL in 1% BSA in PBS for 60 min at 4°C, and then with species-specific secondary antibodies conjugated to an Alexa fluor dye (e.g. Alexa-488).
5. Wash the bacteria three times with cold 1% BSA in PBS (5,000 × *g*, 3 min).
6. Analyze by flow cytometry gating on the bacterial population using forward scatter/side scatter characteristics and determine the MFI of labelling after subtraction of MFI of the bacteria stained with the isotype-control antibody, and the proportion of bacteria labelled with the mucin.

3.3.6. Assessment of Mucin-Bacterial Binding by Confocal Microscopy

1. Prepare stained bacteria as in Subheading 3.4.5, steps 1–5, and smear bacteria onto charged glass slides.
2. Stain with DAPI (0.1 μg/mL) for 15 min, wash with PBS, and mount in fluorescence mounting medium.

3. Examine binding of mucins to bacteria using a confocal microscope with multitracking detecting DAPI (excitation 405 nm, detection 420–480 nm) and Alexa-488 (excitation 488 nm detection, LP 505 nm) fluorescence separately.

3.4. Gastrointestinal Infection of Mice to Assess Mucin Function

There is a range of mouse models available that can be used to assess the function of mucins in infection. The following mucin gene-manipulated mice have been published to date: $Muc1^{-/-}$ (8), $Muc2^{-/-}$ (9), $Muc5ac^{-/-}$ (10), and $Muc13^{-/-}$ (11), strains with overexpression of mucins or expression of human mucins (e.g. human MUC1 transgenic mice (12)), strains with disease causing single-point mutations in Muc2 (e.g. *Winnie* and *Eeyore* (13)), and strains either lacking glycosyl transferases responsible for mucin glycosylation or mice expressing human glycosyl transferases (14–16).

3.4.1. Bacterial Infection of Mice

1. Culture bacteria as in Subheading 3.4.1 and harvest into appropriate broth for that species (see Note 13).
2. Determine concentration by OD and dilute with broth to appropriate concentration. This needs to be determined empirically for individual bacteria and mouse strains, but suggested inoculums are as follows: *H. pylori* strain SS1 10^7 CFU/mouse; *C. jejuni* 10^4–10^8 CFU/mouse; and *C. rodentium* 10^6–10^8 CFU/mouse.
3. Prepare the inoculums in broth at 100 μL/mouse, keep at 37°C, and gavage mice twice with 50 μL inoculum 30 min apart.
4. Bacterial gastrointestinal colonization can be monitored by collecting faeces and plating serial dilutions of homogenized faeces on agar with antibiotics suited to growth of the pathogen but not the normal flora under appropriate oxygen conditions for the bacteria (selective culture is not possible for some bacteria). If luminescent bacteria are available, then serial imaging of infection using a whole-animal imager and/or imaging of the dissected gastrointestinal tract at the end of the experiment is useful. *C. jejuni* can be grown at 42°C because it is adapted to growth in birds and this helps limit growth on non-pathogenic microbes from faeces.
5. At the end of the experiment, collect gastrointestinal tissues for analysis. For histological analysis of the stomach, longitudinal sections of the stomach should be made and aligned for sectioning. For histological analysis of the intestine, "Swiss rolling" the entire region of the tract affected by the pathogen is recommended (see Note 14). To preserve the mucus layer, see methods detailed in Chapter 13.

3.4.2. Parasitic Infection of Mice

1. Harvest *T. muris* eggs as described in Subheading 2.2 and determine the number of embryonated eggs by using a stereo microscope in 200 μL (inoculation volume). To obtain >200

eggs, concentrate or dilute eggs as required in dH_2O. For inoculation with <12 eggs, withdraw <12 eggs using a stereo microscope with a gel-loading tip and dilute in total volume of 200 μL for inoculation.

2. *T. muris* provides a unique model in which acute and chronic infection can be assessed depending on the strain of the mouse or the dose of eggs administered to the mice. BALB/c and C57BL/6 mice are resistant to chronic infection when inoculated with >200 eggs; however, these mice are susceptible to chronic infection when inoculated with <12 eggs. The immunocompetent AKR mice and immunocompromised mice, such as Severe Combined Immuno-Deficient (SCID) mice, are susceptible to chronic infection at any dose of *T. muris* eggs (17).
3. Mice can be sacrificed at different time points to assess the worm burden; worm burden is assessed on day 12 of infection to determine establishment of infection. Caecal tissue can be collected for histological and gene expression analysis. Mesenteric lymph nodes can be collected, cultured, and stimulated with parasitic exo-products to determine *T. muris*-specific immune response generated by the mice using standard leukocyte/T-cell culture conditions (10, 18).

3.5. Assessing the Effect of Mucins on Parasite Metabolism

ATP assays can be used to assess the health of parasites removed from mammalian hosts or exposed to potentially anti-parasitic factors, such as mucins in vitro. Intestinal mucins have been shown to decrease ATP production in *T. muris* (10). To assess the effect of mucins on parasites, obtain parasites from immunodeficient mice as described in Subheading 2.2. This assay has been used for assessing health of the nematode *T. muris* but can be adapted to other parasites and bacteria.

1. Longitudinally cut and segment the intestine of the infected mice.
2. Incubate with 0.1 M NaCl for 2 h at 37°C with frequent shaking to obtain parasites (a method previously developed for *Caenorhabditis elegans*). This does not affect the ATP production of the parasites.
3. Separate parasites from debris and epithelial cells using a 0.7-μm filter and keep in RPMI-1640/FBS.
4. Count the number of parasites present in the filtrate and either homogenize immediately or expose to purified mucins (see Chapter 2, 0.1–0.5 μg/mL) for 24–48 h prior to homogenization.
5. Homogenize live worms in RPMI-1640/FBS at room temperature using FastPrep® (MP Biomedicals) for 2 min. Confirm homogenization by observing under a stereo microscope.

Live worms can be kept at 37°C in RPMI-1640 for 48 h before homogenization without affecting the ATP levels.

6. Add homogenized sample to opaque-walled multiwall plates compatible with the luminometer and equilibrate at room temperature for 2–5 min.
7. Add equal volume of Luminescence Detection Reagent (Cell-Titre Glo) and mix on orbital shaker for 5 min to stabilize the luminescence signal.
8. Record luminescence. In parallel, generate and analyze an ATP standard curve by serial dilution of ATP from 1 μM to 10 nM in RPMI-1640/FBS. RPMI-1640/FBS plus Cell-Titre Glo substrate are used as blank controls. Homogenates from parasites heated to 100°C for 10 min should be used as negative controls.
9. Calculate relative light units (RLUs) per parasite after subtraction of the blank control luminescence: RLU = (sample light units − blank light units)/number of parasites in volume of homogenate used.

4. Notes

1. For the detection of bacteria in microtitre-based assays, it is useful to biotinylate them. However, the drawback with biotinylation is that too high levels of biotinylation may alter the surface properties of the bacteria to the extent that the bacteria does not exhibit its normal binding properties. Therefore, if one already knows what the bacteria bind to, one can analyze binding level to this structure before and after biotinylation as a control to ensure that binding is not lost by biotinylation (for example, *H. pylori* has known adhesins recognizing sialyl-Le^x and Le^b, so the level of binding can be detected by fluorescent sLe^x and Le^b conjugates) or, if antibodies against the bacteria are available, perform an assay in parallel with un-biotinylated bacteria and use ELISA-based antibody detection in parallel to the streptavidin-based detection system.
2. This protocol is optimized for *H. pylori* and should be optimized for any other species used. Other bacteria may need a different centrifugation time to allow the bacteria to settle but without pelleting them too hard, as different types of bacteria have differential sensitivity to centrifugation damage.
3. Many buffers are compatible with binding to plastic in ELISA plates, but this should be optimized empirically. It is suggested that a sample of an unrelated protein not expected to bind the bacteria is used as a negative control, for example bovine serum albumin.

4. The number of bacteria needs to be optimized for each species of bacteria and for individual labelling reactions. The irrelevant antibody controls for non-specific mucin binding to the plate and background bacterial binding to antibody.
5. The in-plate capture with specific mucin antibodies allows use of crude extracts of mucins to be used, for example detergent extraction lysates of cultured cells.
6. Either individual bacteria or combinations of bacteria that grow under the same oxygen concentration conditions can be utilized in this protocol.
7. To determine whether enzymatic activity within the exo-products is responsible for degradation, incubations with mucins can be performed at 4°C or specific protease and glycanase inhibitors can be added to the incubation mixes. To determine whether exo-products affect the oligomerization of mucins or degrade the mucin core itself, unreduced and reduced mucins can be treated with parasitic exo-products before being analysed by agarose gel electrophoresis (see Chapter 2). PAS staining or Western blotting can be used to identify the degraded mucin products and changes to the electrophoretic mobility can indicate the changing size of mucins. Lectins binding or mass spectrometry could be utilized to explore changes to glycan structures due to pathogen glycanases.
8. Do not allow the sides of the jar to become wet. The wet paper towel keeps the atmosphere in the jar humid which greatly helps the *H. pylori* resuscitate. Ensure that Campygen sachet is not in contact with the wet paper towel or any other moisture (water quenches the gas generating reaction).
9. The *H. pylori* will be healthier if subcultured again in broth culture. Use BHI for washing if the harvested *H. pylori* are to be subcultured, and use PBS for washing if the harvested *H. pylori* are for co-culture with mammalian cells.
10. The most effective siRNA transfection reagent needs to be selected for each individual cell line using fluorescently labelled siRNA and examining any effects on growth or apoptosis.
11. The level of knock-down required to see a functional deficit is difficult to predict, but knock-down >80% is generally achievable within 3 days of mucin siRNA treatment. It can be useful to titre knock-down by titre of the siRNA concentration, evaluating functional significance at several differing mucin concentrations.
12. The number of bacteria per mammalian cell should be not excessively beyond that found in infection and should take into consideration potential exponential expansion of the bacteria during the course of the co-culture. Although MOIs of 1:100 are often used, caution is warranted and researcher should

reflect on whether the co-culture reflects true (patho)-physiology at this concentration. MOIs over 1:5 tend to cause mammalian cell death. Co-culture of microaerophilic bacteria in microaerobic conditions can lead to greater expansion of bacteria during co-culture.

13. Different strains of bacteria adapted for growth in mice may have to be used for some pathogens. For example, for *H. pylori*, the SS1 strain has been adapted for growth in mice, whereas other strains commonly used for in vitro work do not establish infection in mice. It should also be noted that mucin carbohydrates differ somewhat between mice and humans and this can affect the utility of murine infection experiments with some pathogens.
14. To "Swiss roll" the intestine, open the intestine carefully along the longitudinal axis, gently remove faecal material, and either carefully roll it into a Swiss roll and pin it in place as a roll in fixative or pin the opened intestine flat in fixative for several hours prior to rolling.

Acknowledgements

Development of these techniques was supported by NHMRC project grant 543704 and the Swedish Research Council (grant K2008-58X-20693-01-4). Michael McGuckin is supported by an NHMRC Senior Research Fellowship.

References

1. Linden, S. K., Sutton, P., Karlsson, N. G., Korolik, V., McGuckin, M. A. (2008) Mucins in the mucosal barrier to infection. *Mucosal Immunol.* **1**, 183–197.
2. Thornton, D. J., Rousseau, K., McGuckin, M. A. (2008) Structure and function of the polymeric mucins in airways mucus. *Ann. Rev. Physiol.* **70**, 459–486.
3. Larsson, J. M., Karlsson, H., Sjovall, H., Hansson, G. C. (2009) A complex, but uniform O-glycosylation of the human MUC2 mucin from colonic biopsies analyzed by nanoLC/MSn. *Glycobiology* **19**, 756–766.
4. McGuckin, M. A., Linden, S. K., Sutton, P., Florin, T. H. (2011) Mucin dynamics and enteric pathogens. *Nat. Rev. Microbiol.* **9**, 265–278.
5. Thornton, D. J., Khan, N., Sheehan, J. K. (2000) Separation and identification of mucins and their glycoforms. *Methods Mol. Biol.* **125**, 77–85.
6. Linden, S. K., Driessen, K. M., McGuckin, M. A. (2007) Improved in vitro model systems for gastrointestinal infection by choice of cell line, pH, microaerobic conditions and optimization of culture conditions. *Helicobacter* **12**, 341–353.
7. Cottet, S., Corthesy-Theulaz, I., Spertini, F., Corthesy, B. (2002) Microaerophilic conditions permit to mimic in vitro events occurring during in vivo *Helicobacter pylori* infection and to identify Rho/Ras-associated proteins in cellular signaling. *J. Biol. Chem.* **277**, 33978–33986.
8. Spicer, A. P., Rowse, G. J., Lidner, T. K., Gendler, S. J. (1995) Delayed mammary tumor progression in Muc-1 null mice. *J. Biol. Chem.* **270**, 30093–30101.
9. Velcich, A., Yang, W., Heyer, J., Fragale, A., Nicholas, C., Viani, S., Kucherlapati, R., Lipkin, M., Yang, K., Augenlicht, L. (2002) Colorectal cancer in mice genetically deficient in the mucin Muc2. *Science* **295**, 1726–1729.

10. Hasnain, S. Z., Evans, C. M., Roy, M., Gallagher, A. L., Kindrachuk, K. N., Barron, L., Dickey, B. F., Wilson, M. S., Wynn, T. A., Grencis, R. K., Thornton, D. J. (2011) Muc5ac: a critical component mediating the rejection of enteric nematodes. *J. Exp. Med.* **208**, 893–900.
11. Sheng, Y., Lourie, R., Linden, S. K., Jeffery, P. L., Roche, D. K., Tran, T. V., Png, C. W., Waterhouse, N. J., Sutton, P., Florin, T. H., McGuckin, M. A. (2011) The MUC13 Cell Surface Mucin Protects Against Intestinal Inflammation by Inhibiting Epithelial Cell Apoptosis. *Gut* **60**, 1661–1670.
12. Peat, N., Gendler, S. J., Lalani, N., Duhig, T., Taylor-Papadimitriou, J. (1992) Tissue-specific expression of a human polymorphic epithelial mucin (MUC1) in transgenic mice. *Cancer Res.* **52**, 1954–1960.
13. Heazlewood, C. K., Cook, M. C., Eri, R., Price, G. R., Tauro, S. B., Taupin, D., Thornton, D. J., Png, C. W., Crockford, T. L., Cornall, R. J., Adams, R., Kato, M., Nelms, K. A., Hong, N. A., Florin, T. H., Goodnow, C. C., McGuckin, M. A. (2008) Aberrant Mucin Assembly in Mice Causes Endoplasmic Reticulum Stress and Spontaneous Inflammation Resembling Ulcerative Colitis. *PLoS Med.* **5**, e54.
14. An, G., Wei, B., Xia, B., McDaniel, J. M., Ju, T., Cummings, R. D., Braun, J., Xia, L. (2007) Increased susceptibility to colitis and colorectal tumors in mice lacking core 3-derived O-glycans. *J. Exp. Med.* **204**, 1417–1429.
15. Magalhaes, A., Gomes, J., Ismail, M. N., Haslam, S. M., Mendes, N., Osorio, H., David, L., Le Pendu, J., Haas, R., Dell, A., Boren, T., Reis, C. A. (2009) Fut2-null mice display an altered glycosylation profile and impaired BabA-mediated *Helicobacter pylori* adhesion to gastric mucosa. *Glycobiology* **19**, 1525–1536.
16. Fu, J., Wei, B., Wen, T., Johansson, M. E., Liu, X., Bradford, E., Thomsson, K. A., McGee, S., Mansour, L., Tong, M., McDaniel, J. M., Sferra, T. J., Turner, J., Chen, H., Hansson, G. C., Braun, J., Xia, L. (2011) Loss of intestinal core 1-derived O-glycans causes spontaneous colitis in mice. *J. Clin. Invest.* **121**, 1657–66.
17. Cliffe, L. J., Grencis, R. K. (2004) The Trichuris muris system: a paradigm of resistance and susceptibility to intestinal nematode infection. *Adv. Parasitol.* **57**, 255–307.
18. Hasnain, S. Z., Wang, H., Ghia, J. E., Haq, N., Deng, Y., Velcich, A., Grencis, R. K., Thornton, D. J., Khan, W. I. (2010) Mucin gene deficiency in mice impairs host resistance to an enteric parasitic infection. *Gastroenterology* **138**, 1763–1771.

Chapter 19

Assessing Mucin Expression and Function in Human Ocular Surface Epithelia In Vivo and In Vitro

Pablo Argüeso and Ilene K. Gipson

Abstract

Mucins of the corneal and conjunctival epithelia are necessary for the protection of the ocular surface against desiccation, pathogen access, and injury. Detection and quantification of mucins is important for the understanding of ocular surface diseases that cause impaired vision and, in advanced stages, blindness. Advances in the field of molecular biology have made it possible to study membrane mucins and their associated O-glycans in established cell culture models of human ocular surface epithelia. This chapter discusses procedures to detect and quantify mucin RNA and protein in biological samples, as well as methods to experimentally manipulate the epithelia in culture by shRNA, to understand the function of specific mucins. Example protocols are provided to evaluate the role of ocular surface mucins in mucosal barrier function and bacteria–host interactions.

Key words: Ocular surface, Mucin, shRNA knockdown, Mucosal barrier, Bacterial adhesion

1. Introduction

All wet-surfaced epithelia of the body, including those on the ocular surface, express both secreted and membrane-tethered mucins. Of the wet-surfaced epithelia, the corneal and conjunctival epithelia are the most exposed to the outside world and, therefore, especially subject to desiccation, pathogen access, and injury. Thus, the secreted mucins and the membrane mucins are especially necessary for protection of the corneal surface, the major refractive surface of the eye, and indeed for vision itself (1).

At the ocular surface, goblet cells intercalated within the stratified conjunctival epithelia express and secrete MUC5AC, which is then moved over the surface of the eye by lid blinking. The apical cells of the stratified epithelium of both cornea and conjunctiva

Michael A. McGuckin and David J. Thornton (eds.), *Mucins: Methods and Protocols*, Methods in Molecular Biology, vol. 842, DOI 10.1007/978-1-61779-513-8_19, © Springer Science+Business Media, LLC 2012

express the membrane-associated mucins MUC1, 4 and 16, albeit central corneal epithelium lacks MUC4 and is especially rich in MUC16 (2, 3). The membrane mucins are major components of the surface glycocalyx and are especially prominent on surface membrane folds termed microplicae (4). Surface diseases such as dry eye syndrome and other drying cicatrizing surface diseases caused by Sjögren's syndrome, ocular cicatricial pemphigoid, and Steven's Johnson Syndrome cause impaired vision and can, in advanced stages, cause blindness. In these diseases, alterations in both secreted (loss of goblet cells) and membrane-tethered mucins have been documented (5).

Since the mucin-expressing cells of the ocular surface are exposed to the external environment, they and the fluids covering their surfaces are readily accessible for analysis of mucin content and mucin gene expression. Thus, it has been feasible to sample human tear fluid as well as conjunctival epithelium in normal subjects and patients with different ocular pathologies to compare mucin protein and O-glycan content, as well as mucin and glycosyltransferase mRNA expression. Although human ocular surface tissue is accessible, obviously it is not possible to experimentally manipulate the epithelia to understand function of specific mucins and their regulation. Additionally, major differences in mucin gene expression profiles are apparent between humans and experimental animals (e.g., mice do not express the MUC16 homologue on their ocular surface, and they express two secreted mucins, Muc5ac and Muc5b). Thus, it has been useful to establish in vitro human epithelial cell culture models of ocular surface epithelia, particularly for study of the membrane mucins and their associated O-glycans (4, 6–12).

This chapter details methods employed for assay of human ocular surface epithelial mucins in health and disease, as well as cell culture models that can be used to study membrane mucin function and regulation relevant to corneal and conjunctival epithelia.

2. Materials

2.1. Collection of Human Samples (see Note 1)

1. Sterile transfer pipet (1 mL) with extended fine tip, individually wrapped.
2. Sterile saline (0.9% (w/v) NaCl).
3. Artificial tears (e.g., Tears Naturale®, Refresh Tears®).
4. Micro BCA Protein Assay Kit (Pierce, Rockford, IL).
5. Topical anesthetic (0.5% proparacaine HCl, Alcaine®).

6. 100% pure nitrocellulose membranes (Protran BA83, Whatman; Sanford, ME).
7. Topical ophthalmic antibiotic formulation (e.g., 0.3% tobramycin, Tobrex®; 0.5% moxifloxacin HCI, Vigamox®).
8. TRIzol reagent (Gibco; Grand Island, NY).
9. RNeasy Mini Kit (Qiagen Inc., Valencia, CA).

2.2. Cell Culture

1. Corneal and conjunctival epithelial cell lines (e.g., HCLE, HCjE, hTCEpi) (9, 13).
2. Neutralization Medium: Add 55 mL newborn calf serum (final concentration 10%) and 5.5 mL 100× Penicillin–Streptomycin (pen/strep) (see Note 2) to 500 mL DMEM/F12 (50/50 mix containing L-glutamine and 15 mM HEPES; Cellgro Mediatech, Washington, DC).
3. Keratinocyte serum-free medium (K-sfm). Add 1.25 mL of 10–14 mg/mL bovine pituitary extract (corresponding to half the vial supplied with the medium; final concentration 25–35 μg/mL), 3 μL of 30–40 μg/mL EGF (final concentration 0.2 ng/mL), 0.5 mL 0.3 M $CaCl_2 \cdot H_2O$ (final concentration 0.4 mM; the medium comes with 0.09 mM $CaCl_2$) (see Note 3), and 5.1 mL 100× pen/strep to 500 mL of Gibco Keratinocyte SFM (Gibco-Invitrogen Corp., Rockville, MD). The medium remains good for at least 1 month at 4°C. All media deteriorate from exposure to light, so protect from light as much as possible.
4. Stratification medium: Add 55 mL newborn calf serum (final concentration 10%), 0.55 mL 10 μg/mL EGF (see Note 4) (final concentration 10 ng/mL), and 5.5 mL 100× pen/strep to 500 mL DMEM/F12.
5. 2× Freezing medium: Add high purity DMSO to chilled DMEM/F12 medium supplemented with 20% calf-serum so that the final DMSO concentration is 20% (v/v). Mix well and let stand for at least 2 h at 4°C. Filter-sterilize with a 0.2-μm filter. Aliquot to plastic tubes. Place in a −80°C freezer to freeze solid, after which the tubes can be stored at −20°C, if desired.

2.3. Stable Gene Knockdown by Short Hairpin RNA

1. DNA oligonucleotides for *MUC16* hairpin RNA expression (4) (see Note 5).

sh*RNA-MUC16-1*:

Sense 5′-AGCCACCTCATCTATTACCTTCAAGAGAGGTAAT-AGATGAGGTGGCT-3′

sh*RNA-MUC16-2*:

Sense 5′-CTGCATGTACTCCCATCTCTTCAAGAGAGA GATGGGAGTAGATGCAG-3′

sh*RNA-MUC16-3*:

Sense 5′-TAACCATCACCACCCAAACTTCAAGAGAGTT TGGGTGGTGATGGTTA-3′.

2. Negative control shRNA with the same nucleotide composition as the *MUC16* shRNA but lacking significant sequence homology to the human genome.
3. Vector system for expression of RNAi: pSUPER.retro.puro (linear) (Oligoengine, Seattle, WA).
4. Annealing buffer: 100 mM NaCl, 50 mM HEPES, pH 7.4.
5. Chemically competent *Escherichia coli* strain: One Shot® MAX Efficiency® DH5α™-T1R cells (Invitrogen Corp., Carlsbad, CA).
6. SOC medium: 2% (w/v) Tryptone, 0.5% (w/v) Yeast Extract, 10 mM NaCl, 2.5 mM KCl, 10 mM $MgCl_2$, 10 mM $MgSO_4$, 20 mM glucose. Stir to dissolve, autoclave, and cool to room temperature. Pass the complete medium through a 0.2-μm filter. Store at 4°C.
7. LB-Amp: Dissolve 10 g bacto-tryptone, 5 g bacto-yeast extract, and 10 g NaCl in 1 L dH_2O. Adjust to pH 7.0 with 1N NaOH and autoclave at 121°C for 15–20 min. Allow solution to cool to room temperature. Add 500 μL of 100 mg/mL ampicillin (final concentration 50 μg/mL). Store at 4°C.
8. Qiaprep Spin Miniprep kit and Qiagen Plasmid Plus Midi kit (Qiagen).

2.4. Preparation of Bacteria for Adhesion Assay

1. Make sure when working on the bench top that glassware and pipet tips are autoclaved and that aseptic techniques (e.g., sterilization by flaming on a Bunsen burner) are used while handling rims of jars and tubes, as well as insides of caps before replacing onto jars. When autoclaving bottles, screw cap on loosely and tape on with autoclave tape. Do not put caps etc. down onto counter or they must be reflamed before use.
2. *Staphylococcus aureus* strains, either clinical derivatives (e.g., RN6390) or mutant strains (e.g., ALC135) (4, 11).
3. Brain heart infusion (BHI) broth, dried powder.
4. Inoculation loops (see Note 6).
5. Fluorescein isothiocyanate (FITC).
6. HEPES buffer: 0.05 M HEPES, 0.15 M NaCl, 1 mM $CaCl_2$, 1 mM $MgCl_2$, pH 7.4.
7. One-well culture chamber slides.

3. Methods

3.1. Assay of Mucin Content and Expression in the Human Ocular Surface

3.1.1. Analysis of Mucin Content in Tear Fluid

1. Using a sterile transfer pipet, instill 60 μL of premeasured sterile saline onto the inferior fornix of an unanesthetized eye.
2. Ask the subject to look left, then right, up and down, four times without blinking, to mix the tear fluid content.
3. Collect the tear wash from the inferior fornix using the previous transfer pipet. Repeat this process with the contralateral eye using a new transfer pipet. Wash out the ocular surface with one or two drops of artificial tears after collection.
4. Remove cellular debris from the tear samples by centrifugation at 18,000 × *g*, in a microfuge, for 30 min at 4°C. Collect and measure volume of supernatant using a micropipet (see Note 7).
5. Determine protein concentration of the samples using a 2 μL-aliquot of collected tears using the Micro BCA Protein Assay Kit.
6. Determine mucin content in tears by western blot (14) or ELISA (15). Mucin carbohydrate content can be analyzed by fluorometric high-performance liquid chromatography (16), lectin blot (16), or electrospray mass spectrometry (17).

3.1.2. Impression Cytology for Assay of Mucin Gene Expression

1. Apply a drop of topical anesthetic to the eye and wait for 1 min.
2. Place a sterile disk of nitrocellulose membrane, 10 mm in diameter, on the temporal bulbar conjunctiva using sterile forceps (see Note 8). Gently fold the edge of the disk for easy handling during application and removal.
3. Apply gentle pressure to the disk with the forceps for 15–30 s.
4. Carefully remove the disk from the eye and transfer into an Eppendorf tube containing 1 mL TRIzol reagent.
5. Repeat the procedure with the contralateral eye using a new set of forceps and nitrocellulose membrane. Combine the two nitrocellulose disks from each subject, placing the empty side of the disks facing each other in the Eppendorf tube.
6. Apply one drop of topical ophthalmic antibiotic formulation onto the ocular surface.
7. Promptly freeze the samples at −80°C until the time of extraction.
8. After thawing, vortex the samples gently for 1 min. Zip spin and transfer the TRIzol solution to a new Eppendorf tube, discarding the nitrocellulose membranes.
9. Extract total RNA with TRIzol reagent as recommended by the manufacturer (see Note 9) and analyze mucin transcripts

by real-time PCR (15). RNA samples can be further purified using RNeasy columns and analyzed using gene expression microarrays (18).

3.2. Induction of Cell Differentiation and Mucin Biosynthesis in Culture

1. Thaw cryovial containing corneal or conjunctival cells quickly at 37°C in a water bath (see Note 10).
2. Transfer cells to a 15 mL centrifuge tube. Add 7 mL of neutralization medium dropwise. Pipet up and down to mix.
3. Spin cells down at 1,500 × *g* for 4 min. Resuspend cells in K-sfm medium and plate at 10^4 to 10^5 cells in 15 mL per 75 cm^2 flask.
4. Set flasks in incubator (37°C, 5% CO_2) where they will not be disturbed for 24 h. Change medium next day.
5. Feed cells every 2 days with K-sfm until they reach 100% confluence (see Note 11).
6. Thereafter, switch cell cultures to stratification medium for 7 days (change media on days 0, 2, 4, 5, and 6) to induce stratification, cell differentiation, and mucin biosynthesis.

3.3. Stable Knockdown of Membrane-Associated Mucin

1. Dissolve oligonucleotides in sterile, nuclease-free H_2O to a concentration of 3 mg/mL.
2. Assemble the annealing reaction by mixing 1 μL of each oligonucleotide (sense + antisense) with 48 μL annealing buffer.
3. Incubate the mixture at 90°C for 4 min, and then at 70°C for 10 min. Cool the annealed oligonucleotides to 37°C within 30 min, then at room temperature. The annealed oligonucleotide inserts can be used immediately in a ligation reaction, or cooled further to 4°C. For longer storage, keep at −20°C until needed.
4. Assemble the cloning reaction by adding 2 μL of the annealed oligos to 1 μL of T4 DNA ligase buffer. Add 1 μL pSUPER.retro.puro vector (linear), 5 μL nuclease-free H_2O, and 1 μL T4 DNA ligase.
5. Incubate overnight at room temperature. A negative control cloning reaction should be performed with the linearized vector alone and no insert (see Note 12).
6. Briefly centrifuge the vials containing the ligation reactions and place on ice.
7. Thaw, on ice, one 50 μL vial of One Shot® cells for each ligation.
8. Pipet 2 μL of each ligation reaction directly into the vial of competent cells and mix by tapping gently. Do not mix by pipetting up and down. The remaining ligation mixtures can be stored at −20°C.

9. Incubate the vials on ice for 30 min.
10. Heat shock cells for exactly 30 s in the 42°C water bath. Do not mix or shake.
11. Remove vials from the 42°C bath and place them on ice for 15–30 min.
12. Add 250 μL of prewarmed SOC medium to each vial; sterile technique must be practiced to avoid contamination.
13. Shake the vials at 37°C for exactly 1 h at 225 rpm, in a shaking incubator.
14. Spread two volumes of 20 and 200 μL from each transformation vial on separate, labeled LB-amp agar plates (prepoured plates are commercially available). The vector contains an ampicillin resistance ORF at 4,558–5,415.
15. Invert the plates and incubate at 37°C overnight.
16. Identify positive clones containing the shRNA insert by picking and growing colonies overnight in 3 mL of LB-amp broth at 37°C at 275 rpm in a shaking incubator (see Note 13). Isolate DNA from colonies using Qiagen's Plasmid Miniprep kit. Check for the presence of positive clones in a 1.5% agarose gel by digesting with EcoRI and HindIII. Positive clones (vector with insert) should be 281 bp, whereas negative clones with no insert should be approximately 227 bp (see Note 14). Make glycerol stocks of positive colonies and maintain at –80°C.
17. Grow large-scale cultures of transformed cells from glycerol stocks of positive colonies to isolate large quantities of pSUPER.retro.puro to be used for transfection into the packaging cell line. Use a Qiagen Plasmid Midi kit to obtain enough plasmid, and store DNA at –20°C.
18. The pSUPER.retro.puro plasmid is transfected into a packaging cell line to produce retroviral supernatants for a higher rate of stable cell integration. Polyfect (Qiagen) has been successfully used with 293-10A1 packaging cells (ATCC; Manassas, VA) to produce retroviral supernatants containing shRNA-MUC16 (4). A standard laboratory protocol for transfection of 293 cells with the pSUPER.retro.puro plasmid is available in the PolyFect Transfection Reagent Handbook (http://www.qiagen.com/literature/).
19. Forty-eight hour posttransfection, filter sterilize (0.45-μm syringe filters are convenient) the virus-containing supernatant to remove any cells in suspension. The virus can now be used directly or stored at –80°C until needed.
20. Plate ocular surface epithelial cells in 24-well plates at 2×10^4 cells per well and grow to 40% confluence in K-sfm medium. Incubate with virus-containing supernatant and 4 μg/mL polybrene for at least 6 h. This procedure can be repeated to

enhance infection, as long as cells are allowed to recover for 24 h between infections with fresh K-sfm medium.

21. Infected clones are selected with puromycin (e.g., 2.5 μg/mL for HCLE cells (see Note 15)) and further expanded to confluence to obtain stably transfected cells. Untransfected cells should die within 5–10 days. When the selection is complete, cells can be grown without antibiotics.
22. Allow stably transfected cells to reach confluence and add stratification medium to induce mucin expression.
23. Seven days later, cells will be ready for functional analysis or extraction of RNA and protein. Quantitative RT-PCR and western blotting can be used to quantify the level of suppression of *MUC16* expression.

3.4. Analysis of Mucosal Barrier Function in Vitro

1. Serum starve stratified ocular surface epithelial cells in one-well culture chamber slides for 2 h with K-sfm. Rinse three times with Ca^{+2}/Mg^{+2}-free PBS, pH 7.4.
2. Add 1 mL of 0.1% rose bengal dye in PBS to cells.
3. Incubate for 5 min at room temperature.
4. Aspirate rose bengal solution from cell culture.
5. Rinse cells once with PBS using a transfer pipet to remove excess rose bengal. Add two drops of PBS to keep cells hydrated.
6. Photograph cell layers (ten images per well) at 10× using an inverted microscope. Ocular surface epithelial cells grown in stratification media will contain areas of stratified cells that exclude the rose bengal dye (Fig. 1).
7. Analyze images using ImageJ software (http://rsb.info.nih.gov/) as follows: (see Note 16).
   (a) File>Open
   (b) Plugins>Colour Deconvolution>Select H&E vector from pull-down menu
   (c) Select color 2 (pink)
   (d) Image>Adjust>Threshold (see Note 17)
   (e) Image>Type>8-bit
   (f) Edit>Invert (see Note 18)
   (g) Analyze>Measure
   (h) Determine Integrated Density

3.5. Bacterial Adhesion Assay

1. Inoculate 5 mL BHI broth (see Note 19) with 20–200 μL of a glycerol stock of *S. aureus* in a sterile 15 mL tube.
2. Place the cap onto the bacterial suspension tube (if using a screw-cap tube, only screw it on loosely so that air can enter the tube) and incubate statically overnight at 37°C.

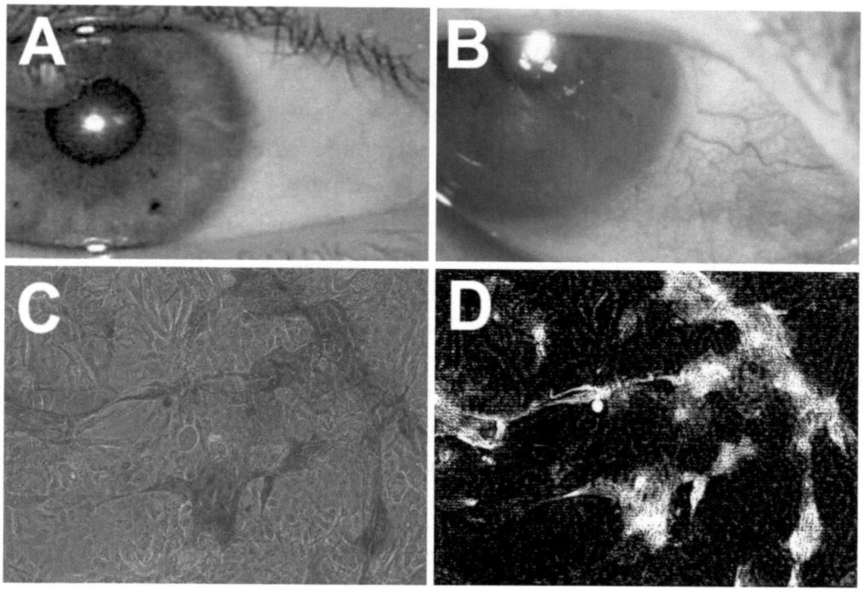

Fig. 1. Rose bengal is an organic anionic dye used to assess damage to the ocular surface epithelium in ocular surface disease. Application of rose bengal onto the ocular surface results in patches of superficial punctate staining in patients with epithelial damage (**b**), but not in normal subjects (**a**). In culture, human corneal epithelial cells grown in stratification media contain areas of stratified cells that exclude the rose bengal dye (**c**). The amount of rose bengal uptake can be determined using ImageJ software (**d**) as described in Subheading 3.4, step 7.

3. At this point, and if unknown, determine the growth curve for each bacterial strain by serial dilution and plate counting as follows:
   (a) Make a 1:100 dilution of the overnight suspension in BHI broth (Subheading 3.5, step 2) to a total volume of 5 mL and place in an orbital shaker at 37°C.
   (b) Incubate for a predetermined period of time that will allow reaching the stationary phase, usually 8 h, taking aliquots (325 μL) approximately every 30–60 min, starting at time point 0. Place aliquots in a 96-well plate or cuvettes and determine the OD_{595}.
   (c) To calculate colony forming units (cfu), take 20 μL of culture at each time point, and dilute in 180 μL of sterile PBS. Make approximately four to five dilutions of the bacterial suspension that is being counted (e.g., 10^{-3}, 10^{-4}, 10^{-5}, 10^{-6} dilutions for low OD values) and spot 5 μL of each dilution across the top of a square agar dish. Prop dish up vertically so drops will run down the dish (making a track) and dry. Then, place plate upside down into the incubator to let grow overnight at 37°C for counts the next day

(count dilution lane with most distinct colonies, usually between 20 and 30) to determine cfu/mL (cfu/mL = (number of colonies × dilution factor)/volume inoculated). Colonies can sometimes overlap; look at regularity of bacterial colonies to determine whether they are actually two or more colonies growing together.

4. Vortex the overnight bacterial suspension to break up the clumps. Dilute 1:200 into 30 mL of prewarmed BHI using a 125-mL sterile Erlenmeyer flask (e.g., 150 μL culture in 30 mL broth). Put flask into an orbital shaker at 200 rpm at 37°C and grow bacteria in 5% CO_2 to the desired exponential growth phase using the OD_{595} reading (use a predetermined growth curve for each bacterial strain).
5. Centrifuge at 5,000 × *g* for 5 min at 4°C and wash the bacterial pellet twice with 10 mL PBS, pH 7.4.
6. Prepare the FITC-labeled bacteria:
   (a) Incubate 5 mL (5×10^9 cfu/mL) *S. aureus* in PBS with 1.25 mL of a stock solution of 500 μg/mL FITC (final concentration 100 μg/mL) for 30 min on ice (see Note 20). Mix the suspension by vortexing every 5 min to avoid cell clumping/aggregation.
   (b) Collect labeled bacteria by centrifugation at 5,000 × *g* for 5 min at 10°C and wash six times with 10 mL PBS, vortexing each time to ensure thorough washing.
   (c) Resuspend the bacterial pellet in DMEM/F12 to a final concentration of 1×10^8 cfu/mL and maintain on ice, protected from light until needed.
7. Perform bacterial adhesion experiments in ocular surface epithelial cells grown on one-well culture chamber slides. Twenty hours prior to assay, switch into an antibiotic-free stratification medium.
8. Wash the slides three times with sterile HEPES or PBS buffer.
9. Incubate each slide with 2 mL FITC-labeled bacteria (1×10^8 cfu/mL) for 1 h at 37°C. Maintain the slides in the dark or covered with aluminum foil.
10. Carefully wash four times with PBS to remove nonadherent bacteria, and fix in 4% paraformaldehyde for 30 min at room temperature.
11. Wash four times with PBS and apply Vectashield mounting medium (Vector Laboratories; Burlingame, CA). Add coverslips to the slides and image them under a fluorescence microscope.
12. Analyze images using ImageJ software (NIH; Bethesda, MD).

4. Notes

1. Collection of human tear fluid and conjunctival impression cytology specimens for research purposes must be performed by an ophthalmologist and requires compliance with good clinical practice, institutional review board regulations, informed consent regulations, and the tenets of the Declaration of Helsinki.
2. Use a 100× solution containing 10,000 U/mL penicillin G sodium and 10,000 μg/mL streptomycin sulfate in 0.85% saline (available from Gibco-Invitrogen Corp.).
3. Dissolve 0.441 g $CaCl_2$ in 10 mL Milli-Q H_2O and filter sterilize with a 0.22 μm filter.
4. Prepare a 1,000× solution: Dissolve 100 μg recombinant human EGF (Invitrogen) in 10 mL of 20 mM HEPES-buffered Earle's salts and 0.1% bovine serum albumin. Store frozen at −20°C.
5. shRNA targets and scramble can be designed using software available from specialized companies (e.g., Oligoengine Inc.; http://www.oligoengine.com/download/). The design tool from Oligoengine displays only the forward oligo, but the second, complementary oligo is automatically designed and ordered. Three distinct regions of the gene should be chosen to find the most effective RNA target. Custom complementary oligonucleotides are synthesized containing a unique 19-nucleotide sequence derived from the gene transcript. Blast the sequences in GenBank to exclude the possible off-target effect to human genes. The loop sequence in shRNA-MUC16 is underscored. The flanking BglII and HindIII are added to the DNA oligonucleotides for insertion into the pSUPER.retro.puro vector.
6. Sterilize the inoculation loop starting at the circular end of the loop, waiting until the wire glows, and then slowly bringing it up through the flame, a little past the actual wire onto the handle. Hold the loop for a few second to let it cool and test on an area of the agar dish without colonies to see if it melts/bubbles.
7. Only minor amounts of protein (insufficient for analysis) are extractable from the pellets obtained after centrifugation.
8. Prepare the nitrocellulose disks by making a circular incision on the nitrocellulose sheet with a sterile 10 mm corneal trephine. Introduce each individual disk into a self-sealing sterilization pouch and autoclave.
9. The amount of total RNA obtained by nitrocellulose membrane stripping of human conjunctiva usually varies from approximately 1–10 μg.

10. Drop vial in container of water and shake back and forth. Watch until cells thaw. Rinse outside of vial with 70% ethanol to sterilize.
11. Subculture from this medium before cells reach 2/3 confluence. To cryopreserve cells, add an equal volume of room temperature 2× freezing medium to cells already suspended to an appropriate density in K-sfm medium. Typical freezing density is 5×10^5–1×10^6 cells per cryovial.
12. After cloning and prior to transformation, plasmids should be treated with BglII to reduce the level of background in the transformation. Add 1.0 μL of BglII to the plasmid and incubate for 30 min at 37°C. As the BglII site is destroyed upon successful cloning of the oligo pair, those vectors cut by the enzyme will not contain the insert fragment.
13. Choose at least ten colonies from each plate for screening.
14. The presence of the correct insert within the recombinant pSUPER.retro.puro vector can also be confirmed by sequencing prior to transfection in mammalian cells.
15. Since optimal concentrations vary for individual cell types, an antibiotic kill curve should be established for each cell line.
16. Analysis of multiple images is facilitated by the use of macro recording to automate the series of ImageJ commands.
17. The threshold is selected after randomly opening five images and closely reproducing the staining observed in the original images.
18. This allows measurement of dye uptake. To measure dye exclusion, omit this step.
19. To make BHI broth, dissolve 37 g of powder in 1 L purified water; mix with a magnetic stir bar; remove stir bar and autoclave at 121°C for 15 min in bottle before use. Aliquot 5 mL of BHI broth into sterile 15 mL tubes. Always inoculate a dummy tube with no cells to verify that the medium and tubes are sterile. Use this tube as blank while taking OD_{595} readings.
20. Make a 5 mg/mL stock solution of FITC in DMSO and then dilute to 500 μg/mL stock in H_2O. Filter the FITC suspension through a 0.45-μm Millex-HV PVDF filter (Millipore; Bedford, MA) and use the filtrate for labeling bacteria.

Acknowledgments

Supported by NIH/NEI Grants no. R01EY014847 (PA) and R01EY03306 (IKG)

References

1. Gipson, I. K. (2007) The ocular surface: the challenge to enable and protect vision: the Friedenwald lecture, *Invest Ophthalmol Vis Sci.* **48**, 4390; 4391–4398.
2. Gipson, I. K., and Argueso, P. (2003) Role of mucins in the function of the corneal and conjunctival epithelia, *Int Rev Cytol.* **231**, 1–49.
3. Govindarajan, B., and Gipson, I. K. (2010) Membrane-tethered mucins have multiple functions on the ocular surface, *Exp Eye Res.* **90**, 655–663.
4. Blalock, T. D., Spurr-Michaud, S. J., Tisdale, A. S., Heimer, S. R., Gilmore, M. S., Ramesh, V., and Gipson, I. K. (2007) Functions of MUC16 in corneal epithelial cells, *Invest Ophthalmol Vis Sci.* **48**, 4509–4518.
5. Mantelli, F., and Argueso, P. (2008) Functions of ocular surface mucins in health and disease, *Curr Opin Allergy Clin Immunol.* **8**, 477–483.
6. Albertsmeyer, A. C., Kakkassery, V., Spurr-Michaud, S., Beeks, O., and Gipson, I. K. (2010) Effect of pro-inflammatory mediators on membrane-associated mucins expressed by human ocular surface epithelial cells, *Exp Eye Res.* **90**, 444–451.
7. Argueso, P., Guzman-Aranguez, A., Mantelli, F., Cao, Z., Ricciuto, J., and Panjwani, N. (2009) Association of cell surface mucins with galectin-3 contributes to the ocular surface epithelial barrier, *J Biol Chem.* **284**, 23037–23045.
8. Argueso, P., Tisdale, A., Spurr-Michaud, S., Sumiyoshi, M., and Gipson, I. K. (2006) Mucin characteristics of human corneal-limbal epithelial cells that exclude the rose bengal anionic dye, *Invest Ophthalmol Vis Sci.* **47**, 113–119.
9. Gipson, I. K., Spurr-Michaud, S., Argueso, P., Tisdale, A., Ng, T. F., and Russo, C. L. (2003) Mucin gene expression in immortalized human corneal-limbal and conjunctival epithelial cell lines, *Invest Ophthalmol Vis Sci.* **44**, 2496–2506.
10. Paulsen, F., Jager, K., Worlitzsch, D., Brauer, L., Schulze, U., Schafer, G., and Sel, S. (2008) Regulation of MUC16 by inflammatory mediators in ocular surface epithelial cell lines, *Ann Anat.* **190**, 59–70.
11. Ricciuto, J., Heimer, S. R., Gilmore, M. S., and Argueso, P. (2008) Cell surface O-glycans limit Staphylococcus aureus adherence to corneal epithelial cells, *Infect Immun.* **76**, 5215–5220.
12. Sumiyoshi, M., Ricciuto, J., Tisdale, A., Gipson, I. K., Mantelli, F., and Argueso, P. (2008) Antiadhesive character of mucin O-glycans at the apical surface of corneal epithelial cells, *Invest Ophthalmol Vis Sci.* **49**, 197–203.
13. Robertson, D. M., Li, L., Fisher, S., Pearce, V. P., Shay, J. W., Wright, W. E., Cavanagh, H. D., and Jester, J. V. (2005) Characterization of growth and differentiation in a telomerase-immortalized human corneal epithelial cell line, *Invest Ophthalmol Vis Sci.* **46**, 470–478.
14. Spurr-Michaud, S., Argueso, P., and Gipson, I. (2007) Assay of mucins in human tear fluid, *Exp Eye Res.* **84**, 939–950.
15. Argueso, P., Balaram, M., Spurr-Michaud, S., Keutmann, H. T., Dana, M. R., and Gipson, I. K. (2002) Decreased levels of the goblet cell mucin MUC5AC in tears of patients with Sjogren syndrome, *Invest Ophthalmol Vis Sci.* **43**, 1004–1011.
16. Guzman-Aranguez, A., Mantelli, F., and Argueso, P. (2009) Mucin-type O-glycans in tears of normal subjects and patients with non-Sjogren's dry eye, *Invest Ophthalmol Vis Sci.* **50**, 4581–4587.
17. Argueso, P., and Sumiyoshi, M. (2006) Characterization of a carbohydrate epitope defined by the monoclonal antibody H185: sialic acid O-acetylation on epithelial cell-surface mucins, *Glycobiology.* **16**, 1219–1228.
18. Mantelli, F., Schaffer, L., Dana, R., Head, S. R., and Argueso, P. (2009) Glycogene expression in conjunctiva of patients with dry eye: downregulation of Notch signaling, *Invest Ophthalmol Vis Sci.* **50**, 2666–2672.

INDEX

Michael A. McGuckin and David J. Thornton (eds.), *Mucins: Methods and Protocols*, Methods in Molecular Biology, vol. 842, DOI 10.1007/978-1-61779-513-8, © Springer Science+Business Media, LLC 2012

MIX
Papier aus verantwortungsvollen Quellen
Paper from responsible sources
FSC® C105338

If you have any concerns about our products,
you can contact us on
ProductSafety@springernature.com

In case Publisher is established outside the EU,
the EU authorized representative is:
Springer Nature Customer Service Center GmbH
Europaplatz 3, 69115 Heidelberg, Germany

Printed by Libri Plureos GmbH
in Hamburg, Germany